DEFORMATION ANALYSIS IN SOFT GROUND IMPROVEMENT

GEOTECHNICAL, GEOLOGICAL AND EARTHQUAKE ENGINEERING

Volume 18

For further volumes:
http://www.springer.com/series/6011

Deformation Analysis in Soft Ground Improvement

by

JINCHUN CHAI

Saga University, Japan

JOHN P. CARTER

The University of Newcastle, Callaghan, NSW, Australia

Jinchun Chai
Saga University
Department of Civil Engineering
and Architecture
Honjo 1
840-8502 Saga
Japan
chai@cc.saga-u.ac.jp

John P. Carter
The University of Newcastle
Faculty of Engineering and Built
Environment
University Drive
2308 Callaghan
Australia
John.Carter@newcastle.edu.au

ISSN 1573-6059
ISBN 978-94-007-1720-6 e-ISBN 978-94-007-1721-3
DOI 10.1007/978-94-007-1721-3
Springer Dordrecht Heidelberg London New York

Library of Congress Control Number: 2011931773

Printed on acid-free paper

Springer is part of Springer Science+Business Media (www.springer.com)

Preface

This book describes a variety of analytical and numerical models, based firmly on experimental evidence from observations made in both the laboratory and the field, which can be used in geotechnical engineering to make predictions of soil deformations arising from or associated with ground improvement techniques. The use of such techniques is becoming ever more popular in soft ground engineering. The deformations of the ground addressed in this work are often time-dependent and, given that soft soil behaviour can be highly non-linear, they are often difficult to estimate reliably.

This volume was inspired by the work of many researchers, practitioners and authors who have made significant contributions to this important field of geotechnical engineering. In writing the book it was the intention of the authors to provide a state of the art and practice in this field, at least as they stand at the time of writing. Ground improvement technology and our analytical and numerical capabilities are advancing at such a pace that there is a significant risk that such a volume may date quickly. Nevertheless, the fundamental mechanics of the soft soils to be treated will remain the same. It is our ability to describe and model them more accurately and reliably that may change with time, along with the evolving technologies being developed to improve their mechanical behaviour.

The book is directed towards students of geotechnical engineering as well as geotechnical practitioners. In the main it provides rather complete derivations of most of the important theoretical results, rather than just bland statements of each result. This is deliberate, as the intention was to write a book that could be used as both a teaching text and a reference work. In presenting such material the authors were also keenly aware of their obligation to present critical evaluation and validation of the various analytical and theoretical models presented in this book and therefore proposed for use in engineering practice. Accordingly, the book also contains numerous case histories and comparisons of the predictions of the proposed models with the results from high quality laboratory and field experimental work. The field case histories are from soft soil sites at various locations around the world.

The book starts with a chapter describing some of the various ground improvement technologies that are in current use in soft ground engineering. Not only does it provide brief descriptions of these technologies, but it also makes the argument that predicting the effects on subsequent ground performance of implementing these technologies is important in engineering practice. One important effect concerns the resulting ground deformations.

Since the development of predictive techniques for estimating soil displacements is underpinned by and critically dependent on the constitutive models adopted to represent the mechanical response of soil, it seems only natural to describe some of these fundamental tools that have been used successfully to date for this task. This is the rationale for including Chapter 2, which contains a description of some of the techniques that have been used successfully in modelling the mechanical response of soft clay subsoils. Included in this treatment are some of the fundamental stress-strain models used to describe the behaviour of clays as well as their implementation in the finite element method (FEM). Also presented in this chapter are example analyses to demonstrate the power and significance of these modern numerical techniques

Chapter 3 contains a description of the methods used for modelling soft ground improved by the installation of pre-fabricated vertical drains (PVDs) using plane strain finite element analyses. Also presented are some example analyses using specific case histories.

Chapter 4 presents the theory of vacuum consolidation and describes the major characteristics of vacuum consolidation of soft ground, the methods adopted for calculating the deformations induced by application of vacuum pressure, as well as some example analyses based on case histories.

Chapter 5 describes a consolidation theory for a clayey subsoil improved by the installation of soil-cement columns. It also describes methods for estimating ground settlement (mainly for the case of ground improved by the prior installation of 'floating' columns), a method for predicting the lateral displacement of the ground induced by the column installation, and the analyses of some case histories.

Finally, Chapter 6 presents suggestions and recommendations for further research and development in this area.

It is worth reiterating that the deliberate intent of the authors was that this book should provide up-to-date information on this important subject from both the theoretical and practical points of view.

While this book was inspired by the many developments already in the field of soft ground engineering, it is also our hope that it will in turn inspire others to focus their efforts on the important task of improving the mechanical behaviour of soft ground, striving to find better and more reliable ways to make such ground 'stiffer and stronger', and striving to develop more accurate and reliable models to be used in association with these techniques.

We are indebted to a number of people for their significant assistance in the task of producing this book, some of whom made direct contributions to its preparation and others whose efforts were more inadvertent but no less important. In particular, we would like to thank Ms M. Komoto and Mrs Y. Kanada for their patient and skillful assistance in preparing some of the figures contained in this book. Importantly, we would also like to thank our life-long partners, Yaru Wang and Heather Carter, for their patience, support, love and understanding. It is to them that this contribution is dedicated.

Saga, Japan — Jinchun Chai
Newcastle, New South Wales — John P. Carter

Contents

Notation

A	cross-sectional area of a drainage channel (L^2)
B	measure of the proximity of the kinematic yield and bounding surfaces in the double surface 'bubble' model (dimensionless)
B	half width of a plane strain unit cell (L)
c	cohesion of soil ($ML^{-1}T^{-2}$)
c'	effective stress cohesion ($ML^{-1}T^{-2}$)
C_c	compression index (dimensionless)
C_d	a constant (dimensionless)
C_f	ratio of field hydraulic conductivity over corresponding laboratory value (dimensionless)
c_h	coefficient of consolidation in horizontal direction (L^2T^{-1})
C_k	a constant (dimensionless)
c_v	coefficient of consolidation in vertical direction (L^2T^{-1})
D	coefficient of dilatancy (dimensionless)
d_e	diameter of a unit cell of PVD (L)
d_m	area equivalent diameter of PVD installation mandrel (L)
d_s	diameter of smear zone (L)
d_w	diameter of vertical drain (L)
e	void ratio (dimensionless)
e_0	initial void ratio (dimensionless)
E	Young's modulus ($ML^{-1}T^{-2}$)
E_b	bulk modulus ($ML^{-1}T^{-2}$)
f	yield function
g	plastic potential function
g	gravitational acceleration (LT^{-2})
G'	elastic shear modulus ($ML^{-1}T^{-2}$)
H_c	optimum thickness of unimproved sub-layer
H_L	thickness of a soil layer (L)
i	hydraulic gradient (dimensionless)
$\boldsymbol{I}$	the second-rank identity tensor
I_{r}	rigidity index (dimensionless)
I_{rr}	reduced rigidity index (dimensionless)

k	a constant for tangential modulus (dimensionless)
K	stiffness matrix (MT^{-2})
k	hydraulic conductivity (LT^{-1})
k_0	initial permeability (LT^{-1})
K_a	the active earth pressure coefficient (dimensionless)
k_b	a constant for bulk modulus (dimensionless)
K_F	element stiffness matrix (MT^{-2})
k_h	hydraulic conductivity in horizontal direction (LT^{-1})
k_{hp}	matched horizontal hydraulic conductivity in plane strain condition (LT^{-1})
K_0	at-rest earth pressure coefficient (dimensionless)
k_s	hydraulic conductivity of smear zone (LT^{-1})
k_{sp}	hydraulic conductivity of the smear zone in the plane strain unit cell (LT^{-1})
k_v	hydraulic conductivity in vertical direction (LT^{-1})
k_{ve}	equivalent hydraulic conductivity in vertical direction (LT^{-1})
L	link (or coupling) matrix (L^2)
l	drainage length (L)
m	a constant for calculating bulk modulus (dimensionless)
M	slope of critical state line in q:p' plot (dimensionless)
n	a constant for calculating tangential modulus (dimensionless)
n	radius ratio = R/r_w (dimensionless)
N	specific volume of isotropic compression line at unit mean effective stress (dimensionless)
OCR	over consolidation ratio (dimensionless)
p	initial mean stress ($ML^{-1}T^{-2}$)
p'	effective overburden pressure ($ML^{-1}T^{-2}$)
p'	mean effective stress ($ML^{-1}T^{-2}$)
p'_0	size of yielding locus on mean effective stress axis ($ML^{-1}T^{-2}$)
p'_0	value of the mean effective stress at the intersection of the current swelling line with the isotropic compression line in the double surface 'bubble' model ($ML^{-1}T^{-2}$)
p'_i	initial mean effective stress ($ML^{-1}T^{-2}$)
p_a	atmospheric pressure ($ML^{-1}T^{-2}$)
p'_a	mean effective stress tensor at the centre of the kinematic yield surface ($ML^{-1}T^{-2}$)
q	deviator stress ($ML^{-1}T^{-2}$)
q	flow rate [L^3T^{-1}]
q_{VD}	flow rate from surrounding soil into a PVD (L^3T^{-1})
q_w	discharge capacity of prefabricated vertical drain (PVD) (L^3T^{-1})
q_{wp}	equivalent discharge capacity of a plane strain drain (L^3T^{-1})
r	radial distance (L)
R	the ratio of the size of the kinematic yield surface to that of the bounding surface (dimensionless)
r_e	radius of a unit cell (L)

R_f	failure ratio of hyperbolic model (dimensionless)
R_q	discharge capacity ratio (dimensionless)
r_s	radius of smear zone (L)
r_w	radius of vertical drain (L)
$\mathbf{s}$	deviator stress tensor ($ML^{-1}T^{-2}$)
s	ratio of the diameter of smear zone over the diameter of a drain
S	spacing of PVDs (L)
$\mathbf{s}_a$	deviatoric stress tensor at the centre of the kinematic yield surface ($ML^{-1}T^{-2}$)
$\mathbf{s}_{ij}$	the (i, j) component of the deviator stress tensor ($ML^{-1}T^{-2}$)
$\mathbf{s}_{ij0}$	initial value of $\mathbf{s}_{ij}$ at the end of anisotropic consolidation ($ML^{-1}T^{-2}$)
S_u	undrained shear strength ($ML^{-1}T^{-2}$)
S_x	length of an element in x direction (L)
S_y	length of an element in y direction (L)
S_z	length of an element in z direction (L)
t	thickness of a prefabricated vertical drain (PVD) (L)
t	time (T)
T_h	time factor in horizontal direction (dimensionless)
T_v	time factor in vertical direction (dimensionless)
u	excess pore water pressure ($ML^{-1}T^{-2}$)
$\bar{u}$	average excess pore water pressure ($ML^{-1}T^{-2}$)
u_i	excess pore pressure at the center of ith element ($ML^{-1}T^{-2}$)
U_h	average degree of consolidation due to radial drainage (dimensionless)
$\overline{U}_h$	average degree of consolidation in horizontal direction (dimensionless)
U_{kp}	average degree of consolidation in horizontal direction under plane strain condition (dimensionless)
U_v	average degree of consolidation due to vertical drainage (dimensionless)
U_{vh}	average degree of consolidation of PVD-improved subsoil (dimensionless)
v	specific volume (dimensionless)
v_d	dilatancy (dimensionless)
w	width of a PVD (L)
W_L	liquid limit (dimensionless)
W_n	natural water content (dimensionless)
W_P	plastic limit (dimensionless)
X	material property in double surface 'bubble' model (dimensionless)
X_p	material property in double surface 'bubble' model for deviatoric shape of plastic potential (dimensionless)
Y	material property in double surface 'bubble' model (dimensionless)
Y_p	material property in double surface 'bubble' model for deviatoric shape of plastic potential (dimensionless)
x	Cartesian co-ordinate
y	Cartesian co-ordinate
z	depth (L)
Z	material property in double surface 'bubble' model (dimensionless)

Z_p material property in double surface 'bubble' model for deviatoric shape of plastic potential (dimensionless)
α secondary compression index (dimensionless)
β constant of proportionality for viscoplastic strain (dimensionless)
β_i coefficient of water flow ($M^{-1}L^3T^2$)
β_{VD} coefficient of water flow of a PVD per unit length ($M^{-1}L^3T^2$)
Γ specific volume on the critical state line in the $p' - q$ plane at unit p'
γ' the effective unit weight of soil ($ML^{-2}T^{-2}$)
γ_t the total unit weight of soil ($ML^{-2}T^{-2}$)
γ_w the unit weight of pore water ($ML^{-2}T^{-2}$)
δ_i lateral displacement (L)
δ_{ij} Kronecker delta (dimensionless)
Δ average volumetric strain (dimensionless)
Δ_d incremental nodal displacement (L)
ΔF_1 incremental nodal force (MLT^{-2})
ΔF_2 incremental nodal flow (L^3T^{-1})
Δt time interval or time increment (T)
Δu incremental nodal excess pore water pressure ($ML^{-1}T^{-2}$)
$\delta {\varepsilon_q}^e$ elastic shear strain increment (dimensionless)
$\delta {\varepsilon_p}^e$ elastic volumetric strain increment (dimensionless)
$\delta {\varepsilon_q}^p$ plastic shear strain increment (dimensionless)
$\delta {\varepsilon_p}^p$ plastic volumetric strain increment (dimensionless)
$\Delta\varepsilon_v$ incremental volumetric strain (dimensionless)
$\Delta\sigma_{vac}$ the incremental vacuum pressure ($ML^{-1}T^{-2}$)
ε_v volumetric strain (dimensionless)
$\dot{\boldsymbol{\varepsilon}}_{ij}^{e}$ the (i, j) component of elastic strain rate tensor (T^{-1})
$\dot{\boldsymbol{\varepsilon}}_{ij}$ the (i, j) component of strain rate tensor (T^{-1})
$\dot{\boldsymbol{\varepsilon}}_{ij}^{vp}$ the (i, j) component of viscoplastic strain rate tensor (T^{-1})
$\dot{\varepsilon}_v$ the volumetric strain rate (T^{-1})
η coefficient of viscosity of water ($ML^{-1}T^{-3}$)
η stress ratio q/p' in modified Cam clay model (dimensionless)
η^* stress parameter in Sekiguchi and Ohta model (dimensionless)
θ interface transmissivity (L^2T^{-1})
θ_b the Lode angle (°)
θ_y Lode angle for the kinematic yield surface (°)
κ slope of unloading-reloading line in $e - \ln p'$ plot (dimensionless)
κ^* slope of unloading-reloading line in $\ln v - \ln p'$ plot (dimensionless)
Λ $1-\kappa/\lambda$
λ slope of virgin compression curve in plot $e - \ln p'$ (dimensionless)
λ^* the slope of isotropic normal compression line in $\ln v - \ln p'$ space (dimensionless)
μ_p parameter representing effects of spacing, smear and well resistance in consolidation theory of a plane strain unit cell of a vertical drain (dimensionless)

ν	Poisson's ratio (dimensionless)
ν	angle of dilation (°)
σ	confining pressure ($ML^{-1}T^{-2}$)
$\boldsymbol{\sigma}'$	effective stress tensor ($ML^{-1}\ T^{-2}$)
σ_3	minimum principle stress or confining stress for triaxial compression ($ML^{-1}T^{-2}$)
σ'_h	in situ horizontal effective stress ($ML^{-1}T^{-2}$)
σ'_v	in situ vertical effective stress ($ML^{-1}T^{-2}$)
σ'_{h0}	initial value of in situ horizontal effective stress ($ML^{-1}T^{-2}$)
σ'_{v0}	initial value of in situ vertical effective stress ($ML^{-1}T^{-2}$)
σ'_x	normal effective stress in x-direction ($ML^{-1}T^{-2}$)
σ'_y	normal effective stress in y-direction ($ML^{-1}T^{-2}$)
σ'_z	normal effective stress in z-direction ($ML^{-1}T^{-2}$)
Φ	hydraulic conductivity matrix ($M^{-1}L^4T$)
ϕ	friction angle of soil (°)
ϕ'	effective stress friction angle of soil (°)
Ψ	parameter in hardening function (dimensionless)

Chapter 1
Introduction

Abstract The book begins with a chapter describing some of the various ground improvement technologies that are in current use in soft ground engineering. Not only does it provide brief descriptions of these technologies, but it also makes the argument that predicting the effects on subsequent ground performance of implementing these technologies is important in engineering practice. Clearly, one important effect concerns the ground deformations resulting from their use. This particular issue is examined in detail in the subsequent chapters of this book. The obligation to validate and calibrate the various analytical and numerical solutions presented in this book by comparing their predictions with experimental data is acknowledged. Indeed, the book includes a large number of case studies and field trials used for this very purpose.

1.1 What is Ground Improvement and When and Why Is It Necessary?

The selection of a suitable site for construction is one of the most important considerations in almost all civil engineering projects. From the point of view of geotechnical engineering, a suitable site usually means one where the foundation soil (or rock) has sufficient strength and stiffness to carry the loads that will be imposed on the ground without causing unacceptably large deformations or stability problems, or else the foundation material has drainage properties suitable for the particular construction application.

However, there are many cases in practice where the most suitable site has been determined by social or economic requirements, rather than purely geotechnical or other engineering considerations. For example, for ease of access to sea transport some of the world's largest cities have been located throughout the history of civilization in coastal areas, many of which are typically underlain by soft clay or clay-like deposits. Obvious examples of this type of city include Shanghai in China and Bangkok in Thailand, but there are many others that could also be listed. In these cases the soft soil deposits underlying these cities often have relatively low strength and high compressibility, sometimes so weak or so compressible that major infrastructure cannot be built directly on the ground as it exists in nature. In many cases, the mechanical properties of these soft deposits, e.g., their strength, stiffness and

J. Chai, J.P. Carter, *Deformation Analysis in Soft Ground Improvement*, Geotechnical, Geological and Earthquake Engineering 18, DOI 10.1007/978-94-007-1721-3_1,

hydraulic conductivity, need to be improved by selected engineering methods, i.e., 'ground improvement' techniques, before any major components of infrastructure are constructed.

Furthermore, due to increased urbanization there is now a growing shortage of available land in many cities, so that new facilities such as airports and sea ports have had to be developed on land reclaimed from the sea. Well known examples include Kansai International airport in Japan, Changi International airport in Singapore, Hong Kong International airport and Incheon International airport in Korea. Since the newly reclaimed land of this type is often too weak to adequately support runways and terminal buildings, some form of ground improvement is normally required in these situations.

Another example of applications for which ground improvement is almost inevitably required occurs in railway and highway construction. Because of existing pressures on land use these transportation routes often pass through regions underlain by deposits of soft or weak soils. The mechanical properties of these softer soils often have to be improved in order to adequately support the associated earth structures and to reduce the residual settlements that occur during the useful life of the associated transport infrastructure.

1.2 Techniques of Ground Improvement

Numerous ground improvement techniques have been developed throughout the course of human history. Bergado et al. (1996) classified the various ground improvement methods according to whether or not additives are used to directly enhance the strength and stiffness of the ground. A slightly modified version of their classification scheme for these methods is as follows:

1. Work on the soil only (primarily to reduce the voids content)
   (a) Densification by applying external forces to coarse-grained soils
      (i) Surface compaction
      (ii) Vibro-flotation and vibro-compaction
      (iii) Dynamic compaction
      (iv) Resonance compaction
   (b) Improvement of the drainage of fine-grained soils (often combined with preloading)
      (i) Sand drains
      (ii) Prefabricated vertical drains (PVDs)
      (iii) Horizontal drains
2. Addition of other materials into the soil deposit
   (a) Soil reinforcement

(i) Mechanically stabilized earth (MSE)
(ii) Sand compaction piles (SCP)
(iii) Geo-piers (stone columns and granular piles)

(b) Use of chemical admixtures

(i) Deep mixing method (DMM) using lime and cement
(ii) Chemical piles

Of these methods, preloading in combination with the installation of prefabricated PVDs, and cement deep mixing (CDM), normally in order to form soil-cement columns in the ground, are widely used to improve the mechanical performance of soft clayey deposits (Bergado et al. 1996). Where preloading is used, the loads can be applied either directly to the soft ground as a surcharge (e.g., by the placement of embankment fill) or by applying a vacuum pressure to the soil (Chu et al. 2000; Tang and Shang 2000; Indraratna and Chu 2005; Indraratna et al. 2004; Chai et al. 2006).

1.3 Why Do We Need to Estimate Ground Deformations?

Almost all geotechnical designs require consideration of at least one or possibly a combination of the following three factors:

1. Strength – such as in the bearing capacity of foundations, slope stability, earthquake resistance of the ground, etc.;
2. Deformation – such as surface settlement, differential settlement, and lateral displacement of the ground and any associated geotechnical structures; and
3. Permeability or hydraulic conductivity – which is significant for water retaining structures like dams, and drainage systems, etc.

There are situations where deformations induced in the ground, or the structures constructed on the ground, become the controlling factors in a design, such as, for example, in problems involving the design of shallow foundations on sand, and also in the design and construction of pile foundations. Clearly, ground deformation has a direct influence on the serviceability, or in-service performance, of the affected structures. In the case of road embankments constructed over soft ground, settlement of the embankment, due mainly to vertical but also lateral deformations of the soft underlying ground, may become excessive. If these settlements become so pronounced that the elevation of the road surface is reduced to the point where the roadway becomes flooded during heavy rainfall, then clearly the serviceability and functionality of the roadway itself becomes compromised. Furthermore, differential settlements of embankments and foundations may induce cracks or inclination (tilting) of the supported structures, which may have adverse effects on the functioning of these structures. Therefore, predicting the settlement and lateral movement of soft ground under the influence of applied loads is often an essential design requirement.

Furthermore, under working conditions the deformation of an earth structure can be measured, at least in principle, but also often in practice. Such measurements can then be compared with the original design predictions. The advantage of conducting such comparisons should be obvious. Clearly, one important advantage is that they allow the predictive models and their underlying assumptions to be refined. Such refinement is common and often essential in geotechnical practice and reflects the fact that the calculation of deformations for real engineering projects is often an iterative process.

1.4 What Is This Book All About?

This book deals with a particular class of deformation problems and the methods of analysis that may be applied to such problems. The focus is on the deformation of soft clay deposits or clay-like deposits whose mechanical properties and performance have been enhanced by various forms of ground improvement. Particular emphasis has been placed on predicting the deformation response of ground improved by the installation of prefabricated vertical drains (PVDs) and soil-cement columns formed by deep cement mixing (DCM). The behaviour of ground treated in this way and then subjected to embankment loading and/or vacuum pressure loading is specifically addressed. Various theories and numerical modelling techniques that can be applied successfully to the analysis of this class of geotechnical problem in soft ground are described.

Since the development of predictive techniques for estimating soil displacements is underpinned by, and critically dependent on, the constitutive models adopted to represent the soil response, it seems only natural to describe some of these fundamental tools that have been used successfully to date for this task. This is the rationale for including Chapter 2, which contains a description of some of the techniques that have been used successfully in modelling the mechanical response of soft clay subsoils. Included in this treatment are some of the fundamental stress-strain models used to describe the behaviour of clays as well as their implementation in the finite element method (FEM). Also presented in this chapter are example analyses to demonstrate the power and significance of these modern numerical techniques

Chapter 3 contains a description of the methods used for modelling soft subsoils whose mechanical performance, particularly their drainage characteristics, have been improved by the installation of pre-fabricated vertical drains (PVDs). Included is a description of how a typical problem may be dealt with using an equivalent plane strain finite element analysis. Also presented are some example analyses using specific case histories, and comparisons of the model predictions with the field performance of PVD systems.

Chapter 4 presents the theory of vacuum consolidation and describes the major characteristics of vacuum consolidation of soft ground, the methods adopted for calculating the deformations induced by application of vacuum pressure, as well as some example analyses based on case histories.

Chapter 5 describes a consolidation theory for clayey subsoil improved by the installation of soil-cement columns. It also describes methods for estimating ground settlements, mainly for the case of ground improved by the prior installation of 'floating' columns, a method for predicting the lateral displacement of the ground induced by column installation, and the analyses of some relevant case histories.

Many of the techniques for deformation analysis presented in Chapters 2, 3, 4, and 5 involve approximations of reality, sometimes quite crude approximations. These have been clearly and deliberately identified. In many cases it has been demonstrated, by presenting comparisons of predictions with field data, that these approximations provide a reasonable balance between tractability of the problem at hand and the accuracy and reliability of the resulting deformation predictions. However, there is still a need for considerable further research in this area of geotechnology. Much still needs to be done to increase our predictive capabilities and particularly the accuracy of our predictions of ground deformations associated with common ground improvement techniques. It is anticipated that this need will only continue as further advances are made in the technology of ground improvement. Chapter 6 provides some suggestions for future research directions, which hopefully will address some of the present shortcomings or gaps in our current predictive abilities.

It is the deliberate intent of the authors that this book should provide up-to-date information on this important subject from both the theoretical and practical points of view.

References

Bergado DT, Anderson, LR, Miura N, Balasubramaniam AS (1996) Soft ground improvement – in lowland and other environments. ASCE Press, New York, NY, p 427

Chai J-C, Carter JP, Hayashi S (2006) Vacuum consolidation and its combination with embankment loading. Can Geotech J 43(10):985–996

Chu J, Yan SW, Yang H (2000) Soil improvement by the vacuum preloading method for an oil storage station. Géotechnique 50(6):625–632

Indraratna B, Chu J (2005) Ground improvement – case histories. Elsevier, Oxford

Indraratna B, Bamunawita C, Khabbaz H (2004) Numerical modelling of vacuum preloading and field applications. Can Geotech J 41:1098–1110

Tang M, Shang JQ (2000) Vacuum preloading consolidation of Yaogiang airport runway. Géotechnique 50(6):613–623

Chapter 2
Modelling Soft Clay Behaviour

Abstract Since the development of predictive techniques for estimating soil displacements is underpinned by, and critically dependent on, the constitutive models adopted to represent the soil response, it seems only natural to describe some of these fundamental tools that have been used successfully to date for this task. This is the rationale for including Chapter 2, which contains a description of some of the techniques that have been used successfully in modelling the mechanical response of soft clay subsoils. Included in this treatment are some of the fundamental stress-strain models used to describe the mechanical behaviour of clays as well as their implementation in the finite element method (FEM). Also presented in this chapter are example analyses to demonstrate the power and significance of these modern numerical techniques. Case histories from Malaysia and Japan are discussed in relation to this form of numerical modelling.

2.1 Introduction

In order to simulate accurately the mechanical behaviour of a clay deposit using mathematical techniques, it is most important to select an appropriate constitutive model and correct values of the model parameters, as well as an accurate assessment of the initial effective stress state in the soil deposit. In order to do so, an objective method should be established for first checking whether the chosen soil model together with the corresponding soil parameters and assumed in situ stress state can simulate, to acceptable accuracy, the stress-strain relationships of the deposit.

An important quantity to determine in most soft ground engineering problems is the undrained shear strength of the clay or clay-like soil. In many geotechnical investigations of soft clay sites, the undrained shear strength (S_u) of the soil deposit is often measured using the field vane shear test, or estimated from the results of cone penetration tests, or else measured in an unconfined compression test or a triaxial compression test using undisturbed soil samples. These measurements to determine the profile of undrained shear strength with depth below the ground surface provide a simple way to check whether the adopted stress-strain model, as well as the corresponding parameters and initial effective stress, are appropriate. In particular, they

J. Chai, J.P. Carter, *Deformation Analysis in Soft Ground Improvement*, Geotechnical, Geological and Earthquake Engineering 18, DOI 10.1007/978-94-007-1721-3_2,

provide a means of checking whether together the assumed soil model, parameter values and initial in situ stress state can provide accurate predictions of the profile of S_u. This process may also be adopted to assess whether it is necessary to make allowance for strain rate effects on the undrained shear strength (Bjerrum 1972, 1973).

Embankments are often constructed as major components of road, railway and dike projects, and therefore they are normally regarded as important geotechnical structures. It is now quite common for embankments constructed on soft clay soils to incorporate some form of ground improvement, such as the installation of prefabricated vertical drains to accelerate the rate of consolidation, or deep cement mixing to reduce the settlement and increase the stability of the embankment.

A theory for analyzing the stresses and deformations induced in the ground under an embankment load is available for the special case where it is assumed that the ground is semi-infinite, homogeneous and elastic (Gray 1936, Giroud 1968). However, in most practical cases the subsoil is neither homogeneous nor elastic, especially if ground improvement techniques have been used to treat the soft clayey subsoil. Most practical problems are much more complicated than the ideal cases for which closed-form solutions exist for the stresses and deformations of the embankment and underlying ground.

The finite element method (FEM) can take into account many of the complicated subsoil and boundary conditions. In particular, it can include elastoplastic and elasto-visco-plastic models of the behaviour of soil, provided an appropriate constitutive model is selected. The FEM has been widely used in the analysis of geotechnical boundary and initial value problems including the simulation of embankments constructed on soft subsoils. In addition to selecting a suitable constitutive model to represent the mechanical behaviour of the soil, there are several other factors that may influence the results of a FEM analysis of an embankment on soft subsoil. These include the methods adopted to simulate application of the embankment load and indeed the entire construction process including any ground improvement, as well as the possibility of including large deformations in the analysis.

In the following sections some of the constitutive soil models more commonly used to represent the mechanical behavior of soft clayey soils are described, as well as their corresponding predictions of the values of undrained shear strength, S_u. Some example analyses are also presented to illustrate the importance of checking the simulated S_u profile before launching in to a complex non-linear analysis of embankment construction. Methods used in FEM analysis to apply embankment load, simulating the embankment construction process and techniques to take into account the effects of large deformation phenomena are also described. Finally, further complications such as the effect of the load application method on the foundation response, the effect of foundation deformations during construction on the factor of safety of the embankment and large-deformation-induced buoyancy effects are discussed and illustrated though the use of example analyses.

2.2 Initial Stiffness and Undrained Shear Strength

2.2.1 Constitutive Models for Clay Soils

It is not intended in this section to provide a comprehensive review of all constitutive models for soft clay soils nor to suggest the "best" soil models for clay soils. Rather, the intention is to describe a selection of the constitutive models that are believed to be of use in representing the mechanical response of soft clays and to demonstrate the importance of particular numerical modelling techniques when simulating the behaviour of a soft clayey subsoil. The Modified Cam Clay model (Roscoe and Burland 1968), the Sekiguchi-Ohta model (Sekiguchi and Ohta 1977) and a generalized two-surface "bubble" model (Grammatikopoulou 2004, Grammatikopoulou et al. 2006) are described in some detail, as well as the equations they provide for calculating the undrained shear strength S_u.

2.2.1.1 Modified Cam Clay (MCC)

The MCC model (Roscoe and Burland 1968) is one of the most widely used models for soft clay soils. As with any elasto-plastic model, MCC has four major components and these are:

1. an elastic response and associated elastic properties;
2. a yield surface defining the boundary between elastic and elastoplastic soil behaviour;
3. a plastic potential function which is used to define the relative magnitudes of the plastic strain increments; and
4. a hardening rule which allows the magnitudes of the plastic strain to be calculated.

MCC assumes that the recoverable volumetric strain ($\delta\varepsilon_p^e$) can be expressed as follows:

$$\delta\varepsilon_p^e = \kappa \frac{\delta p'}{(1+e)\,p'} \tag{2.1}$$

where κ = the slope of unloading-reloading curve in e–ln(p') plot, e = voids ratio, p' = the mean effective stress and $\delta p'$ = the mean effective stress increment. It is also assumed that the recoverable shear strain ($\delta\varepsilon_q^e$) is a linear function of deviator stress (q) increment δq:

$$\delta\varepsilon_q^e = \frac{\delta q}{3G'} \tag{2.2}$$

where G' = the elastic shear modulus. Equations (2.1) and (2.2) imply a variation of Poisson's ratio with mean effective stress, but the alternative assumption of a constant value of Poisson's ratio is also possible and often adopted in practice.

For MCC the yield locus in the $p'-q$ stress plane is an ellipse which is described by Eq. (2.3):

$$\frac{p'}{p'_0} = \frac{M^2}{M^2 + \eta^2} \tag{2.3a}$$

where $\eta = q/p'$. The form of this yield surface (*f*) can also be written as:

$$f = q^2 - M^2 \left[p' \left(p'_0 - p' \right) \right] = 0 \tag{2.3b}$$

where $M =$ the slope of critical state line (CSL) in $p'-q$ stress plane, and $p'_0 =$ the size of yield focus on the mean stress axis. Under a state of triaxial compression, *M* is related to the friction angle (ϕ') as follows:

$$M = \frac{6 \sin \phi'}{3 - \sin \phi'} \tag{2.4}$$

Equation (2.3) describes a set of ellipses, all having the same shape (controlled by the value of *M*), all passing through the origin, and having sizes controlled by the value of p'_0. When the soil is yielding, the change in size of the yield locus, p'_0, is linked with the changes in effective stresses p' and $q = \eta p'$, through the differential form of Eq. (2.3):

$$\frac{\delta p'}{p'} + \frac{2\eta\delta\eta}{M^2 + \eta^2} - \frac{\delta p'_0}{p'_0} = 0 \tag{2.5a}$$

or

$$\left(\frac{M^2 - \eta^2}{M^2 + \eta^2} \right) \frac{\delta p'}{p'} + \left(\frac{2\eta}{M^2 + \eta^2} \right) \frac{\delta q}{p'} - \frac{\delta p'_0}{p'_0} = 0 \tag{2.5b}$$

It is assumed that the soil obeys the normality condition, and the plastic potentials (g) are the same as the yield functions (f) in the $p'-q$ plane:

$$g = f = q^2 - M^2 \left[p' \left(p'_0 - p' \right) \right] = 0 \tag{2.6}$$

In the MCC model, a linear relationship is assumed between the specific volume ($v = 1 + e$) and the logarithm of mean effective stress p'_0 during isotropic normal compression of the soil:

$$v = N - \lambda \ln p'_0 \tag{2.7}$$

where $\lambda =$ the slope of the virgin loading curve in the $e - \ln(p')$ plot, $N =$ the specific volume on the isotropic compression line at unit mean stress, i.e., $p' =$ 1 kPa (Fig. 2.1). It follows then that the magnitude of the plastic volumetric strain is given by:

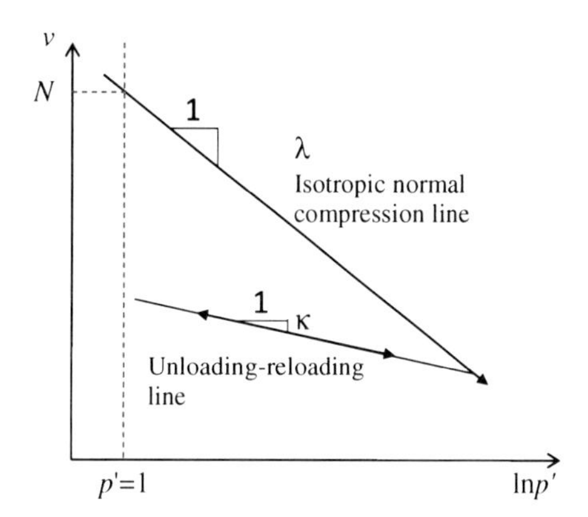

Fig. 2.1 Specific volume and mean effective stress relationship

$$\delta\varepsilon_p^{\mathrm{P}} = \left[(\lambda - \kappa)/v\right] \frac{\delta p'_0}{p'_0} \tag{2.8}$$

and the hardening relationship becomes:

$$\frac{\partial p'_0}{\partial \varepsilon_p^{\mathrm{P}}} = \frac{v p'_0}{\lambda - \kappa} \tag{2.9}$$

In particular, the MCC model predicts the undrained shear strength (S_u) of the soil as follows:

$$S_u = \frac{p'}{2^{1+\Lambda}} M \left(\frac{M^2 + \eta^2}{M^2} \right)^{\Lambda} (OCR)^{\Lambda} \tag{2.10}$$

where OCR = overconsolidation ratio, and $\Lambda = 1 - \kappa/\lambda$.

2.2.1.2 Sekiguchi–Ohta Model

This is an anisotropic Cam clay-type model with an associated flow rule, which also allows for viscosity in the formulation. Sekiguchi and Ohta (1977) introduced a new stress parameter (η^*) to model the shear stress-induced dilatancy of clays as follows:

$$\eta^* = \sqrt{\frac{3}{2} \left(\boldsymbol{\eta}_{ij} - \boldsymbol{\eta}_{ij0}\right)\left(\boldsymbol{\eta}_{ij} - \boldsymbol{\eta}_{ij0}\right)} \tag{2.11}$$

in which

$$\boldsymbol{\eta}_{ij} = \frac{\boldsymbol{s}_{ij}}{p'}, \ \boldsymbol{\eta}_{ij0} = \frac{\boldsymbol{s}_{ij0}}{p'_i} \tag{2.12}$$

where s_{ij} = the (i, j) component of deviator stress tensor, s_{ij0} = the initial value of s_{ij} at the end of anisotropic consolidation, and p'_i = the initial mean effective stress. By definition the stress parameter η^* is non-negative. The dilatancy (v_d) of anisotropically normally consolidated clays is expressed in terms of η^* as follows:

$$v_d = D \cdot \eta^* \tag{2.13}$$

where D = the coefficient of dilatancy which is defined by:

$$D = \frac{\lambda - \kappa}{M\,(1 + e_o)} \tag{2.14}$$

For this model the total volumetric strain, contributed by dilatancy plus isotropic consolidation and creep (ε_v), can be expressed as:

$$\varepsilon_v = \frac{\lambda}{1 + e_0} \ln\left(\frac{p'}{p_i'}\right) + D \cdot \eta^* - \alpha \cdot \ln\left(\frac{\dot{\varepsilon}_v}{\dot{\varepsilon}_{vi}}\right) \tag{2.15}$$

where α = the secondary compression index, $\dot{\varepsilon}_v$ = the volumetric strain rate (a superposed dot denotes the time derivative), and $\dot{\varepsilon}_{vi}$ is the initial value of $\dot{\varepsilon}_v$. The initial state denoted by the subscript (i) is referred to here as the state immediately before the change of loading. Assuming the volumetric strain immediately after the change of loading is elastic, this strain is given by the following equation:

$$\varepsilon_v^e = \frac{\kappa}{1 + e_0} \ln\left(\frac{p'}{p'_i}\right) \tag{2.16}$$

Equation (2.15) can be integrated resulting in the following form:

$$F = \alpha \cdot \ln\{1 + (\dot{\varepsilon}_{v0} t/\alpha) \cdot \exp(g/\alpha)\} = v^{vp} \tag{2.17}$$

where v^{vp} = a strain-hardening parameter, F = a scalar function, t is the elapsed time since the change of loading, and g = a scalar function which represents the plastic potential for the inviscid case, where:

$$g = \frac{\lambda - \kappa}{1 + e_0} \ln\left(\frac{p'}{p'_i}\right) + D \cdot \eta^* \tag{2.18}$$

It can be see that when t is kept constant, Eq. (2.17) describes a surface in effective stress space. It is reasonable therefore to regard the function F as the viscoplastic potential, and its derivatives with respect to any effective stress component define the directions of viscoplastic deformation. It is thus assumed that:

$$\boldsymbol{\varepsilon}_{ij}^{vp} = \beta \cdot \partial F/\partial \boldsymbol{\sigma}'_{ij} \tag{2.19}$$

where $\boldsymbol{\varepsilon}_{ij}^{vp}$ = the (i, j) component of the viscoplastic strain rate tensor and β is a constant of proportionality. Hereunder the superscript (vp) should be read as referring to the 'viscoplastic' component of any physical quantity to which this notation is assigned.

It is noted that the derivative of F with respect to any effective stress component is related to that of g with respect to the same component in the form:

$$\frac{\partial F}{\partial \boldsymbol{\sigma}'_{ij}} = \left\{1 - \exp\left(-\frac{v^{vp}}{\alpha}\right)\right\} \frac{\partial g}{\partial \boldsymbol{\sigma}'_{ij}} \tag{2.20}$$

To determine β, Eq. (2.19) can be expressed in the form:

$$\boldsymbol{\varepsilon}_{ij}^{vp} = \dot{v}^{vp} \cdot \left(\partial F / \partial \boldsymbol{\sigma}'_{ij}\right) / \left(\partial F / \partial p'\right) \tag{2.21}$$

or, in terms of g:

$$\boldsymbol{\varepsilon}_{ij}^{vp} = \dot{v}^{vp} \cdot \left(\partial g / \partial \boldsymbol{\sigma}'_{ij}\right) / \left(\partial g / \partial p'\right) \tag{2.22}$$

where $\dot{v}^{vp}$ is given by:

$$\dot{v}^{vp} = \left\{1 - \exp\left(-\frac{v^{vp}}{\alpha}\right)\right\} \cdot \left(\frac{\lambda - \kappa}{1 - e_0} \frac{\dot{p}'}{p'} + D \cdot \dot{\eta}^*\right) + \dot{\boldsymbol{\varepsilon}}_{vi} \exp\left\{\left(g - v^{vp}\right) / \alpha\right\} \tag{2.23}$$

It is convenient to express the elastic strain rates in the form:

$$\dot{\boldsymbol{\varepsilon}}_{ij}^{e} = \frac{\kappa \dot{p}'}{3\,(1 + e_0)\, p'} \delta_{ij} + \frac{1}{2G'} \dot{\boldsymbol{s}}_{ij} \tag{2.24}$$

where $\dot{\boldsymbol{\varepsilon}}_{ij}^{e}$ = the (i, j) component of elastic strain rate tensor, δ_{ij} = the Kronecker delta. In this way, total strain rates are finally obtained as:

$$\dot{\boldsymbol{\varepsilon}}_{ij} = \dot{\boldsymbol{\varepsilon}}_{ij}^{vp} + \dot{\boldsymbol{\varepsilon}}_{ij}^{e} \tag{2.25}$$

where $\dot{\boldsymbol{\varepsilon}}_{ij}$ = the (i, j) component of the total strain rate tensor.

Sekiguchi and Ohta (1977) demonstrated the capacity of this model by simulating undrained compression and extension under triaxial and plane strain conditions without considering the viscosity effect. In the analysis it was assumed that the coefficient of at-rest earth pressure $K_0 = 0.5$, and the ratio $\sigma'_y / \left(\sigma'_x + \sigma'_z\right)$ remains a constant value of 1/3. Here σ'_x, σ'_y and σ'_z are the effective principal stresses in the x, y, and z directions, respectively. The ratio $\lambda / \{D \cdot (1 + e_0)\}$ was arbitrarily chosen to be equal to $\sqrt{3}$. Computations were made in all cases until an arbitrarily chosen value of $q/p' = 1.269$ was reached.

Presented in Fig. 2.2 are the predicted effective stress paths in a $p'/p'_i - q/p'_i$ plot. This figure clearly shows that the model is capable of simulating well the anisotropic behaviour of clays. Although on the compression side the differences between the predictions for triaxial and plane strain conditions are relatively small,

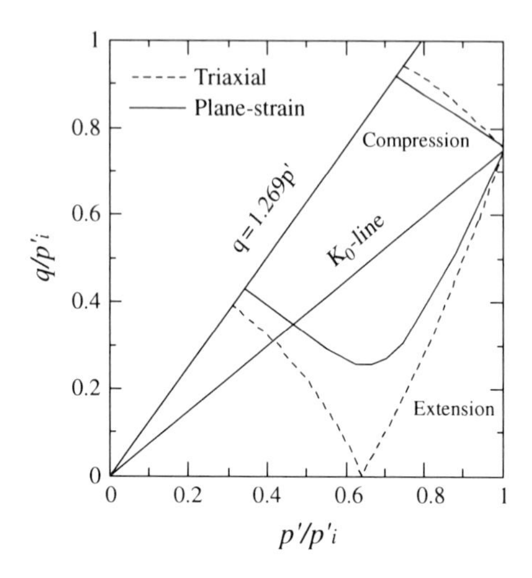

Fig. 2.2 Undrained effective stress paths simulated by the Sekiguchi-Ohta model (modified from Sekiguchi and Ohta 1977)

the differences on the extension side are much more marked. Under triaxial compression conditions, the undrained shear strength (S_u) predicted by this model can be expressed explicitly as:

$$S_u = \frac{p'}{2} M \exp\left(\left(\frac{q}{Mp'} - 1\right)^{\Lambda}\right)(ORC)^{\Lambda} \tag{2.26}$$

2.2.1.3 Generalized Two-Surface "Bubble" Model

This soil constitutive model is an extension of the Modified Cam Clay (MCC) model and it was first proposed by Al-Tabbaa and Wood (1989), and then generalized by Grammatikopoulou (2004) and Grammatikopoulou et al. (2006). It employs a single kinematic yield surface within the MCC bounding surface (Fig. 2.3) and assumes different yield and plastic potential surfaces, i.e., it has a non-associated flow rule. The kinematic yield surface encloses the region within which the soil behaviour is assumed to be elastic. The model can predict anisotropic stiffness and yielding, depending on the position of the kinematic surface, for stress paths that initiate within the bounding surface. However, the strength predicted by the model is always isotropic, since it is controlled by the bounding surface, which is centred on the isotropic stress axis. The model is formulated in general stress space in terms of the mean effective stress and the deviatoric stress tensor **s**:

$$s = \boldsymbol{\sigma}' - p'\boldsymbol{I} \tag{2.27}$$

where $\boldsymbol{\sigma}'$ = the effective stress tensor and $\boldsymbol{I}$ = the second-rank identity tensor. The bounding surface is formulated in general stress space as:

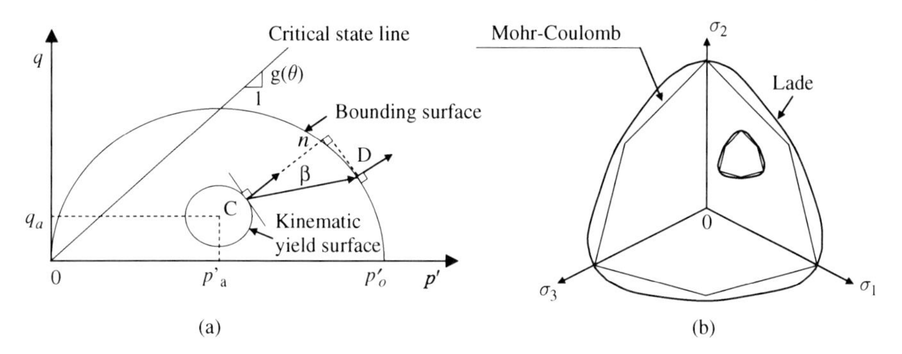

Fig. 2.3 Two-surface "bubble" model: (a) kinematic yield and bounding surfaces in triaxial compression space, (b) deviatoric section through kinematic yield and bounding surface

$$F_b = \left(p' - \frac{p'_0}{2}\right)^2 + \frac{1}{2}\frac{s{:}s}{h^2(\theta_b)} - \frac{{p'_0}^2}{4} = 0 \tag{2.28}$$

where p'_0 = the value of the mean effective stress at the intersection of the current swelling line with the isotropic compression line. The symbol ":" indicates the tensorial contraction. The function $h(\theta_b)$, where θ_b is the value of the Lode angle for the bounding surface, defines the shape of the bounding surface in the deviatoric plane. The Lode angle is defined as follows:

$$\theta_b = -\frac{1}{3}\sin^{-1}\left[\frac{3\sqrt{3}}{2}\frac{\det(s)}{\left\{\left[\frac{1}{2}(s{:}s)\right]^{\frac{1}{2}}\right\}^3}\right] \tag{2.29}$$

The equation of the kinematic yield surface in general stress space is:

$$F_y = \left(p' - p'_a\right)^2 + \frac{1}{2}\frac{(s - s_a):(s - s_a)}{h^2(\theta_y)} - R^2\frac{{p'_0}^2}{4} = 0 \tag{2.30}$$

where p'_a and s_a are the mean effective stress and the deviatoric stress tensor at the centre of the kinematic yield surface respectively, and R = the ratio of the size of the kinematic yield surface to that of the bounding surface. The function $h(\theta_y)$ defines the shape of the kinematic yield surface in the deviatoric plane. The Lode angle θ_y for the kinematic yield surface is calculated by using the centre of the kinematic yield surface as the origin.

Two options are proposed for specifying the function $h(\theta)$. The first gives the shape of a Mohr-Coulomb hexagon in the deviatoric plane (see Fig. 2.3b):

$$h(\theta) = \frac{\sin\phi'}{\cos\theta + \frac{\sin\theta\sin\phi'}{\sqrt{3}}} \tag{2.31}$$

where ϕ' = the friction angle.

The second option gives the family of continuous surfaces in the deviatoric plane proposed by van Eekelen (1980):

$$h(\theta) = \frac{X}{(1 + Y\sin 3\theta)^Z} \tag{2.32}$$

where X, Y and Z are material constants. In particular, $X = \sin\left(\phi'_{\theta=0^\circ}\right)$, where $\phi'_{\theta=0^\circ}$ is the friction angle associated with a Lode angle $\theta = 0$. In order to obtain convex surfaces, certain restrictions apply to the values of Y and Z, and these are discussed in detail by van Eekelen (1980). A major advantage of Eq. (2.32) is that, depending on the values of the constants, different shapes can be obtained in the deviatoric plane, including a Lade shape (Lade and Duncan 1975) (see Fig. 2.3b). A circular shape can be obtained by setting $Y = 0$ and $Z = 1.0$. In the latter case the function $h(\theta)$ is a constant defined by $X = M/\sqrt{3}$, where M is the gradient of the critical state line in the triaxial $q - p'$ plane. It should be noted that the kinematic yield and bounding surfaces always have the same shape in the deviatoric plane, given by either Eq. (2.31) or (2.32), with θ equal to θ_y for the kinematic yield surface and θ_b for the bounding surface.

The plastic potential surface associated with the bounding surface is given by an equation of the form of Eq. (2.28) but with the function $h(\theta_b)$ replaced by the function $h_p(\theta_b)$. Similarly, the plastic potential surface associated with the kinematic yield surface is given by an equation of the form of Eq. (2.30) but with $h(\theta_b)$ replaced by $h_p(\theta_y)$.

The function $h_p(\theta)$ defines the shape of the plastic potential surface in the deviatoric plane, and is given by the family of surfaces proposed by van Eekelen (1980). In the case of the plastic potential, Y and Z in Eq. (2.32) are replaced by the material constants Y_p and Z_p, whereas X is replaced by X_p, which varies such that when the soil is yielding the current plastic potential surface always passes through the current stress state. Depending on the chosen parameters, the deviatoric shapes of the yield and plastic potential surfaces can be different, and hence the flow rule becomes non-associated in the deviatoric plane.

For the case where the kinematic surface moves within the bounding surface the hardening modulus is given by:

$$A = \frac{4}{\lambda^* - \kappa^*}\left\{(p' - p'_a)\left[p'_a(p' - p'_a) + \frac{1}{2}\left(\frac{\partial F_y}{\partial s} : s_a\right) + \frac{R^2 {p'_o}^2}{4}\right] + \left(\frac{B}{B_{\max}}\right)^{\psi}\left(\frac{p'_o}{2}\right)^3\right\} \tag{2.33}$$

where λ^* and κ^* are the slopes of the isotropic normal compression line and the unloading-reloading line in $\ln v - \ln p'$ space ($v = 1 + e$) respectively, B is a measure of the proximity of the kinematic yield and bounding surfaces, $B_{\max} = p'_0\,(1 - R)$ is its maximum value, and ψ is a parameter that controls the variation of the hardening modulus.

The model requires either nine parameters, when a Mohr-Coulomb hexagon is chosen for the shape of the yield surface in the deviatoric plane, or 11 parameters, when the general shape proposed by van Eekelen is adopted instead. These parameters are summarized in Table 2.1.

Under plane strain conditions, if the general shape proposed by van Eekelen is adopted for the plastic potential, then the Lode angle θ at failure can be determined from (Potts and Gens 1984):

$$\tan\theta = -\frac{3Z_p Y_p \cos 3\theta}{1 + Y_p \sin 3\theta} \tag{2.34}$$

The undrained shear strength predicted by this model can be calculated as follows (Grammatikopoulou et al. 2007):

$$S_u = \frac{p'_0}{2} h\,(\theta) \left(\frac{2p'_i}{p'_0}\right)^{\frac{\kappa^*}{\lambda^*}} \cos\theta \tag{2.35}$$

where p'_i = the initial mean effective stress, and θ = the Lode angle at failure.

Table 2.1 Model parameters and values adopted in the finite element analyses

Parameter	Definition
λ^*	Slope of isotropic normal compression line in $\ln v - \ln p'$ space
κ^*	Slope of unloading-reloading line in $\ln v - \ln p'$ space
G'	Elastic shear modulus
ϕ' or X, Y and Z	Internal friction angle (if deviatoric shape of yield surface is assumed to be a Mohr-Coulomb hexagon) or constants $X = \sin\left(\phi'_{\theta=0°}\right)$, Y and Z at critical state (if deviatoric shape of yield surface is assumed to be given by general shape of van Eekelen, 1980)
Y_p and Z_p	Constants Y_p and Z_p for deviatoric shape of plastic potential
R	Ratio of size of kinematic yield surface to that of bounding surface
ψ	Parameter in hardening function
N	Specific volume of isotropic compression line at p' = unity

2.2.2 *Example Analyses*

2.2.2.1 Class-A Prediction of Centrifuge Footing Test Results

In 1994, the former Japanese Society of Soil Mechanics and Foundation Engineering (JSSMFE) (currently the Japanese Geotechnical Society (JGS)) organized a class-A prediction (Lambe 1973) of a footing load test on a clay soil model tested in a centrifuge. The main purpose of this prediction was to assess the ability of numerical analysis when simulating foundation failure and strain localization. The model test conditions and the properties of the soil used were as follows.

The model (centrifuge) container was a rectangular rigid box with inside dimension of 0.6 m in length, 0.15 m in width and 0.4 m in height. The side walls were made of transparent glass. To reduce the side friction, silicon grease was smeared on the inside walls prior to placement of the clay sample. In order to visualize the deformation pattern during the test, target dots for photography were emplaced into the model ground under the transparent side wall in a grid pattern with a spacing of 10 mm. The model ground was made of a mixture of Kobe clay, Toyoura sand and crushed Toyoura sand powder, in the ratios 2:1:1 by dry weight. The mixture was made into a slurry with an initial water content of 70% and de-aired. It was then preconsolidated under a pressure of 39.2 kPa at 1 g conditions to a thickness of about 0.16 m. The model was then set on a centrifuge machine and consolidated under 92 g (with an average arm length of 2.0 m and rotated at 203 rpm). This procedure resulted in an over-consolidated (OC) layer at the surface (about 54 mm in thickness) with a normally consolidated (NC) layer below it. A model footing, 50 mm in width and 30 mm in thickness (the length of the footing was the same as the width of the model box, 0.15 m), was placed on the surface of the model ground at the central location. During the test, the weight of the footing was cancelled by using a counter weight, as shown in Fig. 2.4. The footing test loading was conducted using displacement control. A total of six tests with three different loading patterns were conducted, as listed in Table 2.2. During the tests, the vertical load and displacement were measured at the locations shown in Fig. 2.5 (information from Miyake et al. 1994). Excess pore water pressures within the model ground were measured at six points P1 to P6 (Fig. 2.4). The shear strain distributions in the model ground were analyzed using a digitizer from the enlarged photographs taken during the test.

After each test, the model ground was sliced into 10 mm thick layers and their water contents were measured. The results for Case-1 to 4 are given in Fig. 2.6 (data from Miyake et al. 1994). It can be seen that under the centrifuge loading, the density of the model ground increased with the depth.

The mechanical properties of the model ground were determined by laboratory oedometer, triaxial and plane strain tests, and the results were provided to all predictors to assist them to make their predictions before the model test results were released. Unfortunately, JGS has not retained these property data.

Sakajo and Chai (1994) joined this Class-A prediction activity and, together with others, made their predictions of the footing response. Only the static loading cases indicated in Table 2.2 were simulated by finite element analysis (FEA). The results of all analyses were published (in Japanese) in the journal of Tsuchi-To-Kiso, JSSMFE (Sakajo et al. 1995). Among the submitted predictions, Sakajo and

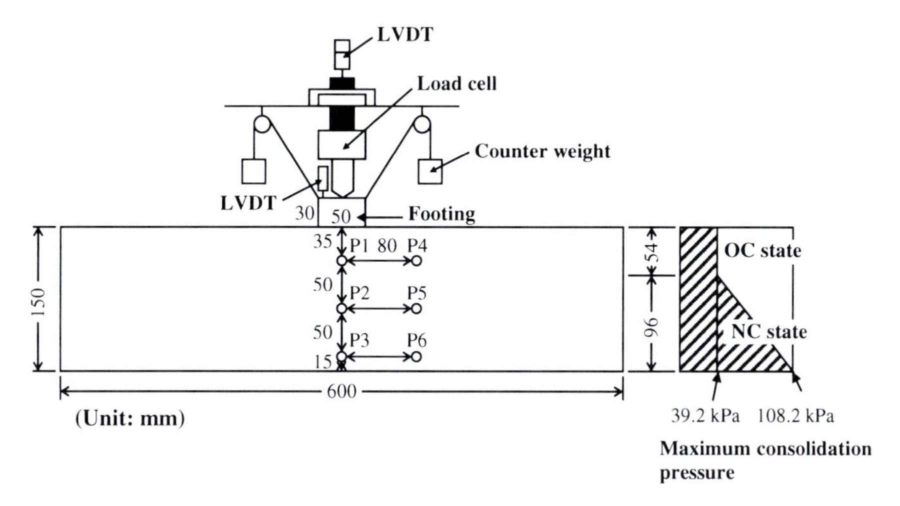

Fig. 2.4 Illustration of centrifuge model test

Table 2.2 Model test and FEA cases

Model test case	FEA case	Pattern
Case-1	FEA-1	Central loading
Case-2		
Case-3	FEA-2	Eccentric loading (loading point: 11 mm from the centerline for model test)
Case-4		
Case-5	–	Dynamic load
Case-6		

Chai's predictions compared best with the test measurements and so their modelling is now described in greater detail. Included in this description is the important initial process of checking the model predictions of the undrained shear strength and the

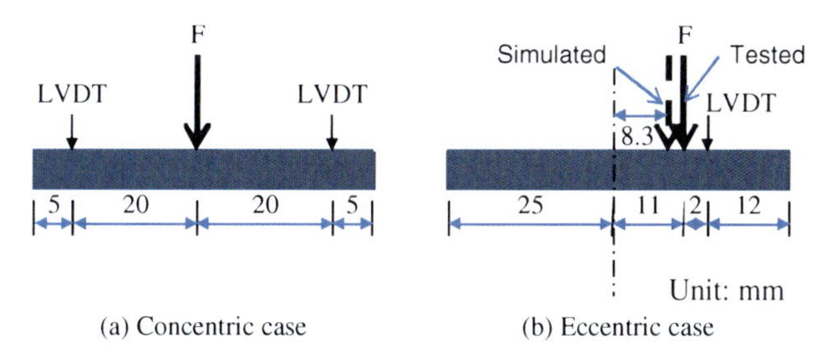

Fig. 2.5 Loading location and displacement measurement points. (**a**) Concentric case (**b**) Eccentric case

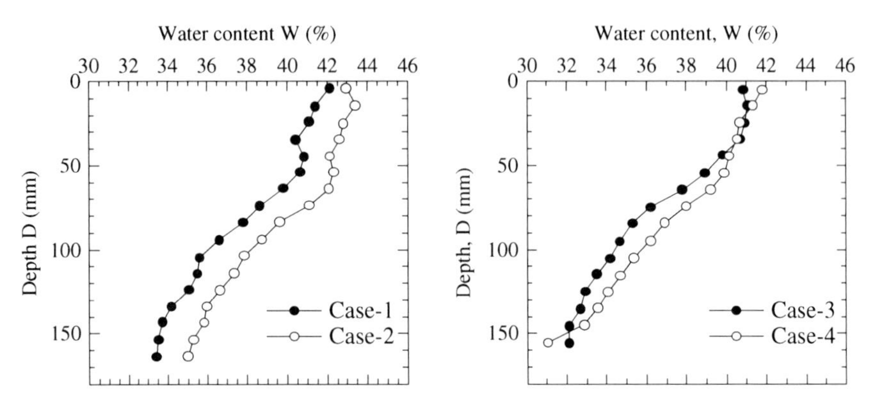

Fig. 2.6 Water content in the model ground after footing loading tests (data from Miyake et al. 1994)

soil stiffness. Also included is a discussion of the specific conditions assumed in the FEA and a comparison of the predicted and measured results.

In the FEA of Sakajo and Chai, the model ground was represented using the Sekiguchi-Ohta constitutive model. Based on the laboratory test results, the deduced and adopted model parameters are as listed in Table 2.3. These values were estimated from the laboratory test results, making use of the empirical methods proposed by Nakase et al. (1988) for estimating the model parameters from the plasticity index (PI) of the soil.

To simulate accurately this kind of model footing test, correct modelling of the shear strength and stiffness of the surface soil layer is an important first step. Adopting the Sekiguchi and Ohta model and the parameters listed in Table 2.3, as well as the initial effective stresses corresponding to the known pre-consolidation condition ($p' = 39.2$ kPa) and the subsequent centrifuge consolidation, the predicted undrained shear strength (S_u) distribution for triaxial compression is shown in Fig. 2.7 as a dashed line, i.e., zero at the surface and increasing non-linearly to about 12 kPa at 5 m depth in the prototype.

For the constitutive model adopted in this study the predicted strengths in plane strain compression and extension are respectively lower and higher than the strength mobilized under triaxial conditions (Sekiguchi and Ohta 1977). The initial value of the constrained modulus for fully drained conditions (D), for point A at a depth of

Table 2.3 Model parameters

Symbol	λ	κ	ν	$\phi'(°)$	e_0	$(K_o)_{NC}$	γ_t(kN/m^3)	Remark
Value	0.16	0.021	0.30	40.3	1.08	0.45	17.6	$PI = 30$

Note: λ = slope of consolidation line in the $e - \ln p'$ plot; κ = slope of unloading-reloading line in the e – ln p' plot; ν = Poisson's ratio; ϕ' = internal friction angle; e_0, initial void ratio; and γ_t, total unit weight

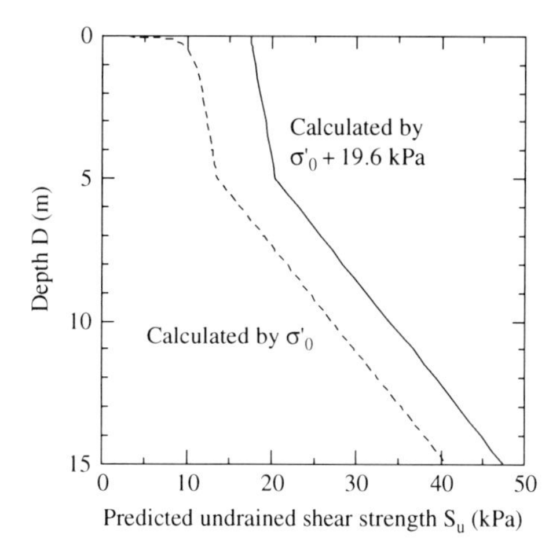

Fig. 2.7 Simulated distribution of undrained shear strength S_u

say 0.05*B* from the surface (about 0.25 m deep for the prototype), the initial effective stress is about 2 kPa assuming the groundwater level is at the ground surface, as adopted in the FEA. For these conditions an initial value of *D* of about 200 kPa can be estimated assuming the soil is in an over-consolidation state.

Although the analyses were conducted assuming undrained conditions, the stiffness of the soil skeleton actually influences the predicted excess pore water pressure in the model ground. From experience, the initial strength and stiffness predicted by the constitutive model and the assumed initial stress conditions appeared to be too low. This may be because the surface of the model ground might have experienced the effects of suction, which will normally increase its strength and stiffness.

In order to increase the initial strength and stiffness predicted for the model ground, it was decided to increase the initial effective stresses in the model ground by applying a distributed load with an intensity of 19.6 kPa over the surface of the soil in the finite element model. The past maximum consolidation pressure was adjusted accordingly to maintain the thickness of the over-consolidated layer. After this adjustment was made, the predicted distribution of S_u for triaxial compression conditions was as indicated by the solid line in Fig. 2.7. In this case the initial value of *D* becomes about 2000 kPa at point A.

FEA was conducted for the prototype condition and the model ground used in the centrifuge tests was represented by 4-noded quadrilateral elements. The footing was simulated by solid elements with a very high elastic modulus. In order to allow the possibility of capturing strain localization, a zone extending to a depth of 0.8*B* under the footing and *B*/3 away from the edge of the footing, was represented by a very fine mesh consisting of elements whose maximum dimension was 0.04*B*. Overall, the finite element mesh extended over a range defined by distances of 3*B* vertically and 5*B* horizontally for one-half of the model in the symmetric loading case, and 3*B* vertically and10*B* horizontally for the eccentric loading case.

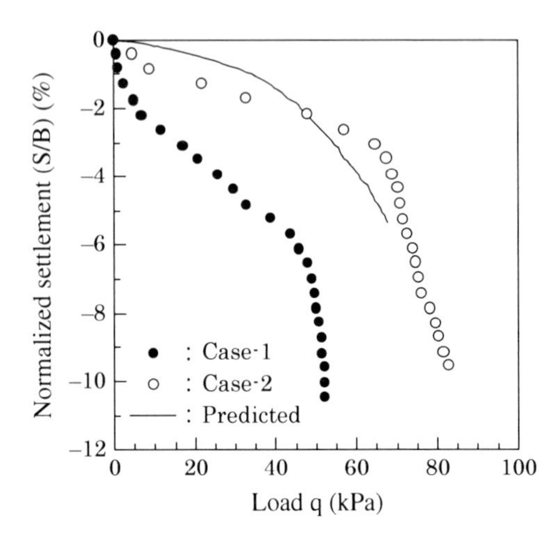

Fig. 2.8 Load-normalized settlement curves for the concentric loading case (modified from Sakajo et al. 1995)

2.2.2.2 Comparison of Results

Symmetric Loading Case

A comparison is given in Fig. 2.8 of the measured and the predicted curves of applied load versus normalized settlement for the case of the concentrically loaded footing. The settlement (S) has been normalized by the width of the footing (B). For test Case-1, obviously the initial contact condition between the footing and the model ground was poor. However, the predictions compared favourably with the measured data for Case-2. Comparisons of the excess pore water pressures at piezometer points P1 to P6 (see piezometer locations indicated in Fig. 2.4) are depicted in Fig. 2.9. Although there are some discrepancies, generally the predicted pore water pressures are quite good, especially given that they are Class-A predictions.

Distributions of shear strains in the ground (ε_s) were deduced from photographs taken during the experiments and these are compared with the predicted shear strain distribution in the model ground. Figure 2.10 compares the results of measurements deduced at $S/B = 8\%$ and predictions corresponding to $S/B = 5\%$. Unfortunately, a comparison at precisely the same normalized displacement was not possible. Considering the fact that the predictions correspond to a smaller value of S/B, the level of agreement with the measured shear strains is quite satisfying.

Eccentric Loading Case

There was a discrepancy between the announced test conditions and the actual test conditions for the eccentric loading case. The announced eccentricity of the loading point was 8.3 mm (50/6 mm) from the centerline of the footing, but in reality it was 11 mm (Miyake et al. 1994), as illustrated in Fig. 2.5. Nevertheless, the predictions, made assuming the incorrect eccentricity, are compared with the experimental measurements in the following paragraphs, in order to provide a reference.

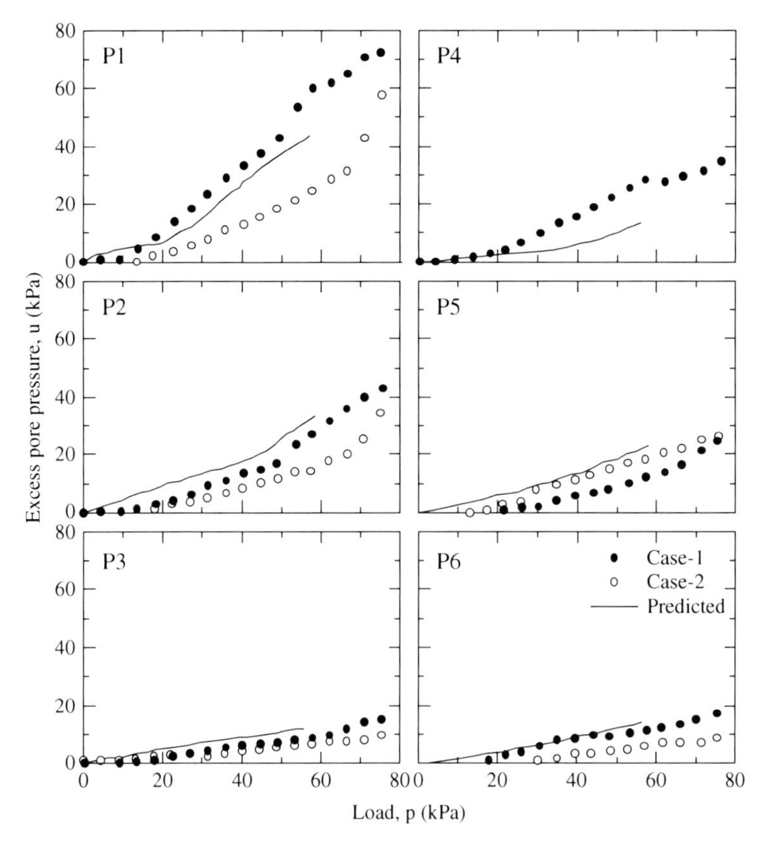

Fig. 2.9 Comparison of excess pore water pressure versus applied pressure for the concentric loading case (modified from Sakajo et al. 1995)

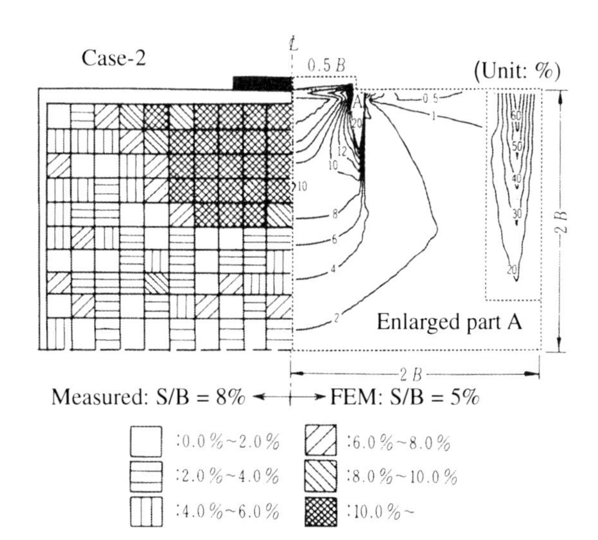

Fig. 2.10 Comparison of shear strain distributions for the concentric loading case (modified from Sakajo et al. 1995)

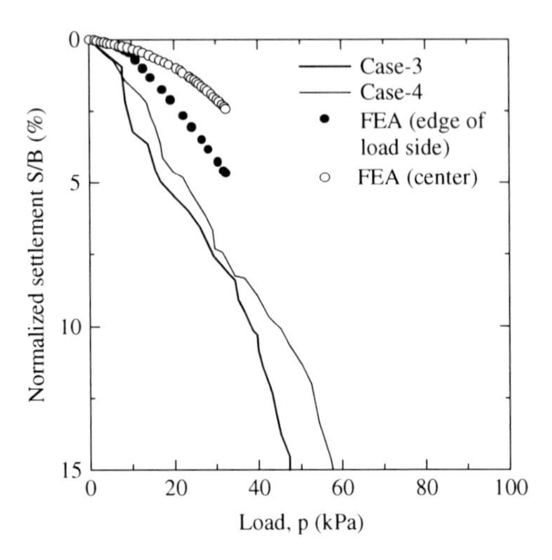

Fig. 2.11 Comparison of load-settlement curves for the eccentric loading cases (modified from Sakajo et al. 1995)

Predicted and experimental curves of the load (in the form of the average footing pressure) versus the normalized settlement are compared in Fig. 2.11. The vertical displacement of the model tests was measured at a point 2 mm outside the eccentric loading point (Fig. 2.5), but the prediction provided values at the centre and the edge (loading side) of the footing. Since the eccentricity of loading was different in the FEA and the experimental test, a definitive comment about the level of agreement between the predictions and the experimental results cannot be made, but it seems that the finite element model has over-predicted the bearing resistance for the eccentric loading case.

In Case-3 the load was applied to the left hand side of the footing, while for Case-4 it was applied to the right hand side. However, the excess pore water pressure gauges P4 to P6 were fixed on the right hand side, as shown in Fig. 2.4. Therefore, for consistency, Cases 3 and 4 are compared together in Fig. 2.12 for the piezometer points located on the centerline. For the piezometer points 80 mm away from the centerline separate comparisons are made in Fig. 2.13. The predicted values at P1 are higher than the measured data and even higher than those observed for the concentric loading case (Fig. 2.9) when compared at the same average load level. The finite element modelling predicted that a shear band develops at a location near P1 (0.7B below the surface) as shown in Fig. 2.14, and the significant shearing that occurred at this location has resulted in the prediction of large excess pore water pressures (u). The measurements show that before the load (p) reached 20 kPa, u did not increase much, but for p greater than about 20 kPa the incremental rate of u measured in the experiments is almost the same as the rate predicted by the model.

At points P2 and P3, the predicted values of u agree well with the measurements and are generally lower than the corresponding measured and predicted values for the concentric loading case (Fig. 2.9).

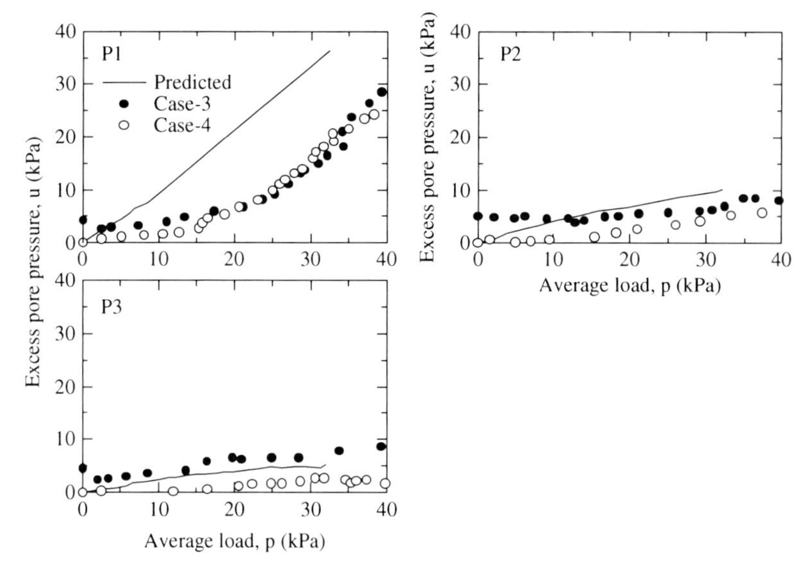

Fig. 2.12 Comparison of excess pore water pressures for the eccentric loading case (measured data from Miyake et al. 1994)

For piezometer points P4 to P6 in Case 3, i.e., where the piezometer points are on the opposite side of the footing to the loading point, the predictions generally compare well with the measurements, as shown in Fig. 2.13. Lower values were predicted for P6, but the measured initial value was not zero and if this zero shift is taken into account then the agreement between measurements and predictions is much better.

For Case 4, the predictions agree very well with the measurements for P4, but the model has generally over-predicted the values of u at P5 and P6.

Figure 2.14 shows a comparison of the shear strain distribution in the ground measured at $S/B = 11\%$ (at LVDT locations shown in Fig. 2.5) and predictions of this strain distribution at $S/B = 5\%$ (as determined at the edge of the loaded side of the footing). Again, considering the fact that the predictions correspond to a lower value of S/B and the fact that there was a significant difference in the location of the loading points, the overall comparison appears to be reasonable.

The comparisons between predictions and measurements described above indicate that in order to analyze geotechnical boundary value problems meaningfully it is important to check whether the soil constitutive model can correctly predict the strength and initial stiffness of the geological medium being considered. The example analyses also indicate that given a reasonably fine mesh, a conventional finite element analysis can predict foundation failure and the possibility of strain localization.

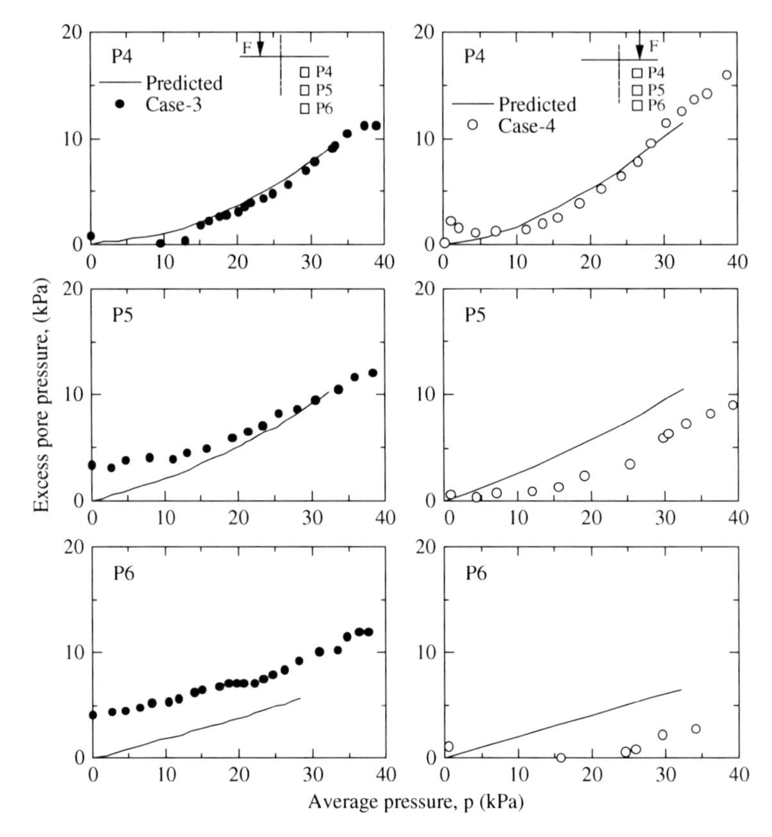

Fig. 2.13 Comparison of excess pore water pressures at P4 to P6 for Case-3 for the eccentric loading case (measured data from Miyake et al. 1994)

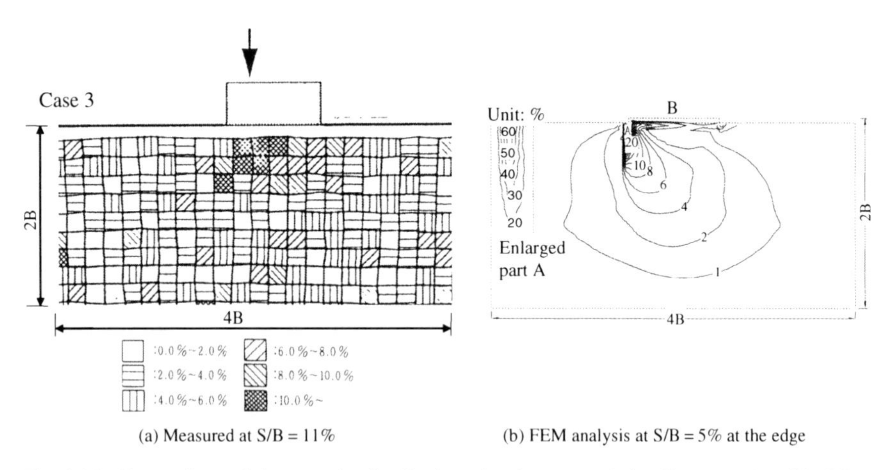

Fig. 2.14 Comparison of shear strain distributions for the eccentric loading case (modified from Sakajo et al. 1995)

2.2.2.3 Effects of Yield Surface and Plastic Potential

Grammatikopoulou et al. (2007) conducted a numerical study on the effect of the yield and plastic potential surfaces on the predicted failure height of an embankment constructed on a deposit of soft Champlain clay at Saint-Alban, Quebec, Canada. The actual test embankment failed at a height of 3.9 m and the geometry of the embankment at failure is shown in Fig. 2.15 (based on information provided by Grammatikopoulou et al. 2007). At the site, a surface crust layer with a thickness of about 2.0 m overlays a soft lightly overconsolidated ($OCR \approx 2$) silty marine clay layer with a thickness of about 7.7 m. Below the clay layer there is a clayey silt and

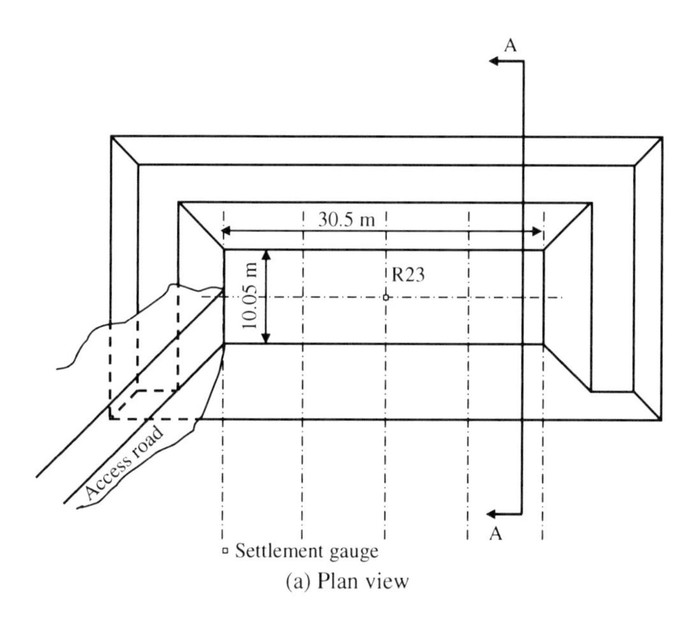

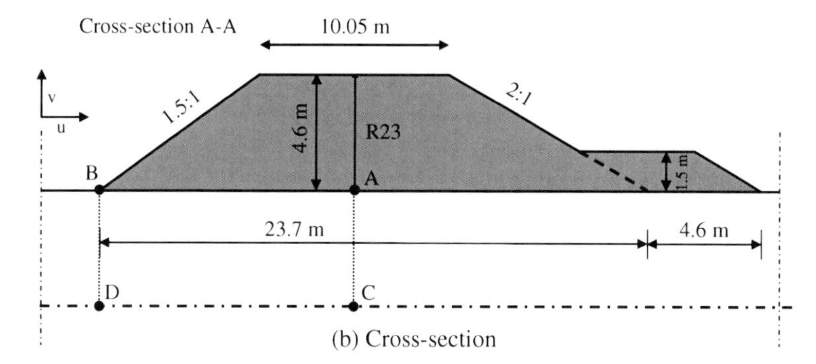

Fig. 2.15 Geometry of the embankment at failure (after Grammatikopoulou et al. 2007). (**a**) Plan view; (**b**) Cross-section

sand layer, which extends to a depth of 13.7 m. Below this layer dense sand extends to a depth of more than 24 m.

The soil model adopted in the numerical simulation was the generalized two-surface "bubble" model (Grammatikopoulou, 2004; Grammatikopoulou et al. 2006). Four different types of yield and plastic potential surfaces were adopted, as follows:

(a) Case A – the yield and plastic potential surfaces were both assumed to have a circular shape in the deviatoric plane;
(b) Case B – the yield surface had the shape of a Mohr-Coulomb hexagon and the plastic potential had a circular shape;
(c) Case C – both the yield and plastic potential surfaces had the shape proposed by Lade and Duncan (1975); and
(d) Case D – the yield surface took the Lade and Duncan shape and the plastic potential took a circular shape.

The values of the parameters for the two-surface model adopted in the analyses are given in Tables 2.4 and 2.5. Values for the parameters N, λ^* and κ^* were derived from the results of oedometer tests, whereas values of the parameters R and ψ were estimated based on experience.

The adopted profile of elastic shear modulus G' was based on independent dynamic measurements reported by Lefebvre et al. (1994), as shown in Fig. 2.16. Grammatikopoulou et al. (2007) explained that these parameters were chosen so that all four cases predicted the same values of ϕ' and $h(\theta)$ (the function that defines the shape of the kinematic yield surface) in triaxial compression (TXC), as listed in Table 2.6.

These parameter values, combined with an assumed coefficient of earth pressure at-rest, K_o, of 0.67, and the OCR profile shown in Fig. 2.17, predict the profiles of undrained shear strength in triaxial compression shown in Fig. 2.18 for all cases. Values of θ and $h(\theta)$ and equivalent values of ϕ' obtained for plane strain (PS)

Table 2.4 Model parameters and the values for all four cases

Parameter	λ^*	κ^*	G'	R	ψ	N
Value	0.215	0.005	See Fig. 2.16	0.06	1.5	7.58

Table 2.5 Model parameters defining shape of yield and plastic potential surfaces in the deviatoric plane

Case	X	Y	Z	Y_p	Z_p	ϕ'
A	0.618	0.0	1.0	0.0	1.0	–
C	0.54	0.416	0.25	0.416	0.25	–
D	0.54	0.416	0.25	0.0	1.0	–
B	–	–	–	0.0	1.0	27°

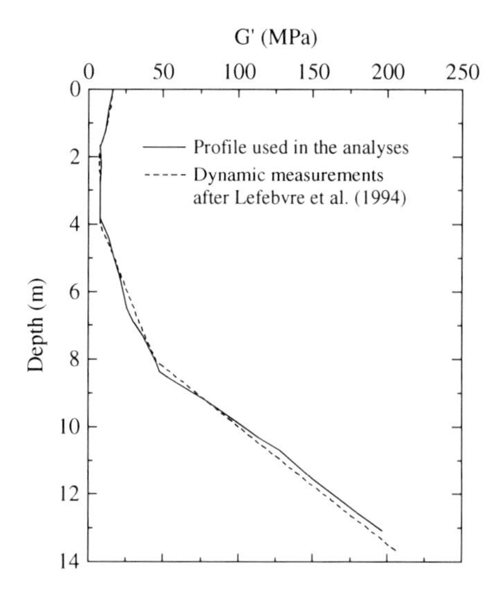

Fig. 2.16 Variation of G' with depth (data from Grammatikopoulou et al. 2007)

Table 2.6 Values of θ, $h(\theta)$ and equivalent values of ϕ' for cases analysed

	Triaxial compression			Plane-strain conditions		
Case	θ	$h(\theta)$	ϕ'	θ	$h(\theta)$	ϕ'
A	–30°	0.618	27°	0°	0.618	38.2°
B	–30°	0.618	27°	0°	0.454	27°
C	–30°	0.618	27°	–16.5°	0.594	31.3°
D	–30°	0.618	27°	0°	0.54	32.7°

conditions for each case can also be seen in Table 2.6. These result in four different predicted undrained strength profiles for plane-strain conditions. The range of field vane shear strength measurements is also included in Fig. 2.18 for reference purposes. The parameter values listed in Tables 2.4, 2.5 and 2.6 and the data points shown in Figs. 2.16, 2.17 and 2.18 are all from Grammatikopoulou et al. (2007).

Figure 2.19 shows the shape of the failure surfaces in the deviatoric plane and the values of $h(\theta)$, which are tabulated in Table 2.6.

The fill material was modelled as a linear elastic-plastic material, with a Mohr-Coulomb yield surface and a non-associated flow rule. The parameter values adopted were: Young's modulus $E = 1000$ kPa; Poisson's ratio $\nu = 0.3$; friction angle ϕ'; and angle of dilation $\psi = 22°$.

With the yield and failure surfaces adopted for the two-surface model and the parameters listed in Tables 2.4 and 2.5, the failure height of the test embankment at Saint-Alban was predicted by FEA using the program ICFEP developed at Imperial College, UK. All the analyses performed assumed small-strain deformation and plane-strain conditions and they were conducted using eight-noded

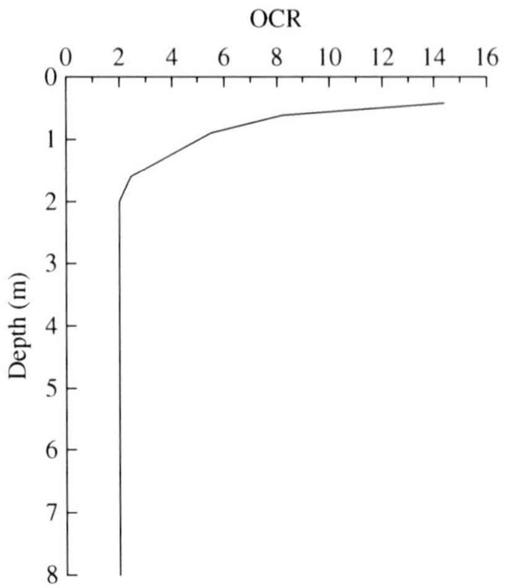

Fig. 2.17 Variation of adopted OCR with depth (data from Grammatikopoulou et al. 2007)

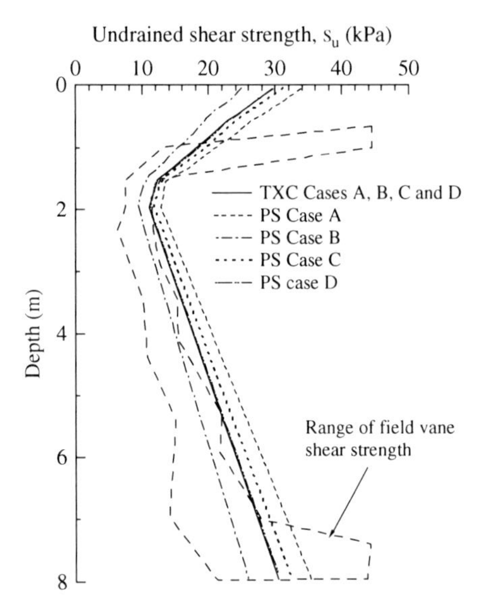

Fig. 2.18 Predicted variations of undrained shear strength with depth (modified from Grammatikopoulou et al. 2007)

isoparametric quadrilateral elements with reduced (2x2) integration. The modified Newton-Raphson scheme, with an error-controlled, sub-stepping stress point algorithm (Sloan 1987) was used as the non-linear solver.

In each finite element analysis the thickness of the simulated embankment was the same in each case up to an overall embankment height 3.3 m, with an individual lift height of 0.3 m. However, for greater embankment heights the thickness of the embankment lifts was gradually reduced to 0.1 m close to failure and each layer

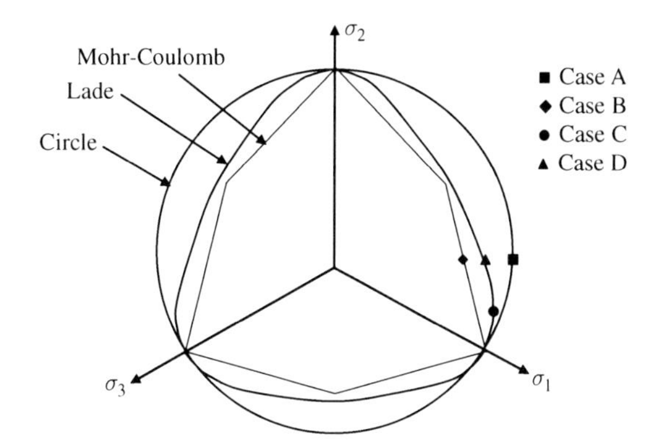

Fig. 2.19 Failure surfaces in the deviatoric plane and values of $h(\theta)$ for 4 cases (after Grammatikopoulou et al. 2007)

was constructed over four load increments. This meant that, close to failure, it was possible to have results for every 0.025 m of fill increment.

In all analyses the clay foundation was assumed to behave in an undrained manner, whereas the fill material was assumed to behave as a fully drained material. It was assumed that at the start of the analyses the kinematic yield surface was centred on the initial stress state corresponding to $K_0 = 0.67$ (Grammatikopoulou et al. 2007).

Curves corresponding to the simulated embankment height versus the settlement of the ground surface under the embankment centreline are shown in Fig. 2.20, while curves of the predicted embankment height versus the predicted horizontal displacements at the toe of the embankment are shown in Fig. 2.21. For Cases B, C and D, the simulations were continued until embankment failure, but the simulation for Case A was carried out until an embankment height of 6.5 m with a crest width of 0.95 m was reached, at which point the calculations ceased. The predicted and measured failure heights of the embankment are listed in Table 2.7. The data points plotted in Figs. 2.20 and 2.21 and the parameter values listed in Table 2.7 are

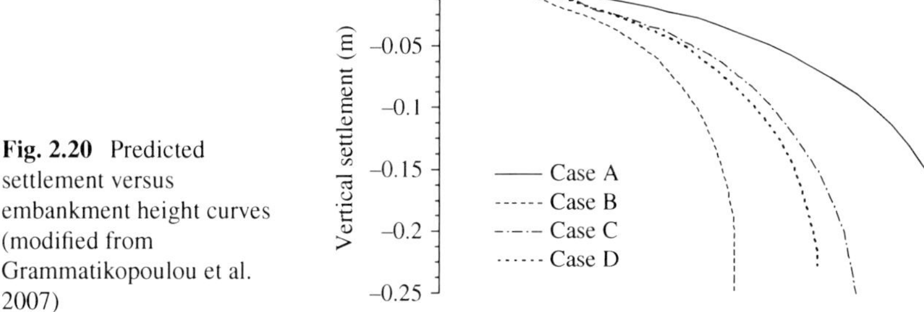

Fig. 2.20 Predicted settlement versus embankment height curves (modified from Grammatikopoulou et al. 2007)

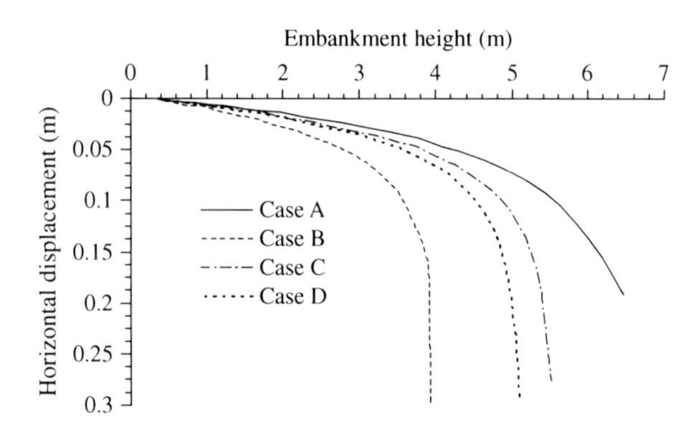

Fig. 2.21 Predicted lateral displacement versus embankment height curves (modified from Grammatikopoulou et al. 2007)

Table 2.7 Predicted embankment failure heights (data from Grammatikopoulou et al. 2007)

Case	Measured	A	B	C	D
Failure height, m	3.9	n/a	3.95	5.55	5.13

taken from Grammatikopoulou et al. (2007). It can be seen that the predicted failure heights follow the same order as the predicted undrained shear strengths for plane strain conditions, as given in Fig. 2.18. Case B provided the closest match with the field measurement of failure height.

This example of the finite element analysis of an embankment clearly demonstrates the importance being able to predict accurately the undrained shear strength profile of the ground using the adopted constitutive model as a necessary pre-cursor for simulating accurately the failure height of an embankment on a soft clay deposit.

2.3 Modelling the Embankment Construction Process

In finite element analysis, the incremental embankment load is normally applied using one of the following methods:

1. applying a distributed surface load (e.g., Asaoka et al. 1992);
2. increasing the gravitational loading applied to all or part of the embankment elements; or
3. placing embankment elements layer by layer (e.g., Britto and Gunn 1987).

If the embankment load is treated as a distributed surface load, the stiffness of the embankment and the resistance to lateral spreading arising from the stiffness of the embankment material are completely ignored. Applying the incremental load by increasing the gravitational force of the entire embankment elements is more realistic than applying a surface loading, but still the sequence by which the load is applied to the soft ground is not closely simulated. Since the behaviour of soft ground is not linear elastic, rather in most situations it behaves elasto-plastically or elasto-visco-plastically, the ground response normally depends on the sequence of load application. Furthermore, unless an appropriate loading sequence is simulated the stiffness of the embankment as well as that of the foundation soils may not be modelled appropriately or accurately, especially if a stress-dependent constitutive model is adopted for the embankment fill material or the foundation soils. Applying the incremental load by placing new layers of elements is much more realistic and is now widely used in geotechnical modelling practice to simulate embankment construction.

2.3.1 Large Deformations

Most finite element programs are formulated on the basis of infinitesimal strain theory, which ignores the effects of the changes in geometry of a structure during loading. If the deformation is small, the error involved in ignoring these changes in geometry is also small. However, in the case of an embankment on soft or very soft subsoil, large foundation deformations are normally experienced during embankment construction. For example, an embankment with a fill thickness of 3.5 m has been known to cause more than 2.0 m settlement during the construction period which corresponded to about 200 days (Chai and Miura 1999). In such cases large deformation phenomena should be considered and a commonly used method involves simply updating the nodal coordinates after each load or time increment. Although this is not a rigorous method (in the applied mechanics sense), provided the rotations of the elements are not significant and the consequent mesh distortions are not too severe, then updating the nodal coordinates can be an effective and pragmatic way of taking into account large deformations.

2.3.2 Embankment Height and Thickness

Normally, the final elevation of a constructed embankment is smaller than the combined thickness of the fill plus the original elevation of the ground surface, due largely to the deformations of the underlying foundation induced by the embankment loading. In finite element analysis the elements representing an embankment are normally pre-specified, as are the positions of the nodes that define them. If the nodal coordinates are not updated during the analysis to incorporate the predicted displacements resulting from the weight of the embankment, the applied embankment load will be exactly the same as pre-specified by the original volume of these elements multiplied by their unit weight. This situation is illustrated schematically in Fig. 2.22a.

However, if large deformations are considered in the analysis, by updating the nodal coordinates after each increment of applied load, then normally the nodal coordinates of the as yet unconstructed elements will not be updated at the end

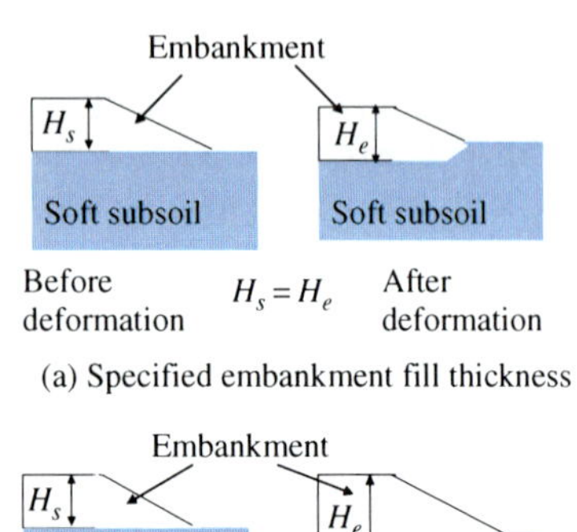

Fig. 2.22 Specified embankment height and embankment fill thickness

of each load increment. In such cases the magnitude of load finally applied to the underlying soil will be larger than might have been anticipated for the case of small deformations, as illustrated schematically in Fig. 2.22b.

To ensure the serviceability of an embankment, in many cases in engineering practice the embankment elevation (at least the final height at the end of construction) is specified and the scenario illustrated in Fig. 2.22b is applicable. However, there are cases where the embankment fill thickness is specified (Fig. 2.22a). In such cases, using the conventional technique of updating the nodal coordinates will lead to the application of more embankment load than desired or intended. If it is desired to include large deformation phenomena in the analysis for cases where the fill thickness is specified, then at least two options are available, as follows:

(a) Include all embankment elements in the finite element calculations from the beginning of the loading stage, but switch on the gravity force loading of the embankment elements layer by layer. Those elements to which the gravity force is not applied are assigned a nominally small value of the elastic modulus, ensuring that the un-constructed elements will 'follow' the deformation of the underlying elements that have been constructed thus ensuring that the thickness of the constructed embankment fill is maintained. However, since the settlement of an embankment is not usually uniform, the upper surface of the embankment may become concave, which is normally inconsistent with conditions desired and achieved at the end of construction.
(b) Adopt the method of Chai and Bergado (1993a), as illustrated in Fig. 2.23, in which the coordinates of the nodes of the embankment elements above the current construction level are corrected at the end of each incremental analysis. In this method the following assumptions are adopted: (i) for the region of the finite element mesh above current construction level, the originally vertical

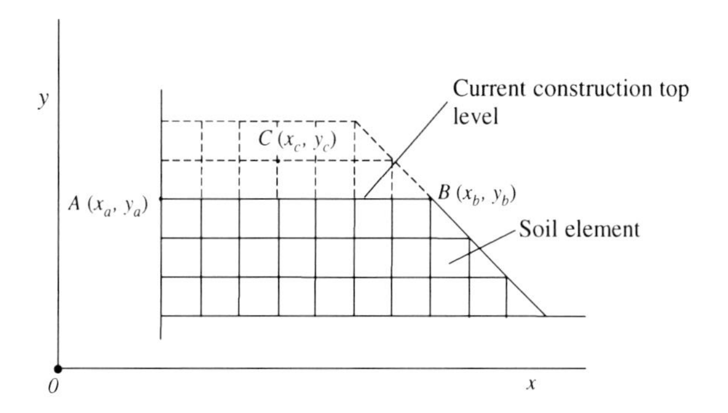

Fig. 2.23 Configuration of embankment construction (after Chai and Bergado 1993b)

element boundaries are kept vertical, and the originally horizontal element boundaries remain straight, and (ii) the incremental displacements of the nodes above the current constructed top surface are linearly interpolated from the displacements of the two end nodes (left and right) of the current top surface, according to their horizontal coordinates.

As shown in Fig. 2.23, Node-C (x_c, y_c) is above current construction top surface. The incremental displacement of Node-C (Δx_c, Δy_c) is calculated by using the incremental displacements of Node-A (Δx_a, Δy_a), Node-B (Δx_b, Δy_b), and their x-coordinates, x_a and x_b as follows:

$$\Delta x_c = \Delta x_a + \frac{x_c - x_a}{x_b - x_a}(\Delta x_b - \Delta x_a) \tag{2.36}$$

$$\Delta y_c = \Delta y_a + \frac{x_c - x_a}{x_b - x_a}(\Delta y_b - \Delta y_a) \tag{2.37}$$

2.3.3 Significance of the Method of Applying Embankment Load

One of the Malaysian trial embankments (Scheme 6/8) (Malaysian Highway Authority, MHA 1989a) was analyzed using different methods of applying the embankment load. The embankment was constructed with a base width of 88 m and length of 50 m, initially to a fill thickness of 3.9 m. Then a 15 m berm was left on both sides and the central embankment was constructed to a final fill thickness of 8.5 m. Two layers of Tensar SR110 geogrid were laid at the base of the embankment with 0.15 m vertical spacing between the layers, and prefabricated vertical drains (PVDs) were installed in the underlying soft clay to a depth of 20 m in a square pattern with 2.0 m spacing (MHA 1989a).

The soil profile at the test site consists of a topmost 2.0 m of weathered crust that is underlain by about 5 m of very soft silty clay. Below this layer lies a 10 m

Table 2.8 Index properties of Muar clay deposit

Depth (m)	Soil	Water content W (%)	Liquid limit W_l (%)	Plasticity index I_p (%)	$\frac{C_c}{1+e_0}$	Clay content (%)	Silt content (%)
0–2	Crust	60–80	–	–	0.30	62	35
2–7	Very soft silty clay	80–110	675–90	40–50	0.50	45	52
7–12	Soft silty clay	70–100	75–90	40–50	0.30	50	47
12–18		55–65	58–65	30–40	0.24	50	47
18 –22	Silty sand	–	–	–	0.10	20	36

thick layer of soft clay, which in turn is underlain by about a 1.0 m thick peat layer with high water content. A thick deposit of medium dense to dense clayey silty sand underlies the peat layer. The groundwater level was 1.0 m below the ground surface. The index properties of the Muar clay deposit at the test site are listed in Table 2.8 (Asian Institute of Technology (AIT) 1988, 1989). The meshes and the boundary conditions used in the finite element analysis are shown in Fig. 2.24. The construction history is given in Fig. 2.25.

The analyses were carried out using the CRISP-AIT finite element program (Chai 1992), which was developed based on the original CRISP program (Britto and Gunn 1987). The constitutive models involved were the Modified Cam Clay (MCC) model (Roscoe and Burland 1968) representing the soft foundation soil and a hyperbolic, nonlinear elastic model (Duncan et al. 1980) representing the embankment fill. Bar and interface elements were used to represent the reinforcement and the soil-reinforcement interface, respectively.

The MCC model parameters for the foundation soil are listed in Table 2.9. Values of all parameters (except the hydraulic conductivity) were directly obtained from

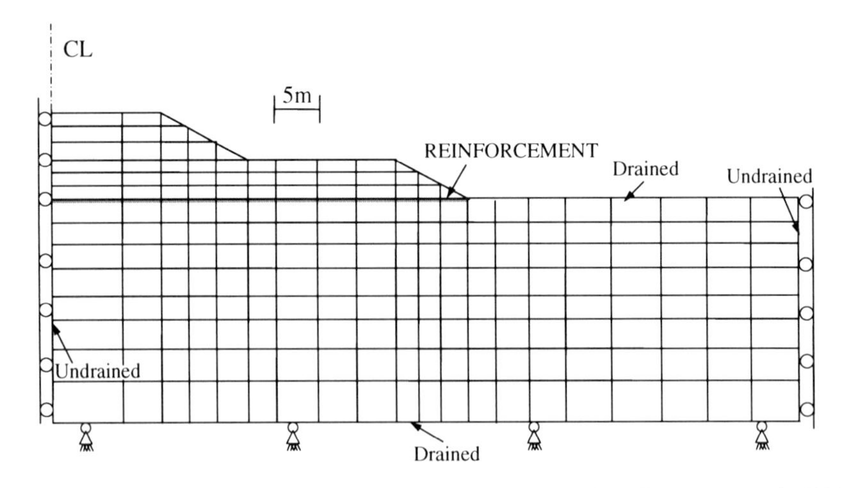

Fig. 2.24 Finite element mesh and boundary conditions (modified from Chai and Bergado 1993a)

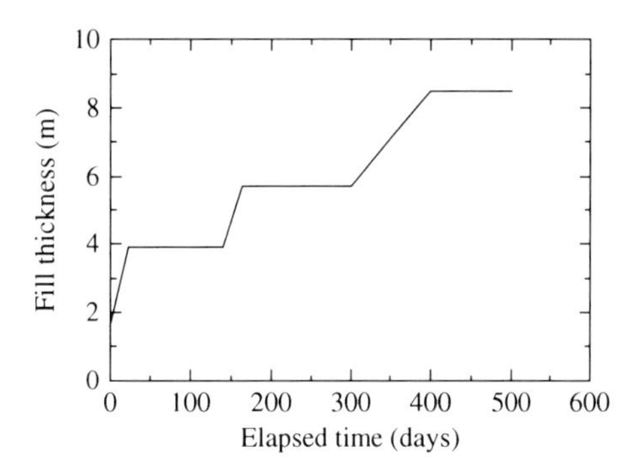

Fig. 2.25 Construction history (after Chai and Bergado 1993a)

Table 2.9 Modified Cam Clay model parameters for Muar clay

Parameter	Soil layers Crust 0–2 m	Very soft clay 2–7 m	Soft clay-1 7–12 m	Soft clay-2 12–18 m	Sand 18–22 m
κ	0.06	0.10	0.06	0.04	0.03
λ	0.35	0.61	0.28	0.22	0.10
M	1.2	1.07	1.07	1.07	1.2
Γ	4.16	5.50	3.74	3.45	2.16
ν	0.20	0.25	0.25	0.25	0.2
γ_t(kN/m^3)	15.5	14.5	15.0	15.5	17.0
k_h(10^{-8} m/s)	2.78	1.40	1.04	0.70	14, 000
k_v(10^{-8} m/s)	1.39	0.70	0.52	0.35	7, 000

laboratory test results (AIT 1989). The fill materials consisted of decomposed granite (sand about 50% and clay about 38%). Values of the parameters of the hyperbolic soil model used to represent the fill material are given in Table 2.10. These parameters were used to calculate the initial tangent modulus (E_i) and bulk modulus (E_b) from the following equations (Duncan et al. 1980):

$$E_i = k \cdot p_a \left(\frac{\sigma'_3}{p_a} \right)^n \tag{2.38}$$

$$E_b = k_b \cdot p_a \left(\frac{\sigma'_3}{p_a} \right)^m \tag{2.39}$$

where σ'_3 = the effective confining stress, and p_a = atmospheric pressure.

For the values of hydraulic conductivity (k), based on existing information (Poulos et al. 1989; Magnan 1989) the initial vertical values were chosen to be twice the laboratory test values (AIT 1989), and the horizontal values were twice the corresponding vertical values. During the consolidation process, values of k were varied according to Taylor's (1948) equation:

$$k = k_0 \cdot 10^{-(e_0 - e)/C_k} \tag{2.40}$$

Table 2.10 Hyperbolic model parameters for sand mat and embankment fill

Parameter	Symbol	Fill material	Sand mat
Cohesion (kPa)	c	19	0
Friction angle (°)	ϕ	26	38
Modulus number	k	320	460
Modulus exponent	n	0.29	0.50
Failure ratio	R_f	0.85	0.85
Bulk modulus number	k_b	270	392
Bulk modulus exponent	m	0.29	0.50
Unit weight (kN/m^3)	γ_t	20.5	20.5

where k_0 = initial hydraulic conductivity, e_0 = initial void ratio, k = current hydraulic conductivity, e = current void ratio, and C_k = a constant given by $C_k = (0.4 \sim 0.5)e_0$, following Tavenas et al. (1986).

The assumed properties of the PVDs (without a filter) are given in Table 2.11. The PVDs were considered as vertical seams that increased the mass hydraulic conductivity of the foundation soil in the vertical direction. The back-fitted value of the hydraulic conductivity was twice the value assumed for the zone without PVDs (Chai and Bergado 1993b). Since the PVDs were installed without a filter, only the holes on the core (which occupy about 8% of the total surface area) acted as drainage paths for pore water into the drains, and so the drainage effect was very limited. For this kind of PVD, the wall of the PVD acts effectively as a part of the "smear zone". Chai et al. (2001) derived an equation for calculating the equivalent vertical hydraulic conductivity in the PVD-improved zone, which can be used to evaluate this "smear" effect. This matter is also considered in more detail in Chapter 3. For the case considered the following values were adopted: the drainage length of the PVDs, $l = 10$ m (two-way drainage deposit), the diameter of the PVDs, $d_w = 48.5$ mm, and the equivalent diameter of a PVD-improved area, $D_e = 2.26$ m. Assuming the discharge capacity of the PVD, $q_w = 50$ m^3/year, and the diameter of smear zone of 0.2 m (actual smear zone and the wall of the PVD), then in order to obtain an equivalent hydraulic conductivity (k_{ev}) equal to twice the original value of the subsoil in the vertical direction, a hydraulic conductivity (k_s) of the smear zone of about 1.5% of the horizontal value (k_h) of the natural subsoil is obtained, e.g., $k_h/k_s = 65$.

Two options were used to apply the embankment load:

Table 2.11 Assumed properties of PVDs

Width (mm)	Thickness (mm)	Discharge cross section (mm^2)	Number of channels	Hole diameter (mm)	Number of holes in a linear meter
95	2	146	24	0.2	24,500

(a) applying a percentage of the self-weight of all the embankment elements; and
(b) the method proposed by Chai and Bergado (1993a), as described previously.

For both methods the adopted loading rate at the embankment centerline closely simulated the actual loading rate. Figure 2.26 (after Chai and Bergado 1993a) shows a comparison of the predicted settlement profiles together with the field measurements of settlement. It can be seen that applying a percentage of the self-weight of the whole embankment yielded a larger predicted settlement under the central point of the embankment and a smaller settlement under the toe.

Figure 2.27 (after Chai and Bergado 1993a) shows a comparison of the predicted maximum lateral displacements at the inclinometer location (see insert to Fig. 2.27) and at about 5 m below the ground surface. It can be seen that applying a percentage of the embankment self-weight as an incremental load results in much larger prediction of lateral displacement, especially at the beginning of construction. This phenomenon coincides with the settlement pattern. The soft foundation soil behaves elastoplastically, in which case the deformation pattern of the soft ground not only depends on the magnitude of the final load, but also on the sequence of applying that load. Although the loading rate at the embankment centreline was assumed to be the same for both methods, when applying a percentage of the self-weight of the whole embankment, the simulated loading rate of the soil under the embankment is slower than the loading rate that actually occurs in practice. In effect, this differential loading rate implies less confinement to the central part of the embankment initially, resulting in larger lateral displacements (Fig. 2.27). Furthermore, as emphasized previously, applying a percentage of the self-weight of the embankment implies that the stiffness of the entire embankment elements exists at the beginning of the analysis, which is not the case in reality.

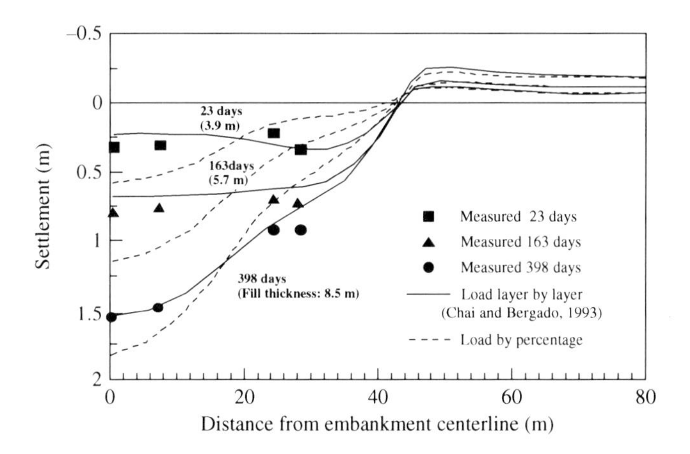

Fig. 2.26 Comparison of surface settlement profiles (after Chai and Bergado 1993a)

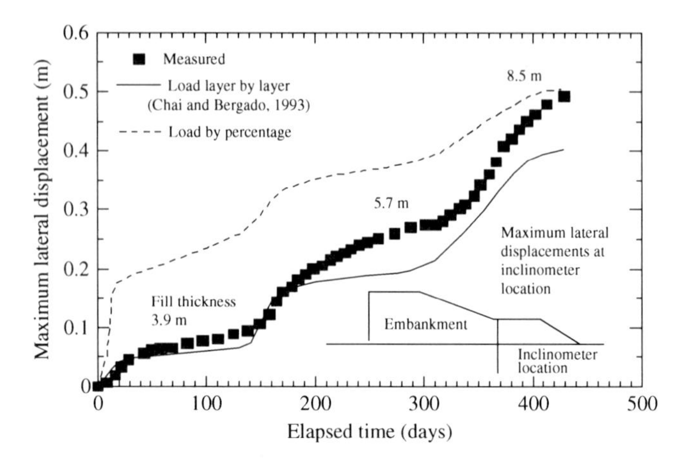

Fig. 2.27 Comparison of maximum lateral displacements (after Chai and Bergado 1993a)

2.4 Effect of Large Deformations on Embankment Stability

2.4.1 General Discussion

The direct effects of subsoil deformation on embankment stability include:

1. the possibility of reducing the surcharge load; and
2. partially replacing the (weaker) soft soil by the (stronger) fill material.

In stability analysis, the net driving moment is derived from the weight of embankment fill above the original ground surface. Therefore, the net driving moment derived from the embankment fill will be reduced by the settlement caused by a fixed thickness of embankment fill. As a first estimate, it can be assumed that the percentage reduction in the net driving moment is the same as the percentage reduction in the embankment height. From the conventional definition of the factor of safety (FS) for a slope stability problem, it follows that approximately the same percentage increase in FS can be expected. In addition, there is an indirect effect of the subsoil deformation. It will tend to shorten the drainage path length in the subsoil which should promote consolidation and hence increase the rate of increase in undrained shear strength of the subsoil. For example, for a deposit with a drainage path length of 10 m, 1.0 m settlement of the surface of the layer can effectively increase the non-dimensional time factor by about 20%, provided the coefficient of consolidation of the foundation soil does not change.

2.4.2 Stability Analysis of an Embankment Constructed to Failure

In order to illustrate quantitatively the influence of construction settlement on the FS of an embankment constructed on soft subsoil, the test embankment built to failure

(MHA 1989b) in Malaysia was analyzed by Chai et al. (1994) using both a finite element approach and the simplified Bishop method. The simplified Bishop slip circle analysis provided numerical predictions of the FS for various construction scenarios.

2.4.2.1 Brief Description of the Test Embankment

The built-to-failure embankment was constructed directly on the natural subsoil. The fill material was decomposed granite, which was compacted in 0.2 m layers with a nominal construction rate of 0.4 m/week until failure occurred. The embankment was constructed with base dimensions of 55 m wide and 90 m long and initially to a fill thickness of 2.5 m. A 15 m wide berm was then left on three sides and the remaining embankment was constructed to failure at an overall fill thickness of 5.4 m (Brand 1991). The index properties of the underlying soil deposit are given in Table 2.8. The groundwater level was about 0.8 m below the ground surface.

2.4.2.2 Finite Element Analysis

The finite element analysis of the test embankment was conducted assuming plane strain conditions and the mesh used and the key field instrumentation points are shown in Fig. 2.28 (after Chai et al. 1994). The construction schedule is shown in Fig. 2.29. Eight-node rectangular and six-node triangular elements were included in the mesh and the program used to conduct the analysis was CRISP-AIT (Chai 1992).

The behaviour of the foundation soil was simulated by the MCC model (Roscoe and Burland 1968) and the compacted fill material by the hyperbolic non-linear elastic model (Duncan et al. 1980). The model parameters are presented in Tables 2.9 and 2.10. The initial stresses assumed in the foundation soil are listed in Table 2.12.

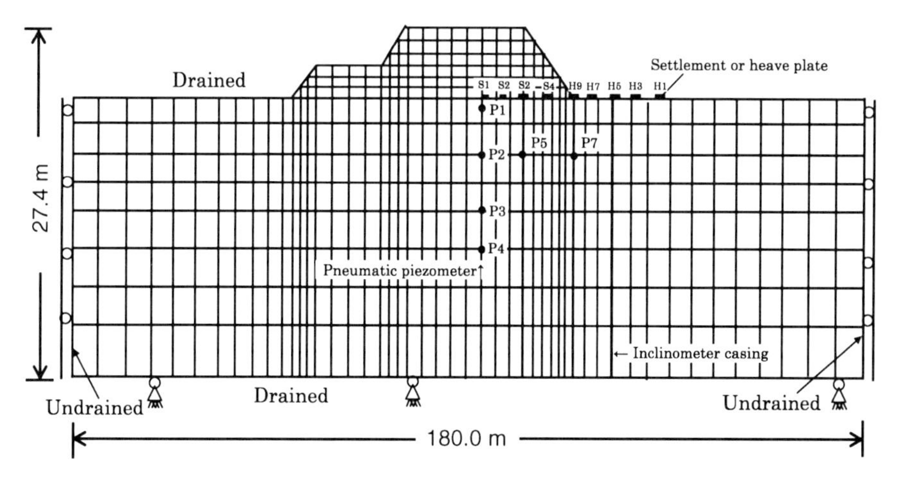

Fig. 2.28 Finite element mesh and key instrumentation points of the Malaysian built-to-failure test embankment

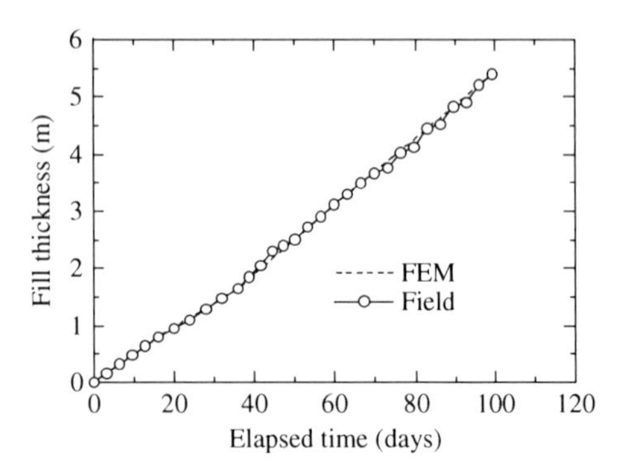

Fig. 2.29 Construction history of the Malaysian test embankment

Table 2.12 Initial stress conditions of Muar clay deposit

Depth (m)	Horizontal stress σ'_{h0} (kPa)	Vertical stress σ'_{v0} (kPa)	Water pressure u (kPa)	Size of yield locus p'_0 (kPa)
0.0	0.0	0.0	0.0	92.7
0.8	12.4	12.4	0.0	92.7
2.0	15.2	19.0	12.0	40.0
3.0	16.2	23.5	22.0	40.0
6.0	22.6	37.0	52.0	47.6
11.0	37.8	62.0	102.0	73.4
18.0	61.0	100.0	172.0	110.0
22.0	78.2	128.0	212.0	130.0

Comparisons of the FEM predictions with the observed data are presented in Figs. 2.30, 2.31, 2.32, and 2.33 (modified from Chai et al. 1994) for the excess pore pressures and displacements. All measured data were from MHA (1989b). It can be seen that the FEM analysis simulated the embankment behaviour reasonably well. Because of this good agreement the deformation and stress conditions predicted by the FEM analysis were then used in a stability analysis of the embankment in a separate investigation of the effects of construction deformations on the FS.

2.4.2.3 Effect of Construction Deformation on the Factor of Safety

In the stability analysis of the Malaysian test embankment conducted using the simplified Bishop method representative values of the undrained shear strength (S_u) of the subsoil were determined using three different methods, viz.: (1) from the results of field vane shear tests after applying Bjerrum's (1972) correction factor; (2) using Ladd's (1991) empirical equation; and (3) as predicted by the MCC model. Ladd's equation is as follows:

$$S_u = S \cdot \sigma'_v \cdot (OCR)^m \tag{2.41}$$

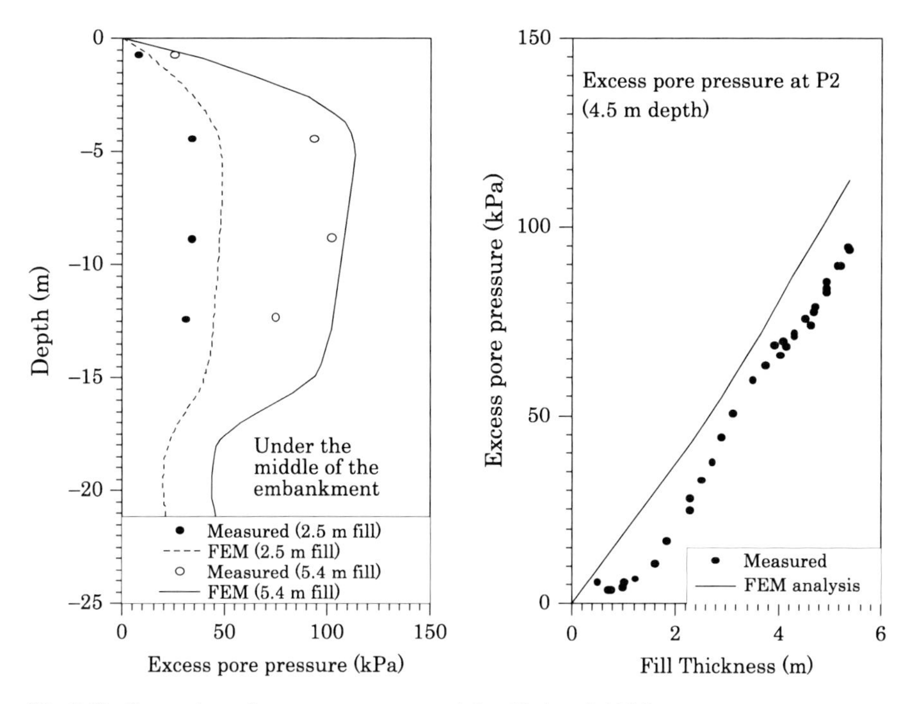

Fig. 2.30 Comparison of excess pore pressures (after Chai et al. 1994)

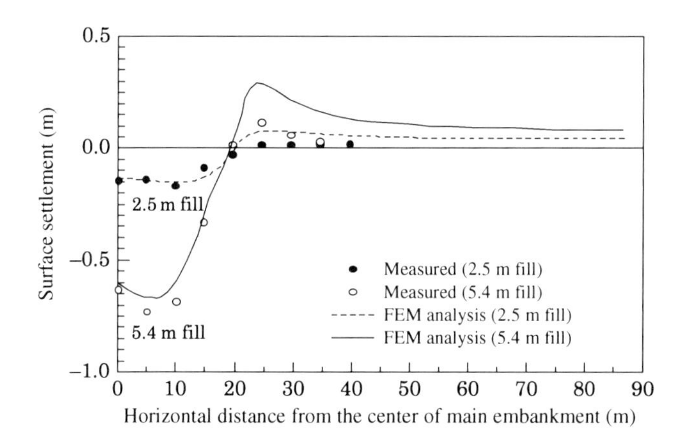

Fig. 2.31 Comparison of surface settlement profiles (modified from Chai et al. 1994)

where σ'_v = effective vertical stress, OCR = over consolidation ratio, and S and m are constants. For the soils investigated by Ladd (1991), the value of S is from 0.162 to 0.25, and m is 0.75–1.0. For the MCC model S_u was calculated from Eq. (2.10).

The calculated values of S_u are summarized in Table 2.13. The values predicted by the MCC model are the average values for the soil elements from the middle

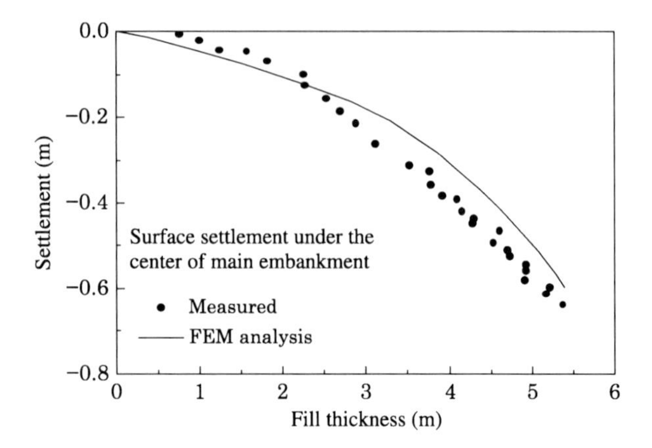

Fig. 2.32 Comparison of surface settlement under the centre of the main embankment (modified from Chai et al. 1994)

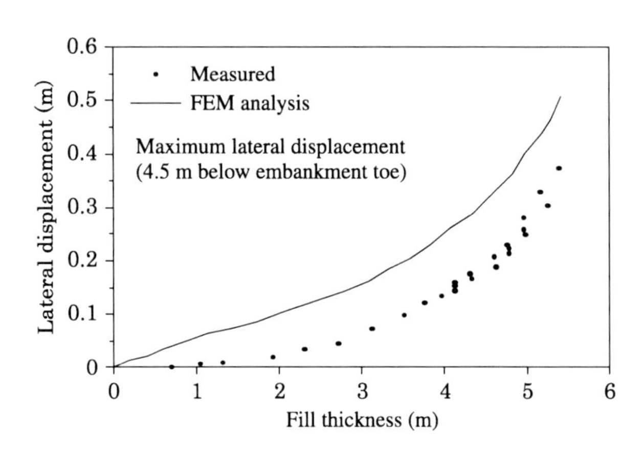

Fig. 2.33 Comparison of maximum lateral displacement (modified from Chai et al. 1994)

Table 2.13 Undrained shear strength of Muar clay deposit (after Chai et al. 1994)

	Initial values (kPa)			Modified Cam clay model predicted at Different fill thickness (kPa)			
Depth (m)	Vane shear	Empirical[a]	Cam clay	Fill 2.5 m	Fill 3.5 m	Fill 4.5 m	Fill 5.4 m
0–2.0	20.0	14.5	20.6	22.9	24.4	25.2	26.6
2.0–4.5	9.5	8.7	10.4	11.0	11.5	12.1	12.8
4.5–7.0	13.5	10.6	12.6	12.9	13.3	13.7	14.0
7.0–9.5	16.5	13.2	15.9	16.2	16.6	17.1	17.5
9.5–12.0	20.0	16.5	19.8	20.1	20.5	21.1	21.5

[a] $S_u = 0.22 \cdot \sigma_v^{'} (OCR)^{0.8}$

of the main embankment (excluding the berm) to 20 m away from the embankment toe. The values of field vane shear strength of the subsoil and the total stress strength parameters of the fill material are from Brand and Premchitt (1989). The cohesion, c, and the friction angle, ϕ, of the fill material were determined from the results of unconsolidated and undrained triaxial tests providing $c = 19$ kPa, and $\phi = 26°$.

As shown in Figs. 2.31 and 2.32, the measured maximum construction settlement was about 0.7 m which corresponds to about 13% of the total fill thickness. The effects of these construction-induced deformations are shown in Fig. 2.34 for the case where the values of S_u were predicted by the MCC model. It can be seen in this figure that foundation settlement increases the value of FS generally by about 0.1. Ignoring the foundation settlement in a stability analysis is generally conservative, i.e., on the safe side. However, in order to understand the mechanism of embankment failure, and for the purpose of back-analyzing accurately the mobilized strength of the soft soil, this effect should probably be considered. For this embankment, the average undrained shear strength of the soft soil, back-calculated assuming the full fill thickness as a surcharge, is about 15% higher than that calculated assuming the surcharge corresponds to the actual embankment height at failure.

Using the various values of S_u listed in Table 2.13 and adopting the full embankment height to estimate the applied surcharge load, the corresponding values of the FS were calculated and these are compared in Fig. 2.35. The results plotted in this figure reveal that the vane shear strength and the initial strength predicted by the MCC model yield almost the same values of FS, while Ladd's (1991) empirical equation provided lower estimates of S_u and hence lower values of FS. The effect of partial drainage resulting in strength increases during construction caused an increased in the FS by about 0.1.

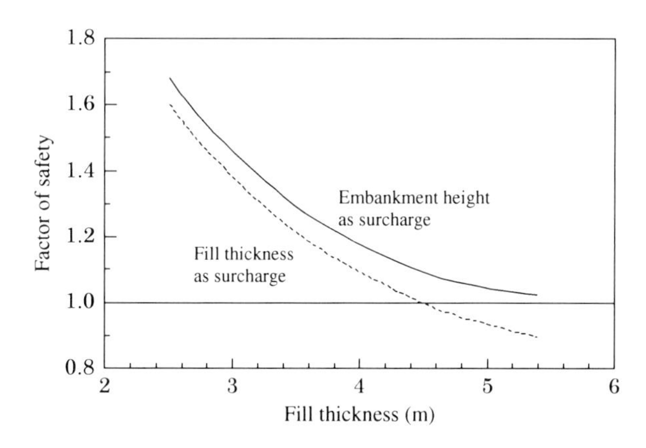

Fig. 2.34 Effect of construction settlement on calculated values of the FS

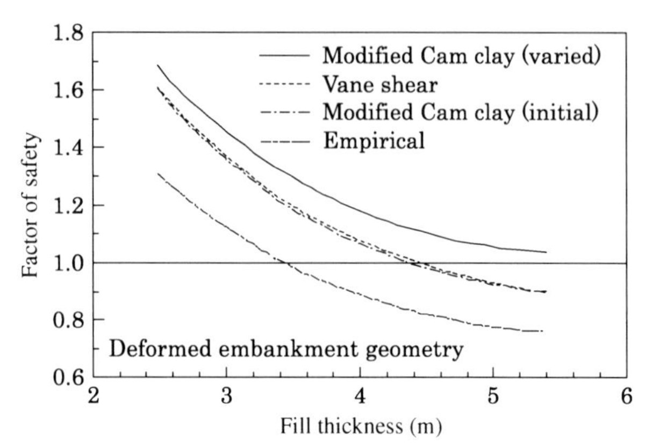

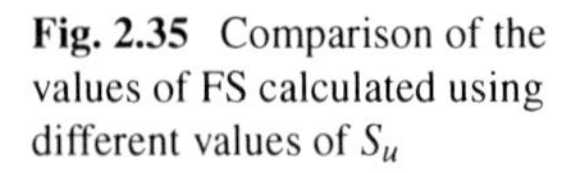
Fig. 2.35 Comparison of the values of FS calculated using different values of S_u

2.5 Buoyancy Effects Due to Large Deformation

2.5.1 General Discussion

In some cases the settlement of the ground surface induced by the weight of the embankment fill may be sufficient to bring the fill material and any subsoil located originally above the groundwater level to a position below the groundwater level. In such cases the effective load on the subsoil may be reduced due to buoyancy effects. Indeed, the effects of subsoil settlement and buoyancy interact strongly. Settlement causes buoyancy of the fill material and buoyancy reduces the further increases in settlement. If the settlement is large, the buoyancy effect may need to be considered in the calculations used to estimate that settlement.

The buoyancy effect described above occurs during both the construction period and afterward as consolidation of the underlying soils proceeds. During construction, the buoyancy force will balance a part of the newly applied embankment load, effectively reducing the load increment applied to the underlying soil. During the consolidation process, excess pore pressure dissipation will increase the effective stress and cause settlement. Consequently this settlement will induce further buoyancy effects, which in turn will compensate for some of the effective stress increase. Therefore, the net effect is that the effective stress increase will be less than would correspond exclusively to the dissipation of the initial excess pore pressure.

The buoyancy effect can be considered in a conventional one-dimensional (1D) settlement analysis by adjusting the final applied load assumed in the settlement calculation. However, under embankment loading, the subsoil not only deforms vertically but also laterally. Furthermore, the behaviour of soil is generally stress path dependent, and so considering only the final value of applied load may not give a correct settlement prediction. In more sophisticated numerical analyses the buoyancy effect should be considered systematically, i.e., by considering the interaction between the applied loading, the excess pore pressure and the buoyancy force. The basic idea is that at the end of each load increment the following steps should be undertaken:

1. consider the total water pressure change due to the incremental settlement and the change in unit weight due to consolidation (i.e., due to the voids ratio change);
2. check equilibrium of the whole system and form the load vector corresponding to the buoyancy force; and
3. iterate until the residual force is less than a given tolerance (Chai et al. 1995a).

During this process, the effect of buoyancy on effective stress, excess pore water pressure, as well as the soil deformations, will be taken into account.

2.5.2 Buoyancy Effects for a Test Embankment at Saga, Japan

In order to demonstrate the effect of buoyancy on embankment settlement, a test embankment on the PVD-improved Ariake clay deposit at Saga Airport in Japan, reported previously by Bergado et al. (1996), was analyzed by the FEM. Saga Airport is located 13 km south of Saga city on reclaimed land close to the Ariake Sea. The deposit mainly consists of soft and highly compressible Ariake clay. The test embankment on the PVD-improved subsoil had a fill thickness of 3.5 m, base dimensions of 71 m by 71 m, and top dimensions of 25 m by 25 m in plan view. The rate of fill application was about 0.03 m/day. The PVDs were installed in a square pattern with a spacing of 1.5 m to around 25 m deep over an area of 45 m by 45 m. Figure 2.36 (after Chai and Miura 1999) shows the geometry of the embankment, the main instrumentation points, and the pattern of PVD installation.

At the test site the soft soil layer is about 25 m deep, consisting of 3 clay layers and 2 sand layers. The top weathered crust (B) is about 1.0 m thick and is underlain

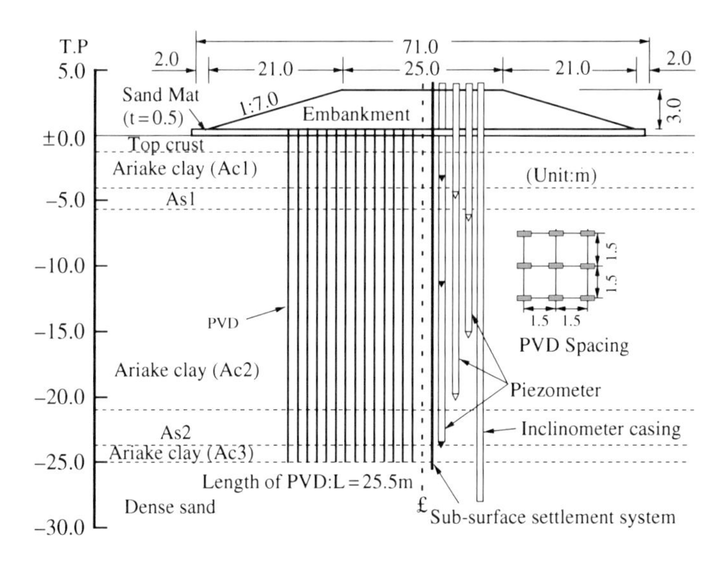

Fig. 2.36 Cross section of test embankment (after Chai and Miura 1999)

by the first soft clay layer, Ac1, with a thickness of about 3.0 m. A sand layer, As1, about 1.5 m thick, underlies Ac1. The main clay layer, Ac2, below As1, is very soft with a thickness of about 15.5 m. Under Ac2 is a sand layer, As2, which has a thickness of about 2.7 m. The third clay layer, Ac3, is soft to medium stiff and has a thickness of about 1.3 m and is underlain by a thick and dense sand layer (DS).

In the finite element analysis plane strain conditions were assumed and the region modelled was 30 m deep and extended horizontally 120 m from the embankment centreline. The finite element mesh and the displacement and drainage boundary conditions are shown in Fig. 2.37. The embankment construction history is shown in Fig. 2.38. For the zone treated with PVDs, one-dimensional drainage elements were adopted to coincide with every other vertical line (Chai et al. 1995b). The mechanical behaviour of the clay layers was represented by the MCC model (Roscoe and

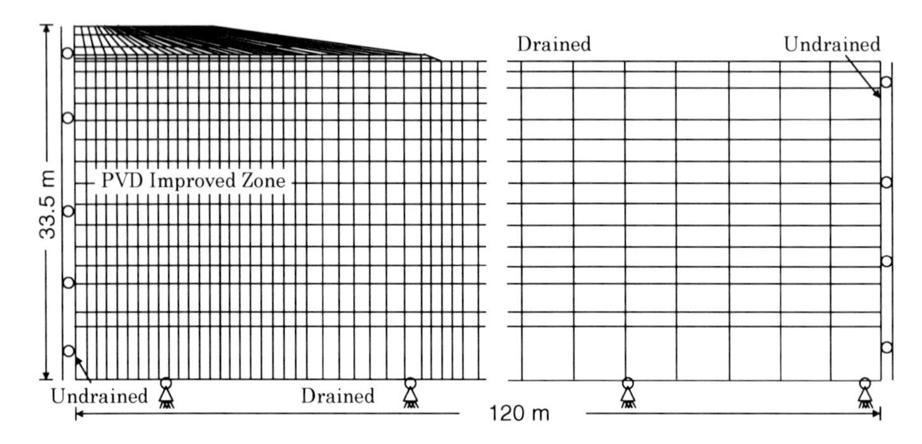

Fig. 2.37 Finite element mesh and boundary conditions assumed for the test embankment at Saga Airport (after Chai and Miura 1999)

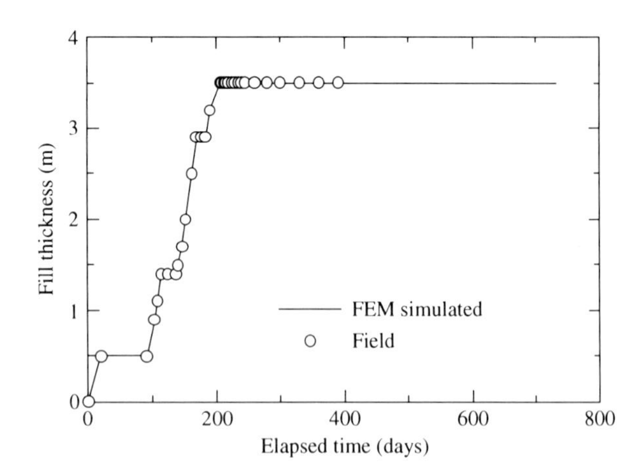

Fig. 2.38 Construction history of the Saga Airport test embankment

Table 2.14 Model parameters for subsoil (after Chai and Miura 1999)

Layer	E kPa	ν	κ	λ	M	e_o	γ_t kN/m^3	k_x 10^{-8} m/s	k_v 10^{-8} m/s
B	–	0.25	0.025	0.25	1.3	2.00	15.0	11.45	7.60
Ac1	–	0.30	0.044	0.44	1.2	2.00	14.5	5.7	3.8
As1	10,000	0.20	–	–	–	–	15.5	290	290
Ac2	–	0.30	0.087	0.87	1.2	2.50	14.5	2.64	1.76
As2	15,000	0.20	–	–	–	–	16.0	290	290
Ac3	–	0.30	0.030	0.30	1.3	1.75	16.0	2.64	1.76
Ds	30.000	0.20	–	–	–	–	19.0	290	290

Note : λ = virgin loading slope in $e - \ln p'$ plot (p' is effective mean stress); κ = reloading/unloading slope in $e - \ln p'$ plot; M = slope of failure line in p' versus q plot (q is deviator stress), E = Young's modulus, ν = Poisson's ratio, γ_t = unit weight

Burland 1968) and the sand layers as well as the decomposed granite fill material were assumed to be elastic materials. The model parameters assumed for the subsoil are listed in Table 2.14. For the clay layers, these parameters were determined from laboratory consolidation tests and triaxial tests conducted on undisturbed samples (Bergado et al. 1996), except for the values of Poisson's ratio and hydraulic conductivity. Selection of the values of Poisson's ratio, ν, was based on empirical experience. For the hydraulic conductivity, first the ratio of horizontal to vertical hydraulic conductivity, k_h/k_v, was determined from laboratory test results as 1.5 (Park 1994). Then values of the hydraulic conductivity in the vertical direction were adjusted to fit the observed field data for an embankment on natural subsoil at the same site (Chai and Miura 1999), which are about 4 times the laboratory values. For the sand layers, Young's modulus was estimated by referring to the results of standard penetration tests (N values). Values of hydraulic conductivity for the sand were simply assumed, based on past experience. The values listed in Table 2.14 correspond to the initial conditions and it was assumed that during consolidation they varied with the voids ratio according to Taylor's (1948) equation (Eq. (2.40)). Parameter C_k in the equation was taken as $0.4e_0$, where e_0 is the initial voids ratio. The subsoils were in a lightly over-consolidated to normally-consolidated state with a maximum over consolidation ratio (OCR) of about 4 for the top crust. The coefficients of earth pressures were calculated using the equation proposed by Mayne and Kulhawy (1982). The groundwater level was about 1.0 m below ground surface. The adopted initial stresses are listed in Table 2.15. The mechanical properties of the fill material were assumed to be a Young's modulus of 15,000 kPa and a Poisson's ratio of 0.2. The unit weight of fill material was assumed to be 20 kN/m^3. The parameters for the PVDs are listed in Table 2.16 (after Chai and Miura 1999).

Figure 2.39 compares settlements at the ground surface and 5.6 m below the ground surface. It can be seen that considering buoyancy effects reduced the surface settlement by about 0.36 m, which is about 13% of the total settlement. The measured data are also included in this figure for comparison purposes. Comparison of the lateral displacement profile at the end of construction is given in Fig. 2.40. There

Table 2.15 Initial stress conditions of Ariake clay deposit at Saga Airport site

Depth (m)	Horizontal stress σ'_{h0} (kPa)	Vertical stress σ'_{v0} (kPa)	Water pressure u (kPa)	Size of yield locus p'_0 (kPa)
0.0	5.0	0.0	0.0	29.0
1.0	9.6	15.5	0.0	29.0
4.0	18.0	29.0	30.0	26.0
5.6	20.8	37.8	46.0	35.0
21.0	53.6	107.1	200.0	95.2
23.7	61.7	123.3	227.0	109.6
25.0	65.6	131.1	240.0	116.5
30.0	88.1	176.1	290.0	151.1

Table 2.16 Parameters relevant to PVD drains

Item	Symbol	Units	Value
Drain diameter	d_w	mm	48.3
Unit cell diameter	d_e	m	1.7
d_e/d_w	N	–	35.2
Smear zone diameter	d_s	mm	300
Hydraulic conductivity ratio	k_h/k_s	–	10
d_s/d_w	S	–	6.2
Discharge capacity	q_w	m^3/yr	85

is no obvious effect of the buoyancy on lateral displacement, but it is noted that lateral displacement is mainly influenced by the magnitude of shear stresses induced in the ground. The method adopted to consider the buoyancy effect involved adjusting the total water pressure and this will not change the magnitude of the shear stress in the ground, and therefore should have little influence on the lateral displacements.

For the test embankment at Saga Airport the 3.5 m thickness of fill material caused about 2.5 m settlement. Figure 2.39 reveals that the inclusion of buoyancy

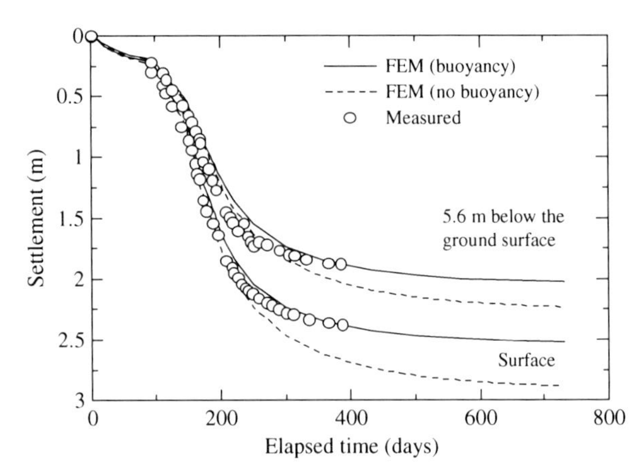

Fig. 2.39 Comparison of settlement-time curves for the Saga embankment on PVD-improved subsoil

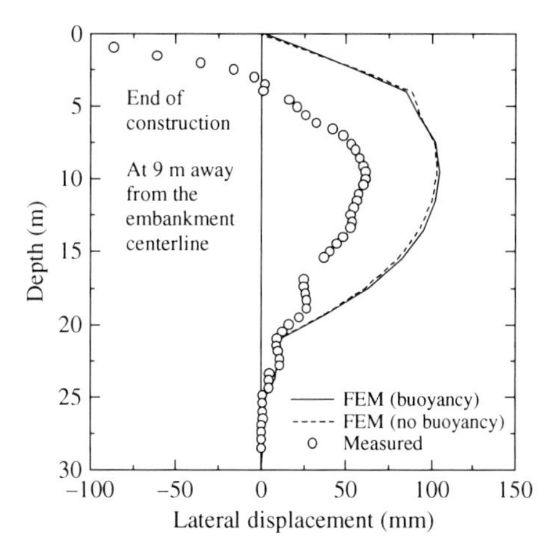

Fig. 2.40 Comparison of lateral displacement profile for the Saga embankment on PVD improved subsoil

effects in the analysis can be significant. In particular, it can make a difference as large as 10% for the predicted total settlement.

2.6 Summary

Some techniques for simulating the mechanical behaviour of a soft clayey subsoil using finite element methods (FEM) have been described. A comparison of the effects of the various techniques for simulating embankment construction has also been provided and a summary of the findings of this chapter is as follows.

2.6.1 Undrained Shear Strength Profile

In order to solve geotechnical boundary and initial value problems accurately and reliably, the use of an appropriate constitutive model to represent the mechanical behaviour of the soil is generally important. However, selection of an appropriate model alone does not guarantee realistic simulations and predictions. The values of the model parameters and the initial stress state of the subsoil are equally important and considerable care is required when selecting them. All these factors together control the operative shear strength and modulus of the modelled soil. Therefore, it has been proposed that comparing the simulated undrained shear strength (S_u) profile with a measured profile (or profiles) provides a useful check on the feasibility and appropriateness of the selected model, its parameter values and the determined initial stress state (including the overconsolidation ratio or OCR). The field values of S_u can be measured by field vane shear tests, laboratory unconfined compression tests or unconsolidated undrained triaxial compression tests using undisturbed soil samples. However, the strain rate adopted during the test is normally different from that experienced in the soil under a real structure and the effect of strain rate on S_u may also have to be considered, with possible corrections applied to allow for the effects of different strain rates (Bjerrum 1972; 1973). Two example analyses, one involving a class-A prediction of a centrifuge model test of a footing on clay and another involving the simulation of a built-to-failure embankment on a deposit of soft Champlain at Saint-Alban, Quebec (Grammatikopoulou et al. 2007), were described in order to demonstrate the importance of first checking the simulated profile of S_u.

2.6.2 Simulating Embankment Loading and Construction

In order to simulate foundation response due to embankment loading using the FEM, the embankment load, corresponding to the self-weight of the embankment fill material, can be applied either by: (1) applying an equivalent distributed surface loading; or (2) increasing the gravity loading of all or some of the elements used to represent the embankment; or (3) placing embankment elements layer by layer, 'switching on' their own stiffness and self weight as the elements are placed or 'constructed'. Method (3) can simulate quite closely the actual construction process. The effect of the embankment load application method on the simulated ground response was demonstrated by analyzing a test embankment on the Muar clay deposit in Malaysia (MHA 1989a).

For an embankment on soft subsoil, normally the embankment load will cause large deformation of the ground resulting in significant changes to the original problem geometry. A practical way to consider these large deformations is to update the nodal coordinates of the FEM mesh frequently during an incremental analysis. In FEM analysis, normally the elements representing an embankment are pre-specified. If the nodal coordinates are not updated during an incremental analysis the applied embankment load will correspond exactly (within calculation tolerance) to the initial volume of the embankment elements multiplied by the unit weight of the fill material. However, if large deformations are considered by updating the nodal coordinates during the course of the incremental non-linear analysis, normally the nodal coordinates of any un-constructed elements will not be updated and consequently the volume of the unconstructed elements may actually change during the analysis. The volume of these elements usually increases so that when they are finally made active during a construction increment the final applied load will be larger than may have been originally anticipated. Hence in FEM analysis, in order to ensure the pre-specified fill thickness is maintained, the unconstructed embankment elements must follow the settlement of the constructed elements and the foundation soil. In order to ensure that this occurs the method proposed by Chai and Bergado (1993a) is recommended, as has been described in this chapter.

2.6.3 Effect of Settlement on Embankment Stability

Soft subsoil deformation during embankment construction will influence the factor of safety (FS) of the embankment. The direct effects include: (1) reducing the surcharge load; and (2) partially replacing the weaker soft soil by the stronger fill material. An indirect effect is the shorting of the effective length of the drainage path within the subsoil, which can increase the partial drainage that takes place during the construction period, and this drainage will usually cause the undrained shear strength of the subsoil to increase. For the test embankment built to failure on the Muar clay deposit in Malaysia (MHA 1989b), results of FEM analysis indicate that taking into account the foundation deformation during construction can increase the computed FS typically by about 0.1.

2.6.4 Effect of Buoyancy on Settlement

For many soft clay or clay-like deposits, the groundwater level is usually close to the ground surface. Large deformations of the foundation soil induced by embankment construction will bring a part of foundation soil and possibly some of the fill material from above the groundwater level to below the groundwater level. In such cases buoyancy effects will tend to reduce the overall effective surcharge load applied to the underlying subsoil. In FEM analysis, the buoyancy effect can be taken into account by: (1) considering the changes in total water pressure due to incremental settlement, and the changes in the unit weight of soft soil and fill due to consolidation; (2) checking equilibrium of the entire mechanical system and forming the load vector corresponding to the buoyancy force; and (3) iterating until the residual force is less than a given tolerance. For the test embankment constructed at Saga Airport in Japan, the results of a finite element analysis of this type indicated that inclusion of the buoyancy effect can reduce the predicted final settlement about 13%.

References

Al-Tabbaa A, Wood DM (1989) An experimental based 'bubble' model for clay. In: Pietruszczak A, Pande GN (eds) Proceedings of international conference on numerical models in geomechanics. Balkema, Rotterdam, pp 91–99

Asaoka A, Nakano M, Matsuo M (1992) Prediction of the partially drained behavior of soft clays under embankment loading. Soil Found 32:41–58

Asian Institute of Technology (AIT) (1988) Laboratory test data on soil samples from the Muar Flats test embankment, Johore, Malaysia. Research report, Division of Geotechnical and Transportation Engineering, Asian Institute of Technology, Bangkok, Thailand

Asian Institute of Technology (AIT) (1989) Laboratory test data on soil samples from the Muar Flats test embankment, Johore, Malaysia. Research report, phase II. Division of Geotechnical and Transportation Engineering, Asian Institute of Technology, Bangkok, Thailand

Bergado DT, Anderson LR, Miura N, Balasubramaniam AS (1996) Soft ground improvement in lowland and other environments. ASCE Press, New York, NY, p 427

Bjerrum L (1972) Embankment on soft ground: SOA report. Proceedings of specialty conference on performance of earth and earth-supported structures, ASCE, Purdue University, Indiana, USA, vol 2, pp 1–54

Bjerrum L (1973) Embankment on soft ground: SOA report. Proceeding of the 8th international conference on soil mechanics and foundation engineering, vol 3, USSR, Moscow, pp 111–159

Brand EW (1991) Predicted and observed performance of an embankment built to failure on soft clay. Geotech Eng, Southeastern Asian Geotechn Soc 22(1):23–42

Brand EW, Premchitt J (1989) Comparison of the predicted and observed performance of the Muar test embankment. In: Hudson RR, Toh CT, Chan SF (eds) Proceedings of the international symposium on trial embankments on Malaysian marine clays, Kuala Lumpur, Malaysia, The Malaysian Highway Authority, Kuala Lumpur, Malaysia, vol 2, pp 10/1–10/29

Britto AM, Gunn MJ (1987) Critical state soil mechanics via finite elements. Ellis Horwood, Chichester, UK

Chai JC (1992) Interaction behavior between grid reinforcement and cohesive-frictional soil and performance of reinforced wall/embankment on soft ground. DEng dissertation, Asian Institute of Technology, Bangkok, Thailand

Chai JC, Bergado DT (1993a) Some techniques for FE analysis of embankment on soft ground. Can Geotech J 30(4):710–719

Chai JC, Bergado DT (1993b) Performance of reinforced embankment on Muar clay deposit. Soil Found 33(4):1–17

Chai JC, Miura N (1999) Investigation on some factors affecting vertical drain behavior. J Geotech Geoenviron Eng, ASCE 125(3):216–226

Chai JC, Sakajo S, Miura N (1994) Stability analysis of embankment on soft ground. Soil Found 34(2):107–116

Chai JC, Sakajo S, Yang CW (1995a) Effect of soft subsoil settlement on embankment behavior. Proceedings of 10th Asian regional conference on soil mechanics and foundation engineering, vol 2, Beijing, China, pp 158–159

Chai JC, Miura N, Sakajo S, Bergado DT (1995b) Behavior of vertical drain improved subsoil under embankment loading. Soil Found 35(4):49–61

Chai JC, Shen SL, Miura N, Bergado DT (2001) Simple method of modeling PVD improved subsoil. J Geotech Geoenviron Eng ASCE 127(11):965–972

Duncan JM, Byrne P, Wong KS, Mabry P (1980) Strength stress-strain and bulk modulus parameters for finite element analysis of stresses and movements in soil. Geotechnical engineering research report, Department of Civil Engineering, University of California Berkeley

Giroud JP (1968) Settlement of an embankment resting on a semi-infinite elastic soil. Highway Research Record No. 223:18–35

Grammatikopoulou A (2004) Development, implementation and application of kinematic hardening models for overconsolidated clays. PhD thesis, Imperial College, University of London, UK

Grammatikopoulou A, Zdravkovic L, Potts DM (2006) General formulation of two kinematic hardening constitutive models with a smooth elastoplastic transition. Int J Geomechan ASCE 6(5):291–302

Grammatikopoulou A, Zdravkovic L, Potts D M (2007) The effect of the yield and plastic potential deviatoric surfaces on the failure height of an embankment. Géotechnique 57(10): 795–806

Gray H (1936) Stress distribution in elastic solids. Proceedings of 1st international conference on soil mechanics and foundation engineering, Harvard University Press, vol 2, p 157

Ladd CC (1991) Stability evaluation during staged construction. J Geotech Eng ASCE 117(4): 541–615

Lade PV, Duncan JM (1975) Elastoplastic stress-strain theory for cohesionless soil. J Geotech Eng Div ASCE 101(GT10):1037–1053

Lambe TW (1973) Predictions in engineering. Géotechnique 23(2):149–202

Lefebvre G, Leboeuf D, Rahhal ME, Lacroix A, Warde J (1994) Laboratory and field determinations of small-strain shear modulus for a structured Champlain clay. Can Geotech J 31(1):61–70

Magnan J-P (1989) Experience-based prediction of the performance of Muar Flats trial embankment to failure. In: Hudson RR, Toh CT, Chan SF (eds) Proceedings of the international symposium on trial embankments on malaysian marine clays, Kuala Lumpur, Malaysia, vol 2. The Malaysian Highway Authority, Kuala Lumpur, Malaysia, pp 2/1–2/8

Malaysian Highway Authority (MHA) (1989a) Chapter 16 Scheme 6/8 Preloading, geogrid reinforcement and vertical drains In: Hudson RR, Toh CT, Chan SF (eds) Proceedings of the international symposium on trial embankments on Malaysian marine clays, Kuala Lumpur, Malaysia. The Malaysian Highway Authority, Kuala Lumpur, Malaysia, pp 16/1–16/12

Malaysian Highway Authority (MHA) (1989b) The embankment built to failure. In: Hudson, RR, Toh, CT, Chan SF (eds) Proceedings of the international symposium on trial embankments on malaysian marine clays, Kuala Lumpur, Malaysia, The Malaysian Highway Authority, Kuala Lumpur, Malaysia, 2:1/1–1/27

Mayne PE, Kulhawy FH (1982) K_0-OCR relationships in soils. J Geotech Eng ASCE 108(GT6):851–872

Miyake M, Akamoto H, Kato Y (1994) Bearing-capacity test of Kobe Marine clay using centrifuge. Proceedings of the symposium on foundation failure and strain localization, Tokyo, pp 9–16 (in Japanese)

Nakase A, Kamei T, Kusakabe O (1988) Constitutive parameters estimated by plasticity index. J Geotech Eng ASCE 114(7):844–858

Park YM (1994) A research on the mechanical properties of lowland marine clay and vertical drain improvement method. Doctor of Engineering Thesis, Saga University, Japan

Potts DM, Gens A (1984) The effect of the plastic potential in boundary value problems involving plane strain deformation. Int J Numer Anal Methods Geomechan 8(3):359–286

Poulos HG, Lee CY, Small JC (1989) Prediction of embankment performance on Malaysian marine clays. In: Hudson RR, Toh CT, Chan SF (eds) Proceeding of the international symposium on trial embankments on Malaysian marine clays, Kuala Lumpur, Malaysia. The Malaysian Highway Authority, Kuala Lumpur, Malaysia 2:4/1–4/10

Roscoe KH, Burland JB (1968) On the generalized stress-strain behavior of wet clays. Proceeding of engineering plasticity, Cambridge University Press, Cambridge, UK, pp 535–609

Sakajo S, Chai J-C (1994) Prediction of load-carrying capacity of strip footing on clay. Proceedings of the symposium on foundation failure and strain localization, Tokyo, pp 31–38

Sakajo S, Chai J-C, Kamei T (1995) Strip footing's load-carrying prediction analysis on cohesive soil. Tsuti-to-Kiso 43(7):28–30 (in Japanese)

Sekiguchi H, Ohta H (1977) Induced anisotropy and time dependency in clays. Constitutive equations of soils. Proceedings special session 9, 9th international conference soil mechanics and foundation engineering, Tokyo, pp 229–238

Sloan SW (1987) Substepping schemes for numerical integration of elasto-plastic stress-strain relations. Int J Numer Methods Eng 24:893–911

Tavenas F, Tremblay M, Larouche G, Leroueil S (1986) In situ measurement of permeability in soft clays. Proceedings ASCE special conference on use of in situ tests in geotechnical engineering, Blacksburg, pp 1034–1048

Taylor DW (1948) Fundamentals of soil mechanics. Wiley, New York, NY

van Eekelen HAM (1980) Isotropic yield surfaces in three dimensions for use in soil mechanics. Int J Numer Anal Methods Geomech 4(1):89–101

Chapter 3
Vertical Drains

Abstract This chapter contains a description of the methods used for modelling soft subsoils whose mechanical performance, particularly their drainage characteristics, have been improved by the installation of pre-fabricated vertical drains (PVDs). Included is a description of how a typical problem may be dealt with using an equivalent plane strain finite element analysis. Also presented are some example analyses using specific case histories, and comparisons of the model predictions with laboratory and field performance of PVD systems. A field case study from China is included in these comparisons.

3.1 Consolidation Theory for Prefabricated Vertical Drains

3.1.1 Introduction

Vertical drains are often installed in soft clayey soils to improve their overall drainage properties and ultimately their strength and stiffness, i.e., to accelerate the consolidation of the soft soil under imposed loading. The types of drains used in this form of ground improvement technology can be loosely divided into sand drains, gravel drains and prefabricated vertical drains or PVDs. The major focus of attention in this chapter will be on PVDs and their influence on the behaviour of the clay layers into which they are installed. A schematic illustration of the major components of a PVD is provided in Fig. 3.1.

Because of their speed of installation and reduced cost PVDs are most commonly used these days to accelerate the consolidation of soft soil deposits. They are usually manufactured from a composite of solid polycarbonate material encased in a geotextile filter. Various proprietary products are available for this purpose. In the field, PVDs are installed by using a mandrel, which is pushed into the subsoil with a PVD inside it. The mandrel is subsequently withdrawn leaving the PVD in the subsoil. This process creates a completely disturbed zone around the PVD, called the smear zone, with an effective radius of r_s (diameter d_s). The hydraulic conductivity in the smear zone, k_s, may be reduced to a very low value. Flow in the drain takes place vertically and is governed by the discharge capacity of the PVD, q_w.

Analytical solutions for the consolidation behaviour of a soil layer penetrated by vertical drains have invariably involved the study of a unit cell, i.e., a cylinder

J. Chai, J.P. Carter, *Deformation Analysis in Soft Ground Improvement*, Geotechnical, Geological and Earthquake Engineering 18, DOI 10.1007/978-94-007-1721-3_3,

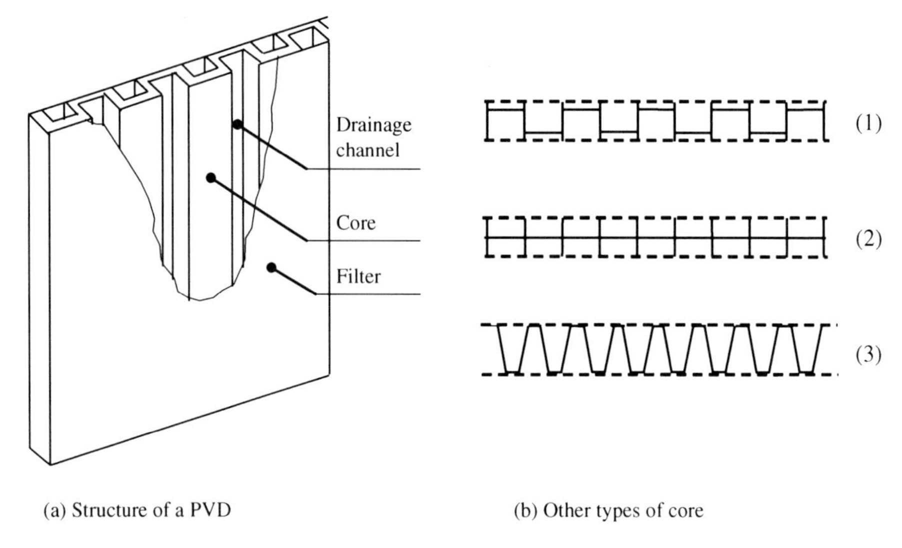

Fig. 3.1 Schematic illustration of a PVD

of soil surrounding a single vertical drain, under simplified boundary conditions and assuming that the soil is uniform in the vertical direction. Figure 3.2 shows a unit cell of fixed external radius r_e (or diameter d_e) and initial length, l, containing a drain of radius of r_w (diameter d_w). The impervious bottom boundary is fixed vertically, while the downward displacement of the pervious upper boundary is permitted. There are two options regarding the displacement pattern assumed at the upper boundary. One is uniform vertical stress in the radial direction (the so called 'free strain' condition), and the other is uniform vertical strain in the radial direction (the so called 'equal strain' condition). The assumption of equal vertical strain is usually adopted in the analysis for mathematical convenience. In obtaining analytical solutions it is also assumed that the soil exhibits uniform linear stress-strain behaviour. Drainage within the soil may take place in both the vertical and radial directions, although it is often considered reasonable to assume that radial flow of the pore water predominates.

In the field, PVDs are installed in square or triangular patterns with a spacing S, and the radius of the unit cell, r_e, is calculated based on the equal area principle:

$$r_e = 0.564S \quad \text{Square pattern} \tag{3.1}$$

$$r_e = 0.525S \quad \text{Triangular patterm} \tag{3.2}$$

The available theories for the consolidation of the unit cell vary in their degrees of rigor and complexity. The most rigorous, but mathematically most complex solutions are based on Biot's (1941) consolidation theory and were derived for equal strain boundary conditions by Yoshikuni and Nakanodo (1974) and Onoue (1988). Both solutions allow for combined vertical and radial drainage in the soil and for

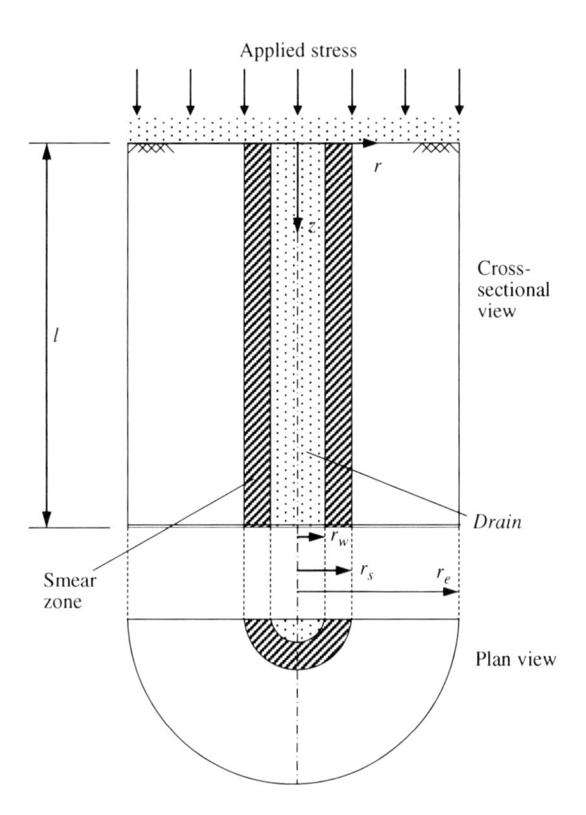

Fig. 3.2 A unit cell of vertical drain consolidation

finite drain discharge capacity (well resistance), but only in Onoue's solution is the smear effect considered. Although the upper boundary is displaced uniformly, variation of vertical strain is generally permitted both radially and vertically. Less rigorous solutions, considering both well resistance and the smear effect but neglecting vertical flow in the soil, have been obtained by Barron (1948), Hansbo (1981) and Zeng and Xie (1989). These are all genuine equal strain analyses where the vertical strain is assumed to be uniform with radius. The solutions of Hansbo (1981) and Zeng and Xie (1989) are relatively easily computed compared to either the Barron (1948) solution or more rigorous solutions. Analytical solutions for free strain boundary conditions have only been obtained in the absence of well resistance by Barron (1948) and, even then, they tend to be complex.

The most commonly used consolidation theories for designing PVD improvement are the Barron (1948) and Hansbo (1981) solutions and these are considered in subsequent sections of this chapter.

3.1.2 Barron's Solution

In the case of equal strains in the drain and the surrounding soil, the differential equation governing the consolidation process of the unit cell is given as:

$$\frac{\partial u}{\partial t} = c_h \left[\left(\frac{\partial^2 u}{\partial r^2} \right) + \frac{1}{r} \left(\frac{\partial u}{\partial r} \right) \right] \tag{3.3}$$

where t = the time after an instantaneous increase of the total vertical stress, u = the excess pore water pressure at any point and at any given time t; r = the radial distance of the considered point from the centre of the soil cylinder; and c_h = the coefficient of consolidation in the horizontal direction. For the case of radial drainage only, the average degree of consolidation (U_h) from Barron's (1948) solution under ideal conditions (no smear and no well resistance) is as follows:

$$U_h = 1 - \exp\left[\frac{-8T_h}{F(n)} \right] \tag{3.4}$$

where

$$T_h = \frac{c_h t}{4r_e^2} \tag{3.5}$$

$$F(n) = \left[\frac{n^2}{(1-n)} \right] \left[\ln(n) - \frac{3}{4} + \frac{1}{n^2} \right] \tag{3.6}$$

and where $n = r_e / r_w$.

3.1.3 Hansbo's Solution

It has been shown (Hansbo 1981; Zeng and Xie 1989; Onoue 1988) that Hansbo's relatively simple theory compares well, not only with the similar solutions of Barron and Zeng and Xie, but also with more rigorous solutions. As Hansbo's solution is also easy to apply, the theory has been widely used in practice. Under an instantaneous step loading, Hansbo's solution for U_h at depth, z, and time, t, is as follows (Hansbo 1981):

$$U_h = 1 - \exp(-8T_h/\mu) \tag{3.7}$$

where μ represents the effect of the spacing between the PVDs as well as the effects of smear and well resistance. It can be expressed as:

$$\mu = \ln(n/s) + (k_h/k_s)\ln(s) - \frac{3}{4} + \pi \cdot z(2l - z)\, k_h/q_w \tag{3.8}$$

For an average well resistance, μ can be expressed as:

$$\mu = \ln(n/s) + (k_h/k_s)\ln(s) - \frac{3}{4} + \frac{2\pi \cdot l^2 \cdot k_h}{3q_w} \tag{3.8a}$$

where $s = r_s/r_w$, l = the drainage length of a PVD, k_h = the horizontal hydraulic conductivity of the native soil containing the PVD, and k_s = the horizontal hydraulic conductivity of the smear zone.

3.2 Parameter Determination

One of the most important parameters affecting the consolidation rate of subsoil improved by the installation of vertical drains is the horizontal coefficient of consolidation (c_h). As discussed later, to date there is no satisfactory laboratory test method to determine this parameter so that the back-analysis of values of c_h from field measurements or field hydraulic conductivity tests is recommended. For given subsoil conditions, the effect of the vertical drains depends on: (1) the drain spacing and the equivalent drain diameter; (2) the well resistance (i.e., the discharge capacity of the drain); (3) the smear effect; and (4) the drainage boundary conditions of the soil layer. Except perhaps for the drain spacing, significant uncertainties exist when attempting to quantify each of these factors. These issues are now described in greater detail.

3.2.1 Equivalent Drain Diameter

The first proposal for determining the equivalent drain diameter (d_w) of a band-shaped PVD was made on the basis of the assumption of an equal drainage perimeter (Hansbo 1979). Subsequent research indicated that due to corner effects the equivalent drain diameter is actually less than the value calculated based on the equal drainage perimeter assumption and a new equation, based on the results of finite element analysis, has been suggested by Rixner et al. (1986) as:

$$d_w = \frac{w + t_d}{2} \tag{3.9}$$

where w and t_d are the width and thickness of the PVD, respectively. Equation (3.9) has been widely used in practice and was also subsequently verified by the independent finite element analyses conducted by Chai and Miura (1999).

3.2.2 Discharge Capacity of a PVD

The discharge capacity of a PVD must be determined experimentally. An ideal discharge capacity test should simulate the drain installation, the confinement of clay on the filter sleeve of the drain, and the deformation of the drain during consolidation. It is obvious that a full-scale test could be expensive, if indeed it is ever possible. For a useful small-scale laboratory test, the important influencing factors, such as the confinement condition, must be considered.

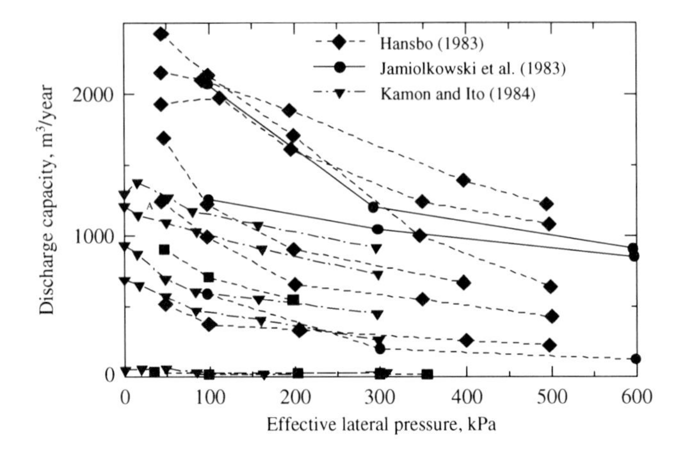

Fig. 3.3 Variation of discharge capacity with confining pressure of several different PVDs

Hansbo (1994) summarized the results of discharge capacity tests for several different PVDs confined in soil, as investigated by Jamiolkowski et al. (1983), Hansbo (1983) and Kamon and Ito (1984). It may be seen from the results plotted in Fig. 3.3 that the discharge capacity is greatly influenced by confining pressure. The reason is that the filter sleeves are squeezed further into the channel system at higher confining pressures, which entails a reduction in the channel areas, or the channels themselves are squeezed together under higher pressures.

Another systematic laboratory test program was carried out to identify the important factors influencing the discharge capacity (Miura et al. 1998). The following factors were investigated:

(a) Confining the drain in clay (compared with confining the drain by a rubber membrane);
(b) The effect of possible air bubbles trapped in the drainage pathways;
(c) The effect of folding of the drain; and
(d) The long-term discharge capacity.

The main findings from these tests are as follows.

(a) For the cases investigated, the discharge capacity observed when confining the drain in clay for one week (i.e., a relatively short term) was only about 20% of the corresponding value when the drain was confined by a rubber membrane. One long-term test, which lasted for about 5 months, indicated that the discharge capacity also reduced significantly with elapsed time.
(b) Air bubbles trapped in the drainage pathways of the drain reduced the discharge capacity by about 20%.
(c) The folding of the drain (with no kinking) combined with a vertical strain up to 20% did not have much effect on the discharge capacity, supporting the conclusion drawn earlier by Hansbo (1983).

These results indicated that the confinement conditions and the test duration are two key factors affecting the discharge capacity of a PVD. To confirm this point and to identify the reasons for a reduction in discharge capacity with time in the case of clay confinement, additional discharge capacity tests with clay confinement were subsequently conducted by Chai and Miura (1999). In order to quantify the creep deformation of the filter under a given confining pressure, creep tests were also carried out for the filter elements of the PVDs tested.

The test device utilized by Chai and Miura is illustrated in Fig. 3.4. The apparatus consists of a cylindrical cell with a diameter of 200 mm and height of 600 mm. The maximum length of PVD that can be tested in this apparatus is 400 mm. The test procedure was as follows:

(a) Install the drain sample on the lower pedestal, which is connected to the inlet water flow system.
(b) Fix the membrane at the lower pedestal, and put the cylindrical mould (2 halves) in position with the membrane inside. The inner diameter of the mould is 100 mm.
(c) Put the remoulded clay (at a water content close to its liquid limit) into the mould layer by layer until the desired height is reached. Each layer is compacted by a stick tamper (about 10 mm in diameter) to avoid air bubbles remaining in the sample. However, the initial density of the sample is basically controlled by its water content.
(d) Install the upper pedestal and connect the PVD sample to the outlet water flow system. Fix the membrane to the upper pedestal.
(e) Apply a suction of about 10 kPa to the sample and remove the mould.

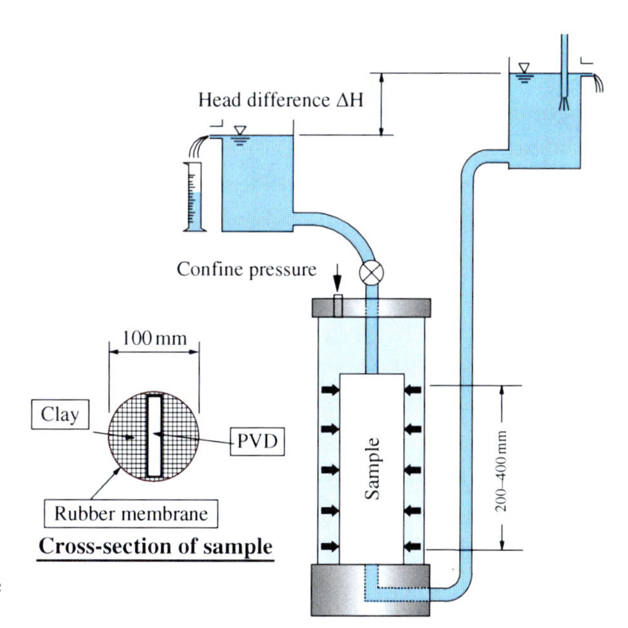

Fig. 3.4 Illustration of discharge capacity test device

(f) Set-up the confining pressure, and gradually release the suction.
(g) Set-up the desired confining pressure (usually in the range from 49 to 392 kPa) and hydraulic gradient (usually in the range 0.08 to 0.8).
(h) Then let the clay sample consolidate under the applied confining pressure. After the clay sample is consolidated, the discharge capacity of the PVD encased in soil is measured. Consolidation of the surrounding clay sample takes about 1 week. Therefore, the discharge measurement after 1 week is defined as the short-term value of the discharge capacity for the test confined in clay.

Slots were cut into the lower and upper pedestals to accommodate the band-shaped PVD. The gap between the drain sample and the slot needed to be sealed by plasticizer to avoid the clay particles entering the drainage pathways of the PVD. It is worth mentioning that this set-up is not intended to be a unit cell test. It merely simulates the clay (smear zone) confinement on the sleeve of the drain.

Table 3.1 gives the properties of the PVDs tested. The remoulded Ariake clay had a liquid limit of about 105.0% and plastic limit of about 42.8%. The soil consisted of 57.0% clay ($< 5\ \mu$m), 41.7% silt and 1.3% sand particles.

The results of two long-term tests are shown in Fig. 3.5. The corresponding values of discharge capacity measured in tests involving confining the PVDs in a rubber membrane are also indicated in the figure for comparison. In these cases the specific test conditions were as follows: the confining pressure was 49 kPa and the hydraulic gradient was 0.08. For both tests in clay the discharge capacities continuously reduced with elapsed time, and the lowest value was about 4% of the value observed when confining the PVDs by a rubber membrane.

On most construction projects a PVD is normally expected to work effectively for at least half a year. Therefore, in design, the longer term behaviour of PVDs should be taken into account. If these test data are linearly extrapolated to conditions where the equivalent hydraulic gradient is 1.0, the lowest discharge capacity would be 75 m^3/year for PVD(A) and 126 m^3/year for PVD(B). PVD(B) had a larger discharge capacity than PVD(A) due to its larger initial cross sectional area of the drainage channel (Table 3.1). The factors considered to cause the reduction in discharge capacity with time under clay confinement are: (1) creep deformation of the filter under a maintained confining pressure, and (2) the clogging effect of fine particles entered the drainage pathways of the PVD drain.

Table 3.1 Physical properties of the PVDs

	Size		Drainage channel				Material	
PVD	t_d[a] (mm)	w (mm)	Depth (mm)	Width (mm)	Number of channels	Unit weight (g/m^2)	Filter	Core
A	2.6	94	1.5	1.8	40	90	Polyester	Polyolefin
B	3.6	97	1.3	2.0–2.4	64–66	120	Polyolefin	Polyolefin

[a] t_d = thickness; w = width

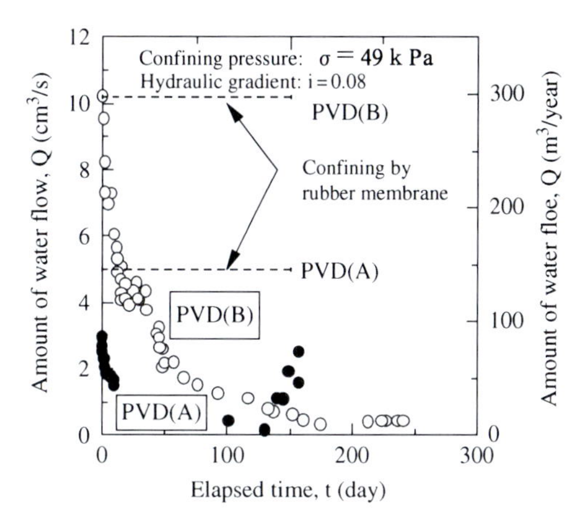

Fig. 3.5 Water flow versus elapsed time for long term discharge capacity tests (after Chai and Miura 1999)

In order to quantify the creep effect, special creep tests on the filter elements of the drains were conducted (Chai and Miura 1999). The test device used in this case is illustrated in Fig. 3.6. The drain filter was tested in the direction corresponding to the transverse direction of the PVD. The size of the sample tested was 200 mm in width and 400 mm in length. To avoid necking of the sample, a clamp was fixed in the middle of the sample. A measuring rule was fixed on the right side of the frame (Fig. 3.6).

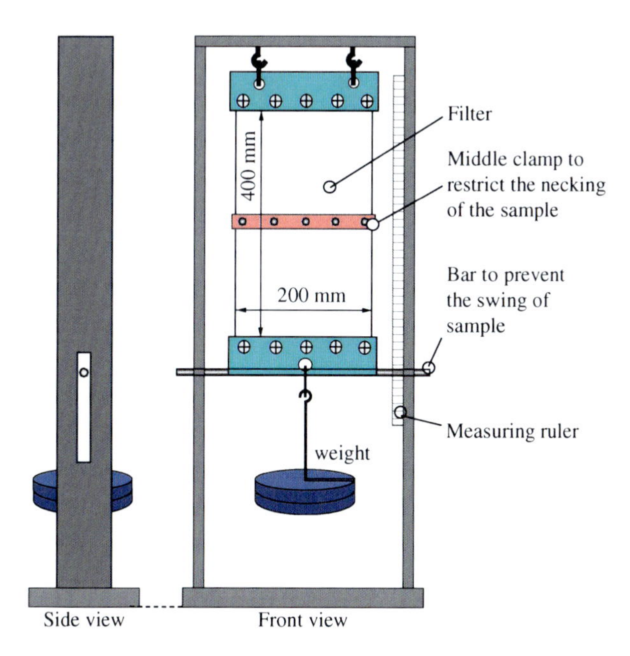

Fig. 3.6 Illustration of setup of creep test (after Chai and Miura 1999)

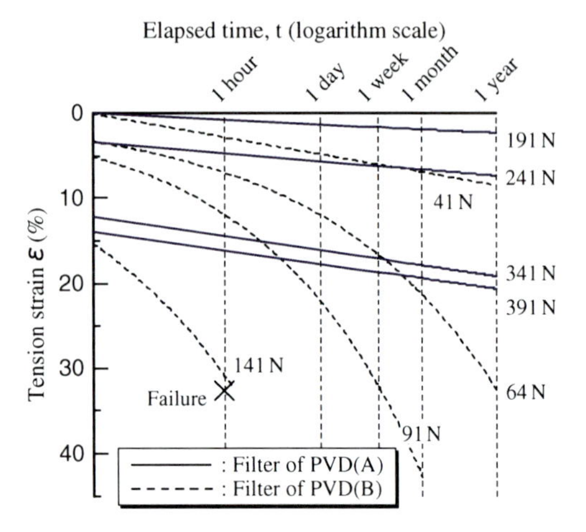

Fig. 3.7 Creep test results (after Chai and Miura 1999)

The results of the creep tests are given in Fig. 3.7. It can be seen that the filter element of PVD(B) is weaker than that of PVD(A). To relate the creep deformation of a filter with the reduction of the cross-sectional area of the drainage channel, the following assumptions are used:

1. As illustrated in Fig. 3.8, the deformed shape of the filter is assumed to be a circular arc.
2. The confining pressure is balanced by the component of the tensile force (T) in the filter in the y direction (Fig. 3.8).

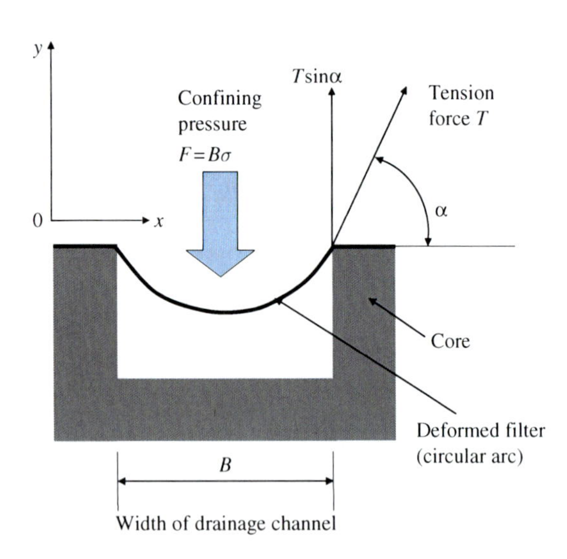

Fig. 3.8 Concept used to calculate the reduction of drainage area (after Chai and Miura 1999)

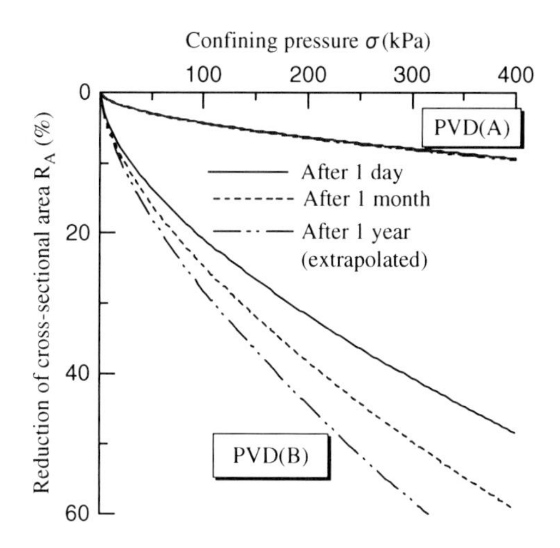

Fig. 3.9 Relationship between reduction of cross-sectional area and confining pressure (after Chai and Miura 1999)

By trial and error, the relationship between the confining pressure and the reduction in the cross-sectional area was obtained, as shown in Fig. 3.9. From this figure it can be deduced that for the two long term tests (Fig. 3.5), the reduction in cross-sectional area due to deformation (including creep) of the filter was 4 and 17% for PVD(A) and PVD(B), respectively.

For the test conducted on PVD(A) and presented in Fig. 3.5, hydraulic shocks were applied after about 130 days, by firmly stepping on the inlet water flow hose, in order to examine the clogging effects. It was observed that some flocculated fine particles were forced out of the drainage channels by these pressure shocks and they were subsequently deposited on the wall of the outlet hose, i.e., some of the PVD clogging was removed by these hydraulic shocks. As shown in Fig. 3.5, the discharge capacities increased after the shock.

Figure 3.10a, b show electron microscope pictures of filter samples obtained from the open channel area. Figure 3.10a shows a sample from the drainage channel side, and Fig. 3.10b shows a sample from the side in contact with the clay. From Fig. 3.10a, it can be seen that there is some bio-film formed on the drainage channel side. On the side in contact with the clay, there is no bio-film but some soil particles are observed in the openings of the filter material. It is considered that the formation of bio-film together with the fine particles inserted in the drainage channel would have caused the partial clogging of the drainage path, resulting in a reduction in the discharge capacity with time.

3.2.3 Smear Effect

Two parameters are needed to characterize the smear effect, namely the diameter of the smear zone (d_s) and the hydraulic conductivity ratio (k_h/k_s), i.e., the value

Fig. 3.10 Electron Microscope pictures of filter after discharge capacity test for PVD(B) (after Chai and Miura 1999). (**a**) Contact with clay side; (**b**) Drainage channel side

(a) Contact with clay side

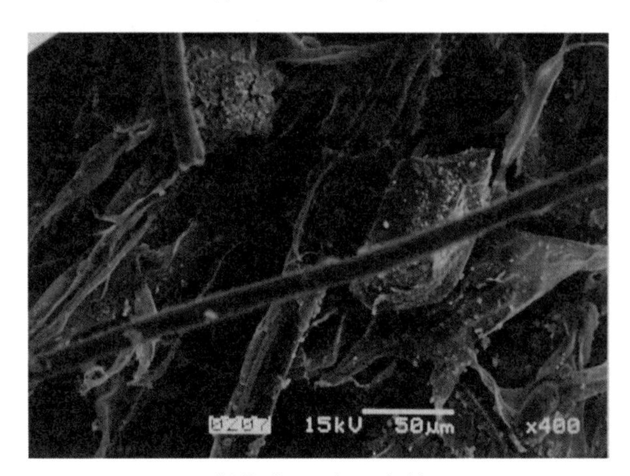

(b) Drainage channel side

of hydraulic conductivity in the undisturbed zone (k_h) divided by that in the smear zone (k_s). Several investigations have been made of these factors (Jamiolkowski and Lancellotta 1981; Jamiolkowski et al. 1983; Hansbo 1987; Miura et al. 1993) and the value of d_s can usually be estimated as:

$$d_s = (2\ to\ 3)\,d_m \tag{3.10}$$

where d_m = the equivalent diameter of the cross-sectional area of a mandrel. In design, if there are no test data available for evaluating the smear zone size, the value of $d_s = 3d_m$ is suggested (Chai and Miura 1999). Since the shape of the mandrel is usually rectangular, the cross-sectional shape of the actual disturbed zone will be close to rectangular or elliptical. In the unit cell theory, a circular shape based on the equal area assumption is used to represent the smear zone. This observation has been checked by independent finite element analysis (Chai and Miura 1999). A plane strain slice (either circular or rectangular) perpendicular to the drain direction was considered for evaluating the effect of the shape of the smear zone. Because of symmetry only a quarter of the slice was required in the finite element analysis (Fig. 3.11). The boundary conditions are also shown in this figure. During the analysis a hydrostatic load of 50 kPa was first applied under undrained conditions, and then the consolidation process was simulated. The soil was assumed to be elastic with a Young's modulus of 10,000 kPa, Poisson's ratio of 0.3, hydraulic conductivity $k_h = 10^{-8}$ m/s, and $k_s = k_h/5$. Figure 3.12 compares the average degree of consolidation of a rectangular smear zone with that of a circular one assuming equal areas. It can be seen that the shape assumed in the model does not make a significant difference to the curves of average degree of consolidation (U) versus the time factor (T). The adoption of a circular smear zone is therefore reasonable.

There are many uncertainties regarding the value of k_h/k_s. Since for most natural deposits the hydraulic conductivity in the horizontal direction is higher than that in the vertical direction, Hansbo (1987) proposed that k_s should be the same as the hydraulic conductivity of the natural soil in the vertical direction, k_v. The value of k_h/k_v can vary from 1 to 15 (Jamiolkowski et al. 1983). It was argued by Chai and Miura (1999) that the assumption of k_s/k_v is based mainly on laboratory test results. Laboratory tests may be an appropriate way to determine the value of k_s, but they generally under-estimate the hydraulic conductivity of a field deposit, mainly because a 20 mm thick soil sample is too small to capture all the effects of stratification of a natural deposit. It is therefore suggested that k_h/k_s can be expressed as:

$$\frac{k_h}{k_s} = \left(\frac{k_h}{k_s}\right)_l \cdot C_f \tag{3.11}$$

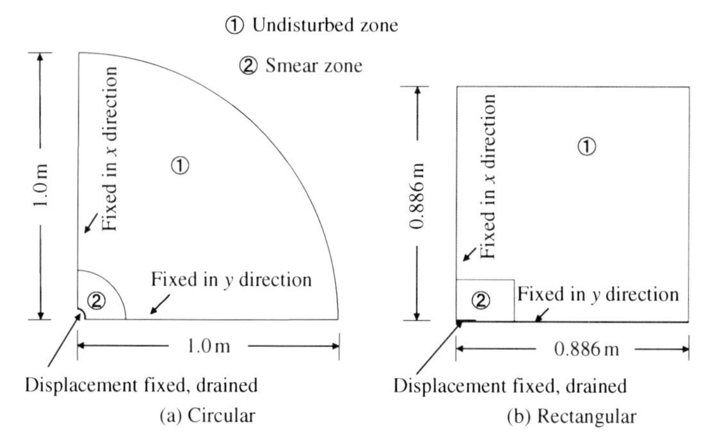

Fig. 3.11 Modelling the smear zone around a drain (after Chai and Miura 1999)

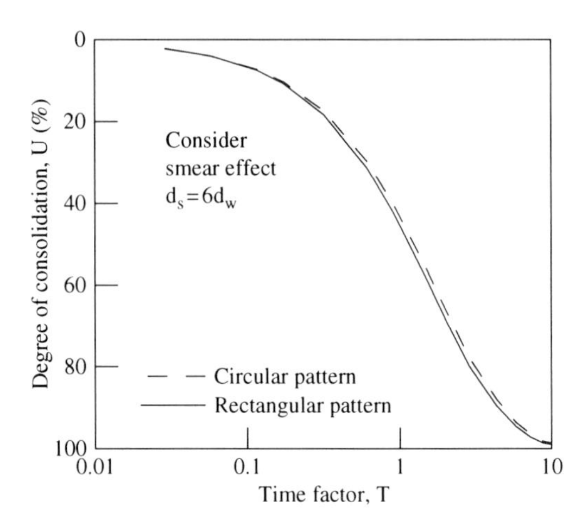

Fig. 3.12 Comparison of average degree of consolidation for different assumption regarding the shape of the smear zone (after Chai and Miura 1999)

where subscript l represents the value determined in the laboratory, and C_f is the hydraulic conductivity ratio between field and laboratory values. In some cases, $(k_h/k_s)_l = (k_h/k_v)_l$. The best way for estimating the value of field hydraulic conductivity is by back-analyzing local case histories, such as measured embankment settlements, or by measuring the hydraulic conductivity in the field using a self-boring permeameter, if such sophisticated equipment is available (Tavenas et al. 1986). It is considered that the most important factor affecting the value of C_f is the stratification of the deposit. For a homogeneous deposit, the value of C_f can be close to 1.0, but for stratified deposits, even those with thin sand layers and sand seams which can not be clearly identified from the borehole record, the value of C_f can be much larger than 1.0. Values of C_f for a few selected clay deposits are listed in Table 3.2 for reference.

Table 3.2 C_f values for various clay deposits

Deposit	C_f	Method	Remarks
Bangkok clay at Asian Institute of Technology campus (about 100 km from sea)	25	Back-analysis	Chai et al. (1995)
Bangkok clay at Nong Hao (close to sea)	4	Back-analysis	Chai et al. (1996)
Malaysia Muar clay deposit	2	Back-analysis	Chai and Bergado (1993)
Ariake clay (close to sea area)	4	Back-analysis	Chai and Miura (1999)
Louiseville (Canada)	About 1[a]	Self-boring permeameter	Tavenas et al. (1986)
St-Alban (Canada)	About 3[a]	Self-boring permeameter	Tavenas et al. (1986)

[a]Laboratory value was determined by direct measurement. For other cases, laboratory values were deduced from c_v value (c_v is coefficient of consolidation)

For the unit cell theory proposed by Hansbo (1981), a single value of hydraulic conductivity is used in the smear zone. However, experimental results indicate that the degree of disturbance, and therefore the coefficient of consolidation in the smear zone, varies with the radial distance from the drain (Onoue 1991; Madhav et al. 1993). Based on this experimental evidence, a unit cell consolidation theory, which considers the hydraulic conductivity in the smear zone to vary linearly or bi-linearly with position has been proposed by Chai et al. (1997). It has been demonstrated that by adopting the average value of hydraulic conductivity in the smear zone, the smear effect will be under-estimated. The value of hydraulic conductivity at the inner smear zone, i.e., in the zone immediately adjacent to the drain surface, has a dominant effect on the consolidation behaviour of a unit cell.

3.2.4 Effect of a Sand Mat

Some or all of the water collected by the PVDs will flow to the ground surface first, and then drain away by the outlet system, usually consisting of a sand mat placed over the ground surface. Since the hydraulic conductivity of sand is considerably higher than that of clay, usually in an analysis it is assumed that there is no hydraulic resistance within the sand mat. If a thick layer (more than 0.5 m) of clean sand, i.e., with the percentage fines less than 5%, is used in the construction of the sand mat, then this assumption may not result in much error. However, in some cases due to material availability as well as cost considerations, lower quality sand, such as clayey sand, may be used to construct the sand mat. In this case, the hydraulic resistance in the sand mat may have some influence on the rate of consolidation of the subsoil containing the vertical drains. The amount of resistance in the sand mat is a function of its hydraulic conductivity as well as the geometry of the embankment constructed over the mat. By analyzing a test embankment at Saga Airport in Japan, Chai and Miura (1999) suggested that a sand mat at least 0.5 m thick with a hydraulic conductivity larger than 10^{-4} m/s, should be sufficient to provide conditions that are close to free draining.

3.3 Optimum PVD Installation Depth

For a deposit with only one-way drainage, PVDs are normally installed so that they fully penetrate the deposit in order to shorten the required preloading period. However, for a deposit with two-way drainage, e.g., in cases where a sand layer underlies soft clay, full penetration of the PVDs into the deposit may not be an economical choice. Retaining a thin clay layer without PVD improvement near the underlying drainage boundary can have advantages. Indeed, almost the same overall rate of consolidation may be achieved with only partial penetration of the clay layer as in the case of full penetration of the layer by PVDs. Therefore, it is possible to design a PVD improvement with partial penetration to achieve approximately the same rate of consolidation as that of a fully penetrated case. This installation depth will be referred to as the optimum depth.

3.3.1 Optimum Thickness of Unimproved Sub-layer (H_c)

The condition assumed in deriving an equation for calculating the optimum thickness of the unimproved soil layer, H_c (Fig. 3.13), is that the average degree of consolidation of the sub-layer of thickness H_c under one-way drainage is the same as the average degree of consolidation of the same deposit with full penetration of the PVDs. Considering both the drainage capacity of the natural deposit and the effect of the PVDs, the average degree of consolidation of a PVD-improved deposit can be calculated as follows (Carrillo 1942):

$$U_{vh} = 1 - (1 - U_v)(1 - U_h) \tag{3.12}$$

where U_{vh} = the average degree of consolidation of the PVD-improved subsoil layer, U_h = the average degree of consolidation of the layer due to radial drainage only (due to the presence of the PVDs), which can be calculated by Hansbo's (1981) solution (Eq. 3.7), and U_v = the average degree of consolidation of the layer due to vertical drainage alone.

In order to obtain a simple explicit expression for H_c, an approximation of Terzaghi's solution for the average degree of vertical consolidation has been proposed (Chai et al. 2001):

$$U_v = 1 - \exp(-C_d T_v) \tag{3.13}$$

where $T_v = \dfrac{c_v t}{h^2}$ is the time factor for vertical consolidation, c_v = the coefficient of consolidation in the vertical direction, h = the vertical drainage path length and C_d = a constant. In order to determine the most appropriate value of C_d, the following factors need to be considered.

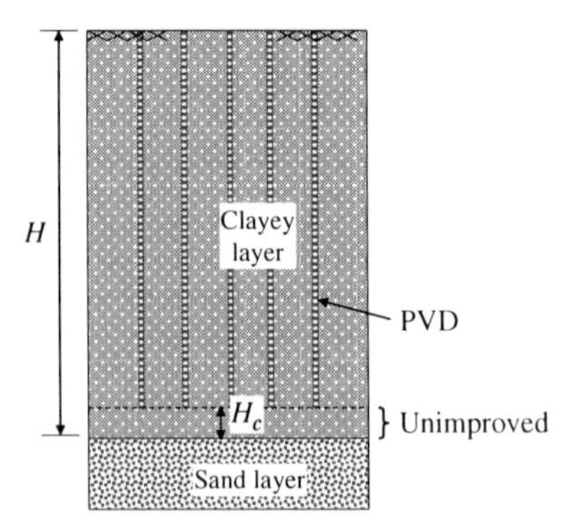

Fig. 3.13 Illustration of improved and unimproved layer

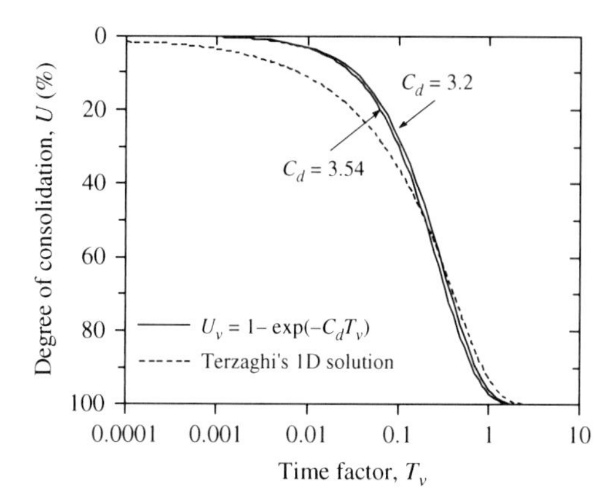

Fig. 3.14 Effect of C_d on the degree of consolidation (after Chai et al. 2001)

Figure 3.14 shows a comparison of Terzaghi's solution for consolidation assuming vertical drainage only as well as the approximate solution given by Eq. (3.13). It may be seen from this figure that the two solutions are almost the same for $U_v = 50\%$, provided C_d is assigned a value of 3.54. In this case, for $U_v < 50\%$, Eq. (3.13) underestimates the Terzaghi solution, while for $U_v > 50\%$ it overestimates the average degree of consolidation, but with a maximum error of less than 10%.

By considering the combined effects of both vertical drainage of the natural subsoil and radial drainage due to the presence of a PVD, the maximum error in the average degree of consolidation will be reduced below that for the case assuming vertical drainage alone. In this case, in order to retain the use of the approximate but simple Eq. (3.13), the optimum value for C_d then becomes a function of the relative importance of the vertical and the radial drainage, which of course may vary from case to case. In order to be definitive, a value of $C_d = 3.2$ has been proposed by Chai et al. (2001) and its adoption is recommended here for most practical cases. Of the cases of combined vertical and radial drainage studied by Chai et al. (2001), the maximum error of the proposed method, in terms of the predicted average degree of consolidation, was less than 5% whenever the value $C_d = 3.2$ was adopted.

For the unimproved layer (H_c), the average degree of consolidation (U_{v2}) can be calculated by Eq. (3.13) with a corresponding time factor, $T_{v2} = c_v t / H_c^2$. By equating the average degree of consolidation of the unimproved layer (of thickness H_c) and the average degree of consolidation of the deposit which is fully penetrated by PVDs and has a vertical drainage length of $H/2$, H_c can be expressed as follows:

$$H_c = \frac{1}{\sqrt{\dfrac{4}{H^2} + \dfrac{2.5}{\mu \cdot D_e^2} \dfrac{c_h}{c_v}}} \tag{3.14}$$

Values of H_c calculated from Eq. (3.14) ensure that the average degree of consolidation of the unimproved layer under one-way drainage is the same as the average degree of consolidation of a similar deposit with full penetration of PVDs. However, with the arrangement shown in Fig. 3.13, the vertical drainage path length for the PVD-improved layer, i.e., $H - H_c$, is longer than that for the case where the PVDs fully penetrate the layer, i.e., $H/2$. Theoretically, this increase of the drainage path length will also result in increased well resistance for the case of a layer with PVDs installed, and it will also reduce the rate of consolidation occurring due to vertical drainage through the natural deposit. For most practical cases, when the discharge capacity (q_w) of a PVD is larger than about 100 m^3/year, the effect of well resistance is not significant (Bergado et al. 1996). In such cases the main effect of an increase in the drainage path length will be to reduce the degree of consolidation due to the vertical drainage through the natural deposit. This relative effect will increase with the increase of PVD spacing, smear effects and the thickness of the deposit in question. To compensate for this effect, the average degree of consolidation of the unimproved layer (H_c) should be higher than that for the case where PVDs fully penetrate the layer. One solution to this dilemma is to calculate H_c by assuming that the average degree of consolidation of the unimproved layer is the same as that of the fully penetrating PVD case, as well as assuming that the drainage path length of the natural deposit in the vertical direction is only $H/8$. This value was determined by trial and error and was found to be applicable for $H < 20$ m. This particular condition results in a modified form of Eq. (3.14), which is now the recommended equation for calculating H_c (Chai et al. 2009):

$$H_c = \frac{1}{\sqrt{\frac{64}{H^2} + \frac{2.5}{\mu \cdot {D_e}^2} \frac{c_h}{c_v}}} \tag{3.14a}$$

The value of μ appearing in Eq. (3.14a) can be calculated from Eq. (3.8a).

It should be noted that Eq. (3.14a) was derived by assuming that the deposit is uniform. Although most natural deposits are non-uniform, it is proposed that Eq. (3.14a) can be used to good effect for non-uniform deposits using weighted average values of the coefficient of consolidation in the horizontal direction (c_h) and in the vertical direction (c_v), where the layer values are weighted by their individual layer thicknesses (Chai et al. 2009). The effectiveness of Eq. (3.14a) has been verified by one dimensional finite element analysis, as reported by Chai et al. (2009).

3.4 Two-Dimensional Modelling of PVD-Improved Soil

For most field projects, the subsoil is not uniform, and also the deformation of PVD-improved soil is not always one-dimensional. Thus numerical methods, typically the finite element method (FEM), are required for designing PVD improvement or for predicting the behaviour of PVD-improved subsoil. Methods for modelling PVD-improved subsoil in two-dimensional (2D) plane strain analysis can be classified into 4 groups. The first group represents the individual PVDs by solid elements and reproduces the axisymmetric (unit cell) and plane strain responses by matching the times for 50% consolidation by adjusting the value of the hydraulic conductivity in the horizontal direction assumed for plane strain (Shinsha et al. 1982; Indraratna and Redana 1997). The second group adopts a macro element in the FEM program to represent the hydraulic behaviour of a PVD (Sekiguchi et al. 1986). The third group employs discrete 1D drainage elements to represent individual PVDs (Hird et al. 1992; Chai et al. 1995), while the fourth group combines the drainage effects of the PVDs and the natural soil in the vertical direction by assuming an equivalent hydraulic conductivity in the vertical direction for a single solid medium (Chai et al. 2001). In the following sections these methods will be described and discussed in chronological order of their appearance in the literature.

3.4.1 Approach of Shinsha et al.

In many cases involving the analysis of a boundary value problem, it is necessary to model the effects of PVDs under plane strain conditions, as shown in plan view in Fig. 3.15. In such cases, matching the degree of consolidation under both axisymmetric and plane strain conditions becomes an essential step in the modelling.

When analyzing the behaviour of subsoils improved by the installation of sand drains, Shinsha et al. (1982) proposed a procedure which can match the average degree of horizontal consolidation that occurs under axisymmetric conditions to that occurring in a plane strain analysis for a given degree of consolidation, normally 50%. The matching equation proposed by Shinsha et al. (1982) is as follows:

$$\frac{T_{hp}B^2}{k_{hp}} = \frac{T_h D_e^2}{k_h} \tag{3.15}$$

where B = the half drain spacing in the plane strain model, k_{hp} = the matched horizontal hydraulic conductivity for the plane strain case, and T_{hp} = the time factor corresponding to the given degree of consolidation for plane strain conditions calculated using Terzaghi's 1D consolidation theory with a drainage path length of B. The value of T_h can be calculated from Eq. (3.5) for the same degree of consolidation. Either Barron's (1948) or Hansbo's (1981) solution is appropriate for the case where vertical drains are present in the subsoil. In such cases the equivalent

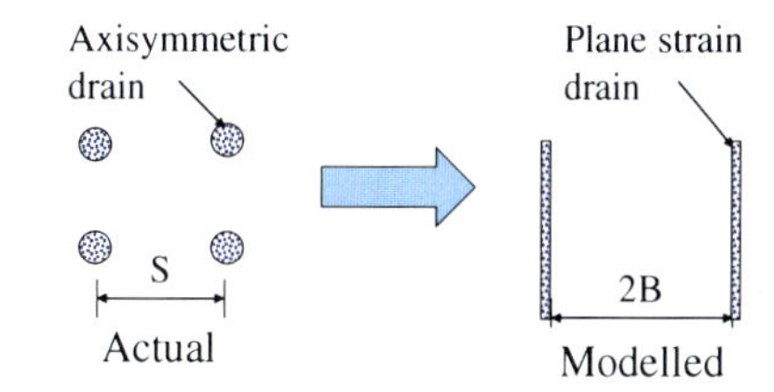

Fig. 3.15 Illustration of modelling the effect of PVDs in plane strain analysis

plane strain drain is represented in the finite element model by solid elements. Since the matching equation was derived on the assumption that Terzaghi's consolidation theory for two-way drainage can be applied to the soil between the two adjacent rows of equivalent vertical drains (effectively assuming horizontal flow of pore water between the adjacent rows), it is automatically assumed that the location of a plane strain vertical drain is indeed a drainage boundary. It is noted that for cases where these equivalent plane strain drains are located at every other vertical line in a finite element mesh, the resulting excess pore water pressure distribution will necessarily be zigzag in form.

3.4.2 Macro-element Approach

A quadrilateral macro-element that includes the beneficial drainage effects of a PVD was formulated by Sekiguchi et al. (1986) in which the degree of freedom corresponding to excess pore water pressure is located at the centre of the element (Fig. 3.16). In formulating the response of this element the following assumptions have been adopted:

(a) the discharge capacity of a PVD is infinite ($q_w \to \infty$);
(b) the smear effect is not considered; and

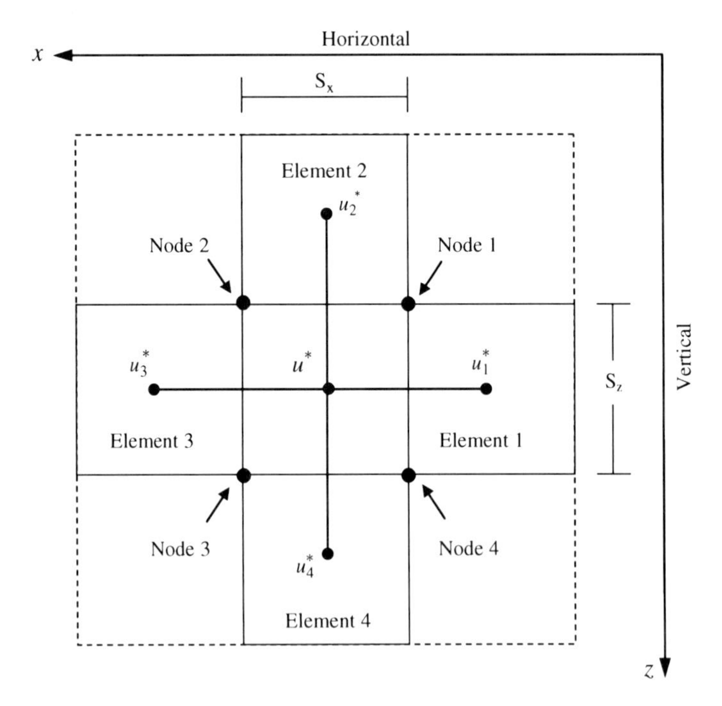

Fig. 3.16 Illustration of excess pore water pressure degree of freedom

(c) the distribution of excess pore water pressure (u) in the radial direction around a PVD is expressed by the eigen-function of an 'equal strain vertical drain consolidation problem' as follows:

$$f(r) = \ln\left(\frac{r}{r_w}\right) - \frac{r^2 - r_w^2}{2r_e^2} \tag{3.16}$$

where r = radial distance. This function satisfies the following boundary conditions: $f(r_w) = 0$, and $\left.\frac{df}{dr}\right|_{r=r_e} = 0$.

3.4.2.1 Equation of Continuity

Assuming that within a time increment Δt the increment of volumetric strain is $\Delta\varepsilon_v$ and that the rate of flow into or out of an element is q, the equation of continuity can be expressed as follows:

$$\int \Delta\varepsilon_v dxdydz = q\Delta t \tag{3.17}$$

As shown in Fig. 3.16, water can flow into or out of an element from the 4 sides of the 2D quadrilateral element. Using Darcy's law, the net change in the amount of water within an element can be expressed as follows:

$$q\Delta t = \left[\left(\sum_{i=1}^{4} \beta_i\right) \cdot u^* - \sum_{i=1}^{4} \left(\beta_i \cdot u_i^*\right)\right] \cdot S_y \tag{3.18}$$

where β_i = the coefficient of water flow (for $i = 1$ to 4), u^* = the excess pore water pressure at the centre of the element being considered, u_i^* = the excess pore water pressure at the centre of ith element adjacent to and surrounding the element being considered, and S_y = the thickness of the element in the direction perpendicular to x-z plane of Fig. 3.16 (normally unity). As an example, the expression for β_i for the line boundary 14 (connecting nodes 1 and 4 in Fig. 3.16) is:

$$\beta_1 = \frac{k_h \cdot \Delta t \cdot S_z/\gamma_w}{d_1 + (l_1 - d_1) \cdot (k_h/k_{h1})} \tag{3.19}$$

where k_h and k_{hl} = the hydraulic conductivities of the central element being considered and the adjacent Element 1, respectively, γ_w = the unit weight of water, d_1 and l_1 = the distances from the centre of the element being considered to the boundary 14 and the centre of Element 1, respectively, and S_z = the length of the line segment 14.

3.4.2.2 Macro-element Formulation

Consider the case where the PVDs are arranged in a square pattern with a spacing $S = S_x = S_y$. For this case the amount of water flowing into or out of a PVD within a time increment of Δt can be expressed as:

$$q_{VD} \cdot \Delta t = 2\pi r_w \cdot S_z \cdot k_h \cdot \Delta t \cdot \left. \frac{\partial (u/\gamma_w)}{\partial r} \right|_{r=r_w} \tag{3.20}$$

where q_{VD} = the flow rate for a section of PVD with a length of S_z. Referring to Eq. (3.18), the following equation can be written:

$$q_{VD} \cdot \Delta t = \beta_{VD} \cdot \bar{u} \cdot S_y \tag{3.21}$$

where

$$\bar{u} = \frac{2\pi \int_0^{r_e} u \cdot r \cdot dr}{\pi {r_e}^2} \tag{3.22}$$

is the average excess pore water pressure, and β_{VD} = the coefficient of water flow for a PVD. Assuming $u = u(r, z, t) = f(r) \cdot g(z, t)$ and using Eq. (3.16) for $f(r)$, from Eq. (3.22), the average excess pore water pressure $\bar{u}$ will be a function of $g(z, t)$. However, by equating Eqs. (3.20) and (3.21) and eliminating $g(z, t)$, an expression for β_{VD} can be obtained for the case of a square pattern of PVDs ($S = S_x = S_y$):

$$\beta_{VD} = \frac{2\sqrt{\pi} \cdot k_h \cdot \Delta t}{\gamma_w} \cdot \frac{S_z}{r_e} \cdot \left[\frac{1 - \frac{1}{n^2}}{\ln(n) - \left(1 - \frac{1}{n^2}\right)\left(3 - \frac{1}{n^2}\right)/4} \right] \tag{3.23}$$

Including the effect of the PVDs on the net water change in Eq. (3.18), replacing u_i^* by $\bar{u}_i$ and equating Eqs. (3.17) and (3.18) results in the following continuity equation for the macro-element:

$$\int \Delta v dx dy dz = (q + q_{VD}) \cdot \Delta t = \left[\left(\beta_{VD} + \sum_{i=1}^{4} \beta_i \right) \cdot \bar{u} - \sum_{i=1}^{4} (\beta_i \bar{u}_i) \right] \cdot S_y \tag{3.24}$$

Replacing the continuity equation in a normal FEM formulation by Eq. (3.24) provides the formulation of a macro-element that includes the effects of the PVDs. Sekiguchi et al. (1986) verified the macro-element they formulated in this way by comparing their numerically simulated results with analytical solutions for selected problems, obtaining good agreement.

3.4.3 One-Dimensional Drainage Elements

3.4.3.1 Formulation of a Drainage Element

Hird et al. (1992) adopted a discrete one-dimensional (1D) drainage element to model the effects of a PVD installed in soil. In the finite element mesh their element has zero cross-sectional area but, in calculating its contribution to the drainage matrix, a unit cross-sectional area can be assumed for convenience. The important quantity defining the behaviour of this element is its discharge capacity, q_w.

This type of element can be used in conjunction with triangular or rectangular continuum elements representing the consolidating soil, i.e., the discrete drainage element can be introduced along one side of a solid or continuum element. Therefore, adding drainage elements into a finite element mesh will not introduce additional nodes or degrees of freedom. Within a drainage element, the variation of excess pore water pressure is assumed to be linear. This type of element is therefore compatible with six noded triangular elements and eight noded rectangular elements, which only have pore water pressure degrees of freedom at the vertex nodes. The formulation of the drainage element is summarised as follows.

With a fully implicit approximation over time, the discretized forms of the equilibrium and continuity equations of Biot's (1941) consolidation theory are:

$$K\Delta d + L\Delta u = \Delta F_1 \tag{3.25}$$

$$L^T\Delta d - \Delta t\Phi\Delta u = \Delta F_2 \tag{3.26}$$

where ΔF_1, ΔF_2, Δd, Δu and Δt = increments of nodal force, nodal flow, nodal displacement, nodal pore water pressure and time, respectively; K = the material stiffness matrix; L = the link (or coupling) matrix and Φ = the hydraulic conductivity matrix. The element 'stiffness' matrix, K_E, for displacement and pore water pressure variables is given by:

$$K_E = \begin{bmatrix} K & L \\ L^T & -\Delta t\Phi \end{bmatrix} \tag{3.27}$$

For a PVD, the relationships between the axial force and displacement and between the axial flow rate and the hydraulic gradient are represented by the matrices K and Φ respectively. Physically, it is assumed that the PVD has negligible cross-sectional area. Hence, no volume change results from displacement and no change of force in the drain results from a change of pore water pressure. Consequently for these elements the link matrices L and L^T are composed entirely of zero terms. Normally in such cases the structural stiffness a PVD can be ignored ($K = 0$), and Φ is a function of q_w.

3.4.3.2 Matching Procedure for Plane Strain Analyses

To get a better matching between plane strain and axisymmetric conditions, Hird et al. (1992) applied Hansbo's (1981) solution to a plane strain vertical drain. Figure 3.17 shows a plane strain unit cell of half width, B, containing a drain with a discharge capacity, q_{wp}, per unit width in the direction perpendicular to the plane under consideration. For matching purposes the drain is assumed to possess negligible thickness and no smear zone. From Hansbo's solution, the average degree of consolidation (U_h) can be expressed as follows:

$$U_{hp} = 1 - \exp(-8T_{hp}/\mu_p) \tag{3.28}$$

where

$$T_{hp} = \frac{c_h t}{4B^2} \tag{3.29}$$

and

$$\mu_p = \frac{2}{3} + \frac{2k_h}{Bq_{wp}}\left(2lz - z^2\right) \tag{3.30}$$

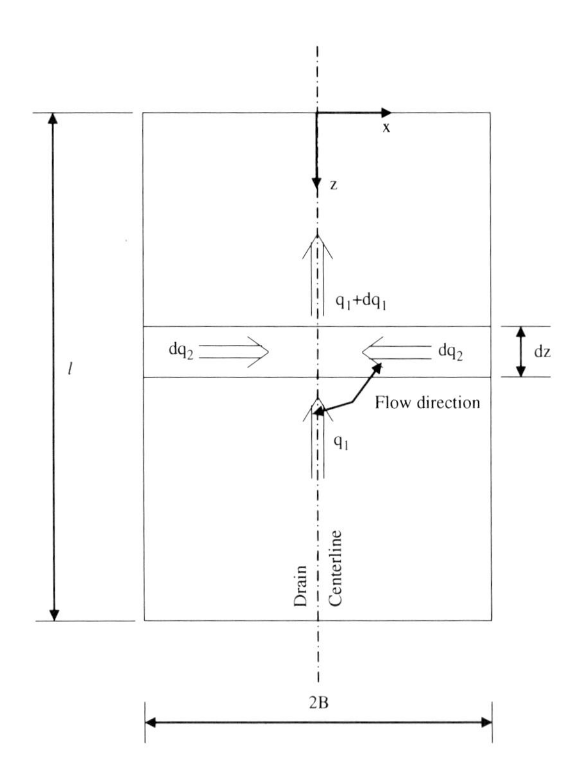

Fig. 3.17 Plane strain unit cell

For the rates of consolidation in a plane strain and an axisymmetric unit cell to be matched, equality of the average degree of consolidation at any time and at every level in the unit cell is required. From Eqs. (3.7) and (3.28), this implies:

$$\frac{T_{hp}}{\mu_p} = \frac{T_h}{\mu} \tag{3.31}$$

Geometry Matching

Assuming c_h (and therefore, k_h) is the same for both axisymmetric and plane strain conditions, Eq. (3.31) can be re-written as:

$$B^2 \mu_p = {r_e}^2 \mu \tag{3.32}$$

The relevant expressions for μ_p (Eq. 3.30) and μ (Eq. 3.8) can be substituted into Eq. (3.32) and the terms rearranged to give:

$$\begin{aligned} \frac{2}{3}B^2 - {r_e}^2\left[\ln(n/s) + (k_h/k_s)\ln(s) - \tfrac{3}{4}\right] = \big[\left(\pi {r_e}^2 k_h/q_w\right) \\ - \left(2Bk_h/q_{wp}\right)\big] \cdot \left(2lz - z^2\right) \end{aligned} \tag{3.33}$$

The condition for geometric matching is obtained by considering the case of negligible well resistance (q_w and $q_{wp} \rightarrow \infty$). Therefore, geometric matching, including the effect of the smear zone, is achieved if:

$$B/r_e = \left\{\frac{3}{2}\left[\ln(n/s) + (k/k_s)\ln(s) - \frac{3}{4}\right]\right\}^{1/2} \tag{3.34}$$

The effect of well resistance is matched independently if:

$$q_{wp} = \left(2B/\pi {r_e}^2\right) q_w \tag{3.35}$$

Hydraulic Conductivity Matching

An alternative procedure can be developed which involves changing the hydraulic conductivity instead of the drain spacing. Adopting $B = r_e$ and following the same logic as before, the hydraulic conductivity and well resistance matching requirements are:

$$k_{hp} = \frac{2k_h}{3\left[\ln(n/s) + (k_h/k_s)\ln(s) - \frac{3}{4}\right]} \tag{3.36}$$

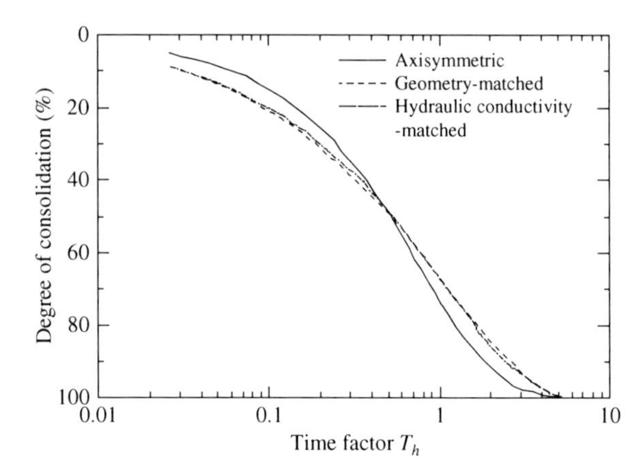

Fig. 3.18 Comparison of the results of axisymmetric and matched plane strain analyses (after Hird et al. 1992)

and

$$q_{wp} = (2/\pi r_e)\, q_w \tag{3.37}$$

Although geometry and hydraulic conductivity matching procedures have been described separately, they can be used in combination by setting B to a desired value and re-deriving the matching requirements for k_{hp} and q_{wp}. This is useful in avoiding the necessity of modelling a small equivalent drain spacing with an excessively large number of finite elements.

Assuming the soil behaves elastically with a Young's modulus $E = 10^4$ kPa, Poisson's ratio $\nu = 0$, and $k_h = 10^{-8}$ m/s, $r_e = 10.0$ m, $r_w = 0.2$ m, $l = 20.0$ m, $s = 5$, $k_h/k_s = 2$, and $L = 8k_h l^2/(\pi q_w) = 2$, Hird et al. (1992) conducted finite element consolidation analyses of axisymmetric and matched plane strain unit cells. The average degrees of consolidation based on predicted settlements are compared in Fig. 3.18 (after Hird et al. 1992). The results indicate that the error introduced from the plane strain matching is less than 11%. The geometry and hydraulic conductivity matching produced almost identical results.

Well Resistance (Discharge Capacity) Matching

While the geometry and hydraulic conductivity matching procedures can match well the average degree of consolidation within a unit cell, the excess pore water pressure distributions within an axisymmetric unit cell and a plane strain unit cell are quite different. Theoretically, excess pore water pressure (u) distributions in the horizontal direction in axisymmetric (Hansbo 1981) and plane strain (Hird et al. 1992) unit cells are as follows:

$$u = \frac{\gamma_w}{2k_h}\left[{r_e}^2 \ln\left(\frac{r}{r_w}\right) - \frac{r^2 - r_w^2}{2}\right]\frac{\partial \varepsilon_v}{\partial t} \quad \text{(axisymmetric)} \tag{3.38}$$

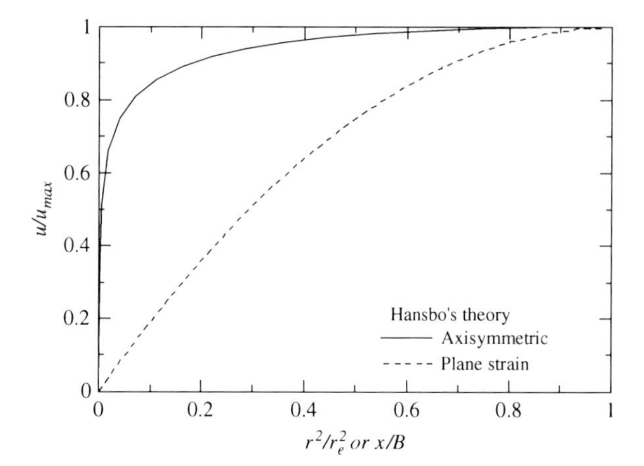

Fig. 3.19 Theoretical excess pore water pressure variation within an axisymmetric and plane strain unit cell

$$u = \frac{\gamma_w}{2k_h}\left(2Bx - x^2\right)\frac{\partial \varepsilon_v}{\partial t} \qquad \text{(plane strain)} \qquad (3.39)$$

where, ε_v = vertical strain, and x = the distance from the centre of the drain for plane strain conditions. It should be noted that for the equal vertical strain assumption, as used elsewhere, the strain rate, $\partial \varepsilon_v / \partial t$, is not a function of r or x. The theoretical excess pore water pressure distribution, normalized by the value at the periphery of a unit cell ($u_{\max}$), is shown in Fig. 3.19. Instead of plotting r or x on the horizontal axes, the ratio of areas, $r^2/{r_e}^2$or x/B is used, which provides a better picture of the excess pore water pressure variation within the complete unit cell area. It can be seen that theoretical excess pore water pressure distribution patterns for the plane strain and the axisymmetric conditions are different, mainly because of different boundary conditions. The stronger variation of excess pore water pressure in the plane strain case implies a more non-uniform response of the soil.

The average degree of horizontal consolidation of a unit cell is not only influenced by drain spacing and the hydraulic conductivity of the soil, but is also influenced by the discharge capacity of a drain. Therefore, the average degree of horizontal consolidation can be matched by fixing both the geometry and the hydraulic conductivity of the soil and varying the well resistance (Chai et al. 1995). In geometry and hydraulic conductivity matching, it can be easily proved that for the condition where $n(n = r_e/r_w)$ is larger than 5 (in practice, normally $n > 20$), either the drain spacing must be increased or the hydraulic conductivity of the soil has to be reduced for the plane strain condition. Therefore, logically, in well resistance matching, the well resistance for plane strain conditions will be higher, and the excess pore water pressure variation in the horizontal direction will be reduced within a unit cell. For the condition of equal average horizontal degree of consolidation in the plane strain and the axisymmetric conditions, based on the unit cell solutions, both axisymmetric (Hansbo 1981) and plane strain (Hird et al. 1992), and using the average well resistances along the drain, the following equation can be obtained:

$$r_e{}^2\left(\ln\frac{n}{s}+\frac{k_h}{k_s}\ln s-\frac{3}{4}+\pi\frac{2l^2k_h}{3q_w}\right)=B^2\left(\frac{2}{3}+\frac{4l^2k_{hp}}{3Bq_{wp}}\right) \tag{3.40}$$

If $B = r_e$ (note that B can be larger than r_e) and $k_{hp} = k_h$, and with some rearrangement, the equivalent discharge capacity of the drain in plane strain conditions can be determined by matching the average degree of horizontal consolidation as follows:

$$q_{wp}=\frac{4k_hl^2}{3B\left[\ln\left(\frac{n}{s}\right)+\frac{k_h}{k_s}\ln(s)-\frac{17}{12}+\frac{2l^2\pi k_h}{3q_w}\right]} \tag{3.41}$$

As illustrated in Hansbo's solution, for a constant discharge capacity of the drain the well resistance increases with depth. Therefore, partly matching the well resistance based on the geometry and smear effects for the axisymmetric case to the well resistance for the plane strain case determined by Eq. (3.41), the variation of excess pore water pressure with depth will not be the same as for the axisymmetric case, although a more realistic excess pore water pressure distribution in the horizontal direction can be obtained. To overcome this drawback, it is reasonable to consider artificially varying the discharge capacity of the drain with depth in the model.

Assuming the average discharge capacity is $(q_w)_{av}$, which may vary with depth (z) from a value of $A_1(q_w)_{av}$ at the top of the PVD and a value of $(A_1 + A_2)(q_w)_{av}$ at the bottom (where $A_1 + A_2/2 = 1.0$, Chai et al. (1995) obtained a unit cell consolidation solution for axisymmetric conditions, and then matched the varying discharge capacity to the plane strain condition. It has been shown that the selection of $A_1 = 0.1$ and $A_2 = 1.8$ can result in an acceptable excess pore water pressure variation with depth under plane strain conditions. In cases where the same average discharge capacity is required, the variable discharge capacity case ($A_1/A_2 > 0$) will yield a higher well resistance, or a lower degree of consolidation for a given time, than that of the constant discharge capacity case. To obtain the same average well resistance, the average discharge capacity for the variable discharge case ($A_1/A_2 > 0$) must be increased. If the ratio between the average discharge capacity for the case of variable discharge to the constant discharge capacity is designated as R_q, then under the condition of the same average well resistance, R_q can be determined as follows (Chai et al. 1995):

$$R_q=\frac{3}{A_2^2}\left\{-\frac{A_2}{2}+(A_1+A_2)\left[\left(1+\frac{A_1}{A_2}\right)\ln\left(1+\frac{A_2}{A_1}\right)-1\right]\right\} \tag{3.42}$$

R_q is the same for both the axisymmetric and the plane strain conditions. For $A_1 = 0.1$ and $A_2 = 1.8$, $R_q = 2.875$. In finite element analysis, the discharge capacity of each PVD element can be determined by its location relative to the top and the bottom of the PVD.

In order to check this approach for matching the well discharge, an ideal drain case was analyzed using FEM. The soil was assumed to be elastic with Young's

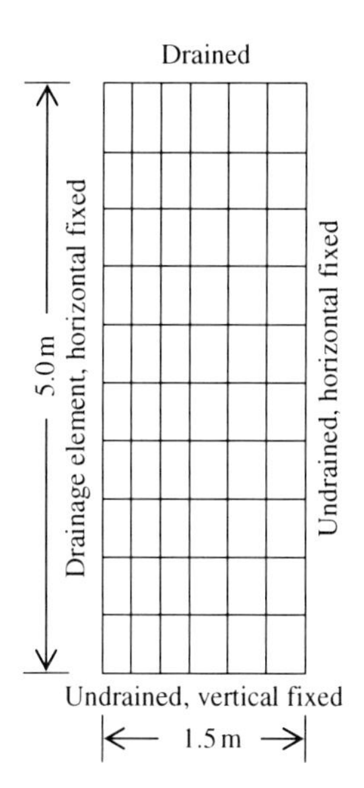

Fig. 3.20 Finite element mesh used in drain matching problem

modulus $E = 10^4$ kPa, Poison's ratio $\nu = 0.3$, and the horizontal hydraulic conductivity was given by $k_h = 10^{-8}$ m/s. The other parameters in the problem were defined as $l = 5$ m, $d_w = 0.013$ m and $r_e = 1.5$ m. The finite element mesh used for this analysis is shown in Fig. 3.20.

For the ideal case ($q_w \rightarrow \infty$), Eq. (3.41) yielded an equivalent discharge capacity in plane strain, q_{wp} of 8.9×10^{-8} m^3/s (or 2.81 m^3/year). Figure 3.21 (modified from Chai et al. 1995) compares the average degree of horizontal consolidation based on the value of the excess pore water pressure at the bottom of the drain. Hansbo's (1981) solution is also shown. It can be seen that the proposed matching procedure for well resistance yielded a satisfactory match in terms of the average degree of horizontal consolidation. The predicted maximum difference between the axisymmetric and plane strain cases is about 2%. A comparison of the normalized excess pore water pressure distributions is shown in Fig. 3.22 (modified from Chai et al. 1995). It can be seen that the well resistance matching procedure has resulted in a more realistic excess pore water pressure distribution in the horizontal direction (compare the distributions plotted in Figs. 3.19 and 3.22).

3.4.4 Modelling PVDs Using Equivalent Solid Elements

Indraratna and Redana (1997; 2000) derived a solution for a plane strain unit cell with a half width of PVD of b_w, and a half width of the smear zone of b_s, as shown in Fig. 3.23. The equation for the average degree of consolidation is the same as Eq. (3.28), but the expression for μ_p is as follows:

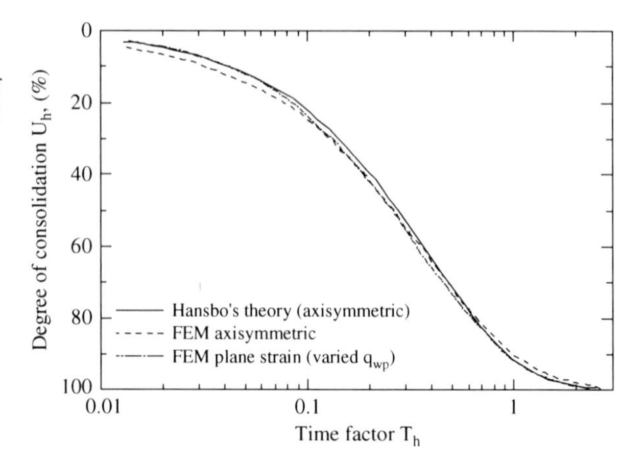

Fig. 3.21 Comparison of the average degree of horizontal consolidation at the bottom of the PVD (modified from Chai et al. 1995)

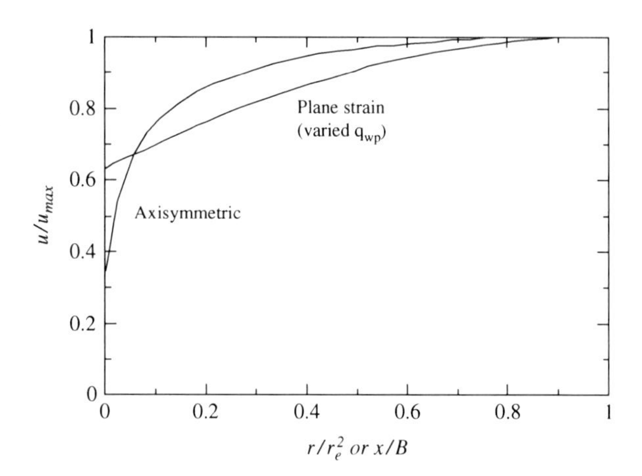

Fig. 3.22 Comparison of normalized excess pore water pressure distribution in the horizontal direction (modified from Chai et al. 1995)

$$\mu_p = \left[\alpha + \beta \frac{k_{hp}}{k_{sp}} + \theta \left(2lz - z^2\right)\right] \tag{3.43}$$

where

$$\alpha = \frac{2}{3} - \frac{2b_s}{B}\left(1 - \frac{b_s}{B} + \frac{b_s^2}{3B^2}\right) \tag{3.44}$$

$$\beta = \frac{1}{B^2}(b_s - b_w)^2 + \frac{b_s}{3B^3}\left(3b_w^2 - b_s^2\right) \tag{3.45}$$

$$\theta = \frac{2k_{hp}^2}{k_{sp}Bq_{wp}}\left(1 - \frac{b_w}{B}\right) \tag{3.46}$$

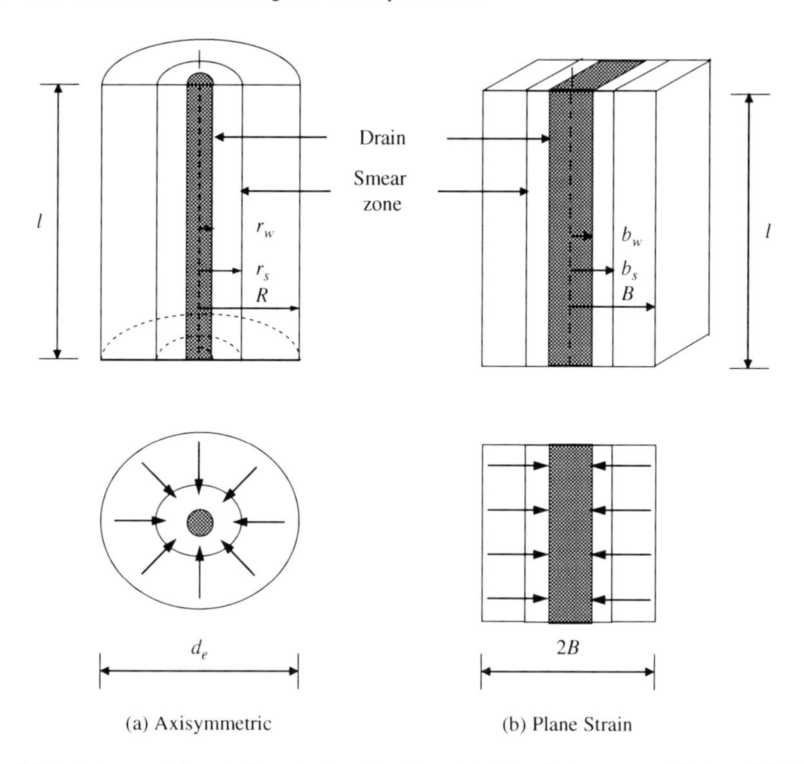

Fig. 3.23 Axisymmetric and plane strain unit cell models (after Indraratna and Redana 1997)

and where k_{sp} = the hydraulic conductivity of the smear zone in the plane strain unit cell.

Assuming the radius of an axisymmetric unit cell is the same as the half width of a matched plane strain unit cell ($r_e = B$), a matching equation for the hydraulic conductivity has been derived as:

$$k_{hp} = \frac{k_h\left[\alpha + \beta\frac{k_{hp}}{k_{sp}} + \theta\left(2lz - z^2\right)\right]}{\left[\ln\left(\frac{n}{s}\right) + \left(\frac{k_h}{k_s}\right)\ln(s) - 0.75 + \pi\left(2lz - z^2\right)\frac{k_h}{q_{wp}}\right]} \tag{3.47}$$

In order to use Eq. (3.47), the values of the discharge capacity (q_{wp}) and the ratio of the horizontal hydraulic conductivity in the smear zone to that in the undisturbed zone (k_{sp}/k_{hp}) of the plane strain unit cell first have to be determined. Indraratna and Redana (1997) proposed that q_{wp} can be calculated by Eq. (3.37), which has been proposed by Hird et al. (1992). The ratio k_{sp}/k_{hp} can be calculated by ignoring the well resistance in Eq. (3.47) and thus can be expressed as follows (Indraratna and Redana 2000):

$$\frac{k_{sp}}{k_{hp}} = \frac{\beta}{\frac{k_{hp}}{k_h}\left[\ln\left(\frac{n}{s}\right) + \left(\frac{k_h}{k_s}\right)\ln(s) - 0.75\right] - \alpha} \tag{3.48}$$

However, the right hand side of Eq. (3.48) also contains the term k_{hp}/k_h, and therefore in order to obtain the value of k_{hp}, either iteration is needed or some kind of approximation is required. Indraratna and Redana (1997) proposed the following approximation of k_{hp}/k_h by ignoring both the smear effect and the well resistance:

$$\frac{k_{hp}}{k_h} = \frac{0.67}{[\ln(n) - 0.75]} \tag{3.49}$$

Finally, if Eqs. (3.37), (3.48) and (3.49) are substituted into Eq. (3.47), the matched horizontal hydraulic conductivity for the plane strain case can be obtained.

In order to use the solution for the plane strain unit cell with smear effect in a plane strain FEM analysis, a plane strain PVD has to be modelled by at least 5 columns of finite elements, i.e., 1 for the PVD itself, 2 for the smear zone and 2 for the remaining zones of the plane strain unit cell. In engineering practice this requirement may not always be convenient to implement.

3.4.5 *Equivalent Vertical Hydraulic Conductivity*

For those methods of finite element analysis that require the use of 1D drainage elements or those adopting a macro-element formulation, conducting the FEM analysis for PVD-improved subsoil can be very time consuming and even inconvenient (Olson 1998). To avoid this shortcoming Chai et al. (2001) proposed a method involving the adoption of an equivalent vertical hydraulic conductivity, in which PVD-improved subsoil can be analyzed in the same manner as unimproved subsoil.

3.4.5.1 Equivalent Vertical Hydraulic Conductivity, k_{ve}

From a macro perspective, the presence of vertical drains increases the mass hydraulic conductivity of the subsoil particularly in the vertical direction. Therefore, it is logical to try to establish a value of the vertical hydraulic conductivity which approximately represents both the effects of the vertical drainage of the natural subsoil and the effects of radial drainage due to the existence of the PVDs. This equivalent value of vertical hydraulic conductivity (k_{ve}) can be derived based on the same average degree of consolidation under 1D conditions (e.g., Chai et al. 2001). To obtain a simple explicit equation for k_{ve}, Eq. (3.13) has been used to approximate Terzaghi's 1D consolidation theory, and Hansbo's (1981) solution (Eq. 3.7) for calculating PVD-induced consolidation has been adopted. Carrillo's (1942) equation is used to combine the vertical and radial drainage effects (Eq. (3.12)). In such cases the equivalent vertical hydraulic conductivity, k_{ve}, can be expressed as (Chai et al. 2001):

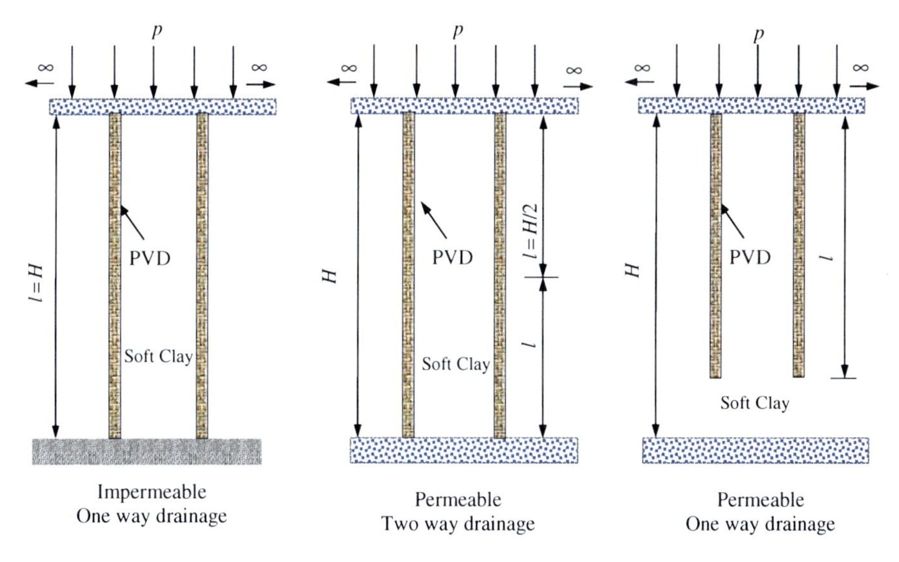

Fig. 3.24 One and two way drainage conditions (after Chai et al. 2001)

$$k_{ve} = \left(1 + \frac{2.5l^2}{\mu d_e^2}\frac{k_h}{k_v}\right)k_v \tag{3.50}$$

where k_v = the hydraulic conductivity of the soil in the vertical direction. The other parameters appearing in this equation have been defined previously.

For multilayer conditions, when using Eq. (3.50) to calculate the value of k_{ve} for each layer, it is simply assumed that the drainage path length l is the same as the total thickness (H) of the PVD-improved zone, i.e., for one-way drainage $l = H$, and $l = H/2$ for two-way drainage. One-way or two-way drainage conditions should be applied based on the conditions in the PVD improved zone only. In the case where a clay deposit has two-way drainage but the PVDs do not penetrate through the whole thickness of the layer, then for the PVD-improved zone, one-way drainage conditions should be assumed, as illustrated in Fig. 3.24.

3.4.5.2 Verification of the Method

For brevity and convenience, the equivalent vertical hydraulic conductivity method will be referred to in the following as the 'simple method'. The simple method was tested by comparing its predicted average degree of consolidation and excess pore water pressure distribution in the vertical direction with values calculated using a combination of Terzaghi's 1D consolidation theory and Hansbo's (1981) solution for radial consolidation. It is possible to obtain a closed form solution for the case of a uniform subsoil. For multi-layer subsoils, the analyses were conducted numerically using the FEM and the results of the simple method were compared with those obtained from separate analyses using drainage elements (Chai et al. 1995). All comparisons were made assuming 1D deformation conditions.

(a) Uniform Subsoil

The assumed 1D configuration is illustrated on the left hand side of Fig. 3.25 (after Chai et al. 2001) and the assumed subsoil and drain parameters are listed in Table 3.3. It should be mentioned that the absolute values of the parameters adopted in this example have no effect on the trends observed in this comparison. With these conditions, Eq. (3.50) yields a value of $k_{ev} = 11.94k_v$.

A comparison of the average degree of consolidation is shown in Fig. 3.26 (after Chai et al. 2001). It can be seen that the maximum difference between the predictions of the simple method and the theoretical solution is about 5%. The predicted excess pore water pressure distribution in the vertical direction is compared in Fig. 3.27 (after Chai et al. 2001) at an average degree of consolidation of 50%, where u_0 is the initial uniform excess pore water pressure. The simple method indicates a faster excess pore water pressure dissipation in the upper part of the

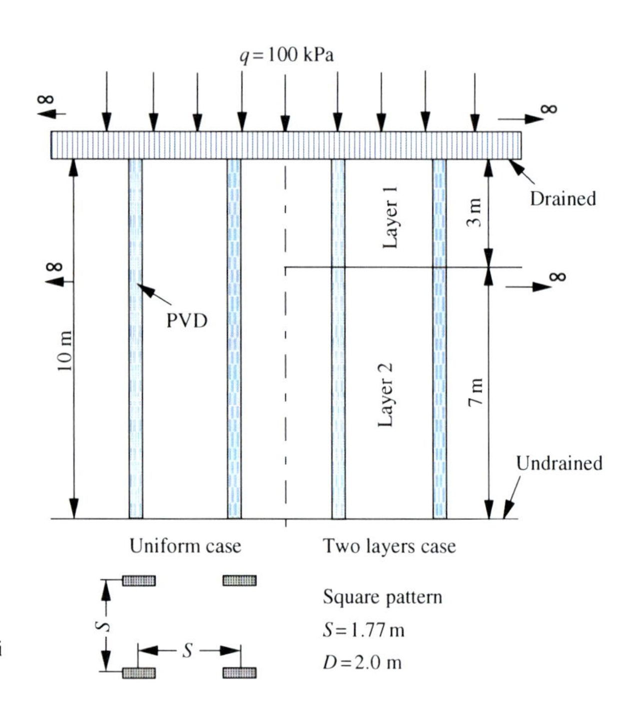

Fig. 3.25 Assumed 1-D subsoil conditions (after Chai et al. 2001)

Table 3.3 Assumed subsoil and drain parameters

Subsoil				
E (kPa)	ν	k_h (10^{-8}m/s)	k_v (10^{-8}m/s)	γ_t (kN/m^3)
4,000	0.3	2.0	1.0	18.0
Drain				
d_w (m)	d_e (m)	d_s (m)	k_h/k_s	q_w (m^3/year)
0.05	2.0	0.3	5.0	100

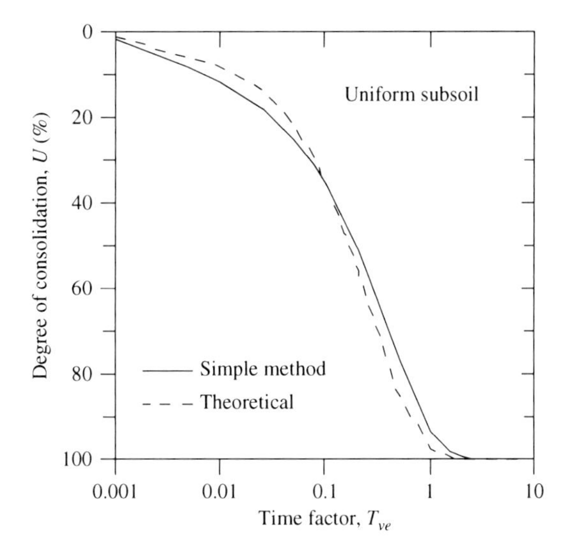

Fig. 3.26 Comparison of the average degree of consolidation of uniform subsoil (after Chai et al. 2001)

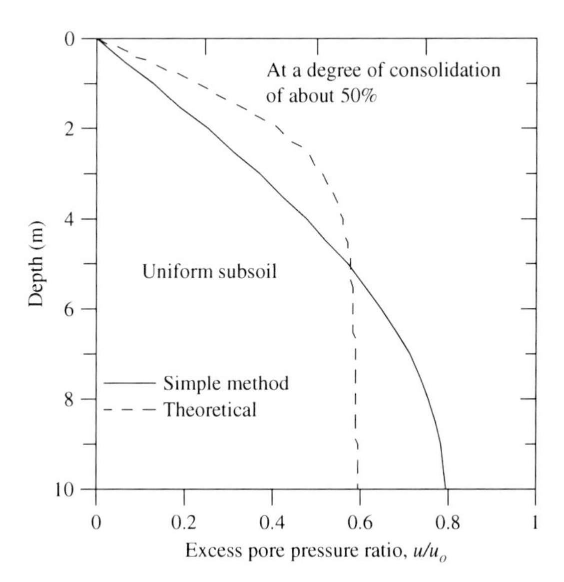

Fig. 3.27 Comparison of excess pore water pressure distribution of uniform subsoil (after Chai et al. 2001)

subsoil and slower dissipation in the lower part. This is because the simple method converts both the effects of vertical and radial drainage into an equivalent vertical drainage. Radial drainage tends to result in almost uniform excess pore water pressure distribution with depth, provided the well resistance of the PVDs is not significant.

(b) Multilayer Subsoil

A uniform subsoil condition is rarely encountered in engineering practice. A more common situation is where a weathered crust with higher hydraulic conductivity is underlain by softer layers. A two layer subsoil model is indicated on the right hand side of Fig. 3.25. The assumed values of the hydraulic conductivities of layer 1 are $k_{v1} = 2 \times 10^{-8}$ m/s and $k_{h1} = 2k_{v1}$, and the elastic modulus of layer 1 (E_1) is 8,000 kPa. The other soil parameters for layer 1 and layer 2, and the drain parameters are the same as those given in Table 3.3. Consequently, the coefficient of consolidation of layer 1 (c_{v1}) is 4 times that of layer 2 (c_{v2}).

A rigorous closed form solution is not available for the case of a multilayered subsoil and so the FEM was used for these calculations. FEM solutions using the simple method were compared with base solutions calculated using the method in which radial drainage due to the presence of the PVDs was modelled by including 1D drainage elements in the finite element mesh (Chai et al. 1995).

Figure 3.28 (after Chai et al. 2001) compares predictions of the average degree of consolidation, which shows that the maximum difference is about 5%. In Fig. 3.28 the time factor T_{ve} was calculated using the weighted average value of the coefficient of consolidation in the vertical direction (weighted by the individual layer thickness). The distribution of excess pore water pressure with depth is given in Fig. 3.29 (after Chai et al. 2001). Due to the higher hydraulic conductivity of layer 1, the differences between the two solutions evident in Fig. 3.29 are smaller than those predicted for the uniform case (Fig. 3.27).

(c) Discussion

Regarding the accuracy of the simple method, it can be concluded that the error in the predicted average degree of consolidation is usually less than 10% for 1D

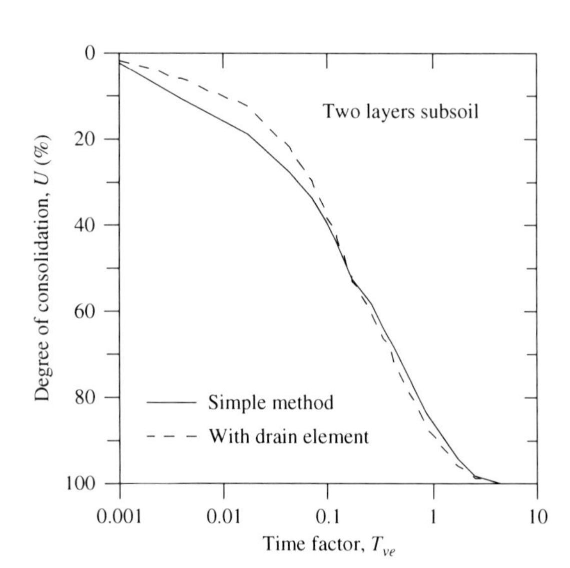

Fig. 3.28 Comparison of the average degree of consolidation for the two layer case (after Chai et al. 2001)

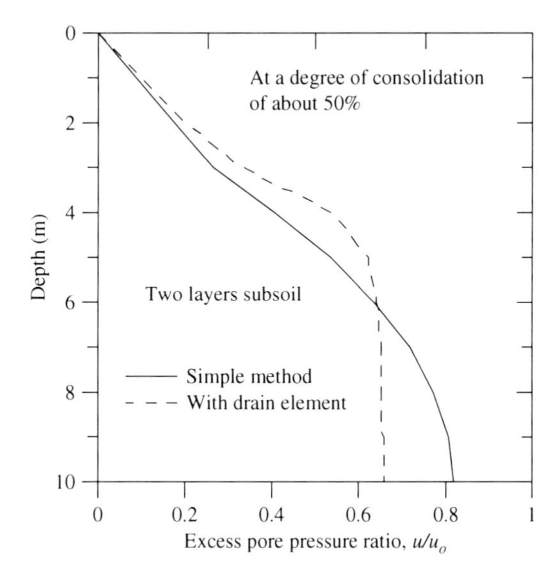

Fig. 3.29 Comparison of excess pore water pressure distribution of two layers case (after Chai et al. 2001)

conditions. However, it should be noted that the comparisons presented previously have been made assuming specific conditions, and strictly speaking the relative errors would only be valid for those particular cases. Generally, it has been found that the relative error is affected by the following factors.

(i) The relative significance of vertical and radial drainage. The smaller the effect of vertical drainage in the natural subsoil, the larger the error will be.
(ii) The effect of well resistance. Well resistance tends to result in a non-uniform excess pore water pressure distribution with depth. Therefore, the larger the well resistance, the smaller the error will be. Field evidence has shown that due to drain clogging the well resistance of a PVD may be quite high and the corresponding discharge capacity therefore may be quite small (Chai and Miura 1999).
(iii) The vertical drainage path length. The larger the vertical drainage path length (l) and the smaller the drain spacing (S), i.e., the larger the value of the ratio l/S, the larger the error will be.
(iv) The uniformity of the subsoil. As shown in the above comparisons, the errors involved in predicting the consolidation of multilayer soils using the simple method, particularly in cases where the topsoil has a higher hydraulic conductivity than the underlying subsoil, can be reduced below those occurring in the case of a uniform soil layer. This is because the higher hydraulic conductivity of the upper layer enhances the vertical drainage. For most natural deposits, the hydraulic conductivity of the top layer is normally higher than that of the underlying layers, as previously noted.

3.5 Modelling a Large Scale Laboratory Test

3.5.1 Large Scale Consolidometer

Tests of PVD materials installed in large scale laboratory samples of clay have been reported by Saowapakpiboon (2010) and Saowapakpiboon et al. (2010). Some of these test details and some relevant findings of the testing program conducted at the Asian Institute of Technology are reported here.

The large scale consolidometer shown in Fig. 3.30 consists of a cylindrical cell of 0.45 m inner diameter and 0.95 m height made of polyvinylchloride (PVC). The upper and lower pedestals have a thickness 40 mm and are connected by eight 12 mm diameter steel rods. The piston system consists of a 40 mm thick piston and a hollow shaft with an outside diameter of 100 mm. To prevent tilting of the piston, a guide is installed on the upper pedestal around the shaft and this guide is fixed to the upper pedestal with eight steel bolts each 15 mm in diameter. 'O' rings lubricated with silicon grease are used to form seals between the upper and lower pedestals and the cylindrical cell, the piston and the cell, the shaft and the upper pedestal and the guide and the shaft. Air pressure is applied through the upper pedestal to the top of the piston. A natural rubber membrane with a thickness of 3 mm was installed in the chamber above the piston to prevent leakage of the pressurized air. The rubber membrane is initially folded in the vertical direction to allow free vertical displacement of the piston during consolidation of the soil sample contained

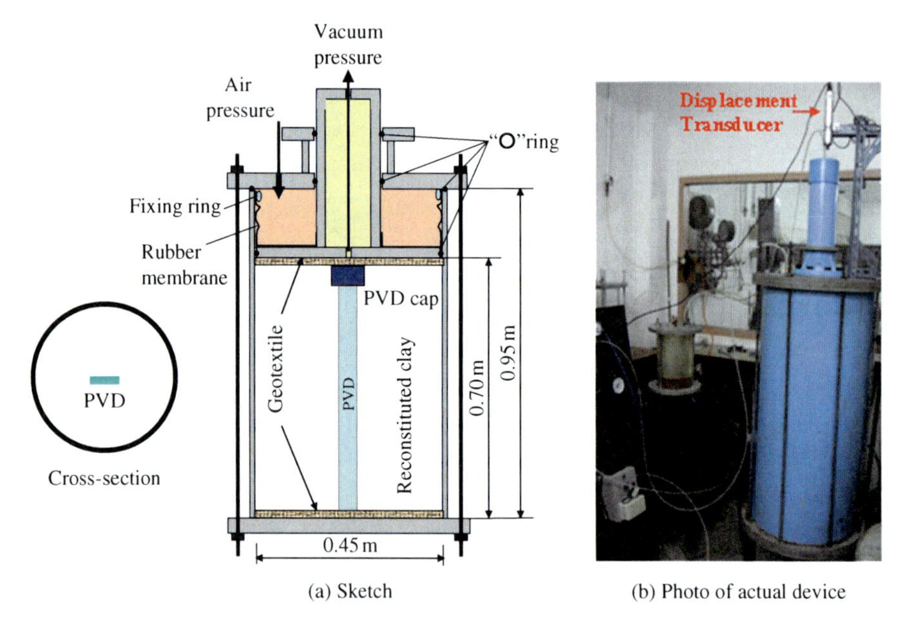

Fig. 3.30 Large scale consolidometer (after Saowapakpiboon et al. 2010). (**a**) Sketch; (**b**) Photo of actual device

within the cell. Geotextiles were placed on the top and bottom of the soil samples to serve as drainage layers. A dial gauge was placed on top of the shaft to allow settlement measurement. A KPD-200 kPa excess pore water pressure transducer (manufactured by Tokyo Sokki Kenkyujo Co, Ltd, Japan) was installed in the soil layer at about its mid height and about 0.11 m away from its central axis. This transducer was connected to a data logger to monitored pore water pressure in the sample during consolidation (Saowapakpiboon 2010).

3.5.2 Soil Properties and Testing Procedures

The soft clay samples tested in the large scale consolidometer were obtained from the site of the Second Bangkok International Airport (SBIA), which is located about 30 km southeast of Bangkok. Samples were recovered from 3.0 to 4.0 m depth below a weathered crust layer. Table 3.4 shows the physical properties of the soft Bangkok clay. The PVD material used in the consolidometer test was the CeTeau CT-D911 drain and its properties are summarized in Table 3.5. In the model tests, the installed PVD was formed from a section of this drain material cut into a half width (50 mm wide).

Before putting the soil sample into the consolidometer, silicone grease was applied to the inside of the large consolidometer to reduce friction. The soil samples were then remoulded by adding a sufficient amount of water until its water content was greater than its liquid limit. The samples were then thoroughly mixed in a mechanical mixer and transferred in layers into the consolidometer. For each layer, air bubbles were eliminated by using a vibrator. A vertical surcharge pressure of 50 kPa was applied under two-way drainage conditions to reconstitute the

Table 3.4 Physical properties of soft Bangkok clay (after Saowapakpiboon et al. 2010)

Properties	Value
Liquid limit (%)	102.24
Plastic limit (%)	39.55
Water content (%)	112.69
Plasticity index	62.69
Total unit weight (kN/m^3)	14.70
Specific gravity	2.66

Table 3.5 Properties of CeTeau CT-D911 drain (after Saowapakpiboon et al. 2010)

Drain Body	Configuration	⊢+++++⊣
	Material	Polypropylene
	No. of channels	44
Filter Jacket	Material	Polypropylene
	Colour	Grey
Weight (g/m)	78	
Width, W (mm)	100	
Thickness, t_d (mm)	3.5	

samples in the large consolidometer,. The consolidation was terminated when the degree of consolidation was more than 90%, as determined by Asaoka's (1978) observational method. The reconstituted sample created in this way had a thickness of approximately 0.7 m (Saowapakpiboon et al. 2010).

A PVD was then installed in the reconstituted soil sample using a mini-mandrel 81.9 mm wide and 18.2 mm thick (a cross-sectional area with an equivalent diameter of 44 mm), and the consolidation test was conducted by applying an overall vertical consolidation pressure of 100 kPa, i.e., a 50 kPa increment in the vertical pressure. During the consolidation test, drainage was allowed only at the top surface and the bottom drainage valve was closed.

3.5.3 Comparison of Measurements and Numerical Simulations

The laboratory model tests were simulated using the finite element method in order to back-analyze the parameters related to the drain performance, e.g., the smear zone parameter, k_h/k_s, etc. (Saowapakpiboon et al. 2010), as well as to evaluate the effectiveness of the various methods for modelling PVD performance. The soil sample was simulated by the Modified Cam clay model (Roscoe and Burland 1968) and the model parameters determined from laboratory test results are listed in Table 3.6. The back-analyzed parameters for the PVD are given in Table 3.7. The value of d_s listed in this table is twice the equivalent diameter of the cross-sectional area of the model mandrel.

In the FEM analysis, two different methods were adopted to model the effects of the PVD. One made use of the axisymmetric unit cell model with the drain at the centre, in which the soil sample with the PVD was modelled in three distinct

Table 3.6 Model parameters for reconstituted soft Bangkok clay (after Saowapakpiboon et al. 2010)

H (m)	e_0	k (m/day)	κ	ν	λ	M	e_{cs}
0.70	2.29	7.51E-05	0.052	0.3	0.569	0.8	4.51

e_0 = initial void ratio; k = hydraulic conductivity; κ = slope of unloading-reloading curve in $e - \ln p'$ plot (e is void ratio and p' is consolidation pressure); λ = slope of virgin loading curve in $e - \ln p'$ plot, M = slope of critical state line (CSL) in $p' - q$ plot (q is deviator stress), and e_{cs} = void ratio corresponding to $p' = 1$ kPa and on CSL in $e - \ln p'$ plot

Table 3.7 Parameters for PVD consolidation (after Saowapakpiboon et al. 2010)

Parameter	Symbol	Unit	Value
Drain diameter	d_w	m	0.0268
Diameter of smear zone	d_s	m	0.087
Discharge capacity	q_w	m^3/year	100
Ratio of k_h to k_s	k_s/k_h	–	3

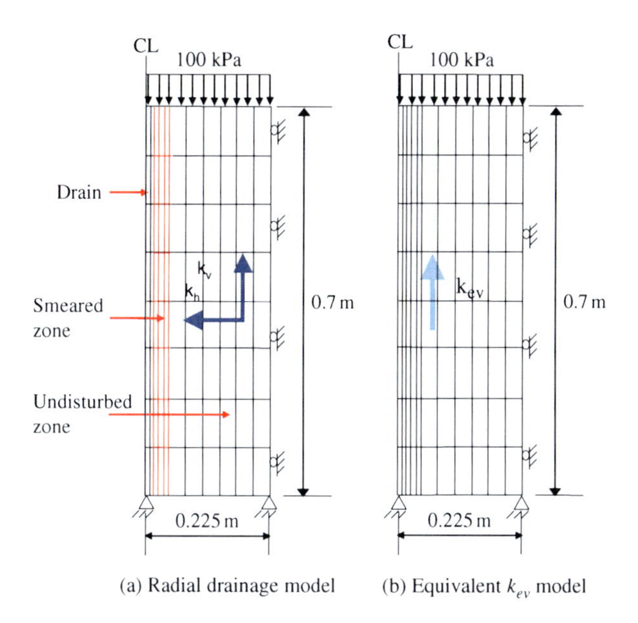

Fig. 3.31 FEM meshes (after Saowapakpiboon et al. 2010). (**a**) Radial drainage model; (**b**) Equivalent k_{ev} model

material zones: the drain zone; a smear zone; and the undisturbed zone, as shown in Fig. 3.31a. The hydraulic conductivity of the drain element was calculated using the discharge capacity of the PVD and the cross-sectional area of the elements in the drain zone. In the other finite element model the equivalent vertical hydraulic conductivity method, i.e., the 'simple method' proposed by Chai et al. (2001), was adopted. In this case there was no zoning of the model ground (Fig. 3.31b). In both cases the soil was represented by 8 noded quadratic displacement elements with only 4 pore water pressure nodes at the element corners and reduced integration was adopted for all calculations. The displacement boundary conditions were, at the bottom both the vertical and horizontal displacements were fixed, and at the right hand boundary (corresponding to the wall of the model) the horizontal displacement was fixed. Only at the top surface was drainage allowed. The program used to compute the numerical simulations was ABAQUS version 6.7 (Abaqus Inc. 2008).

Using the test data from (Saowapakpiboon et al. 2010; Saowapakpiboon 2010), the simulated settlements are compared with the measurements in Fig. 3.32. It can be seen that the simple method resulted in a faster predicted settlement rate at the earlier stage and a slower settlement rate at the later stages of consolidation. However, the differences between the two methods are quite small. Both simulated results compare well with the measurements. The predicted excess pore water pressures (u) are compared in Fig. 3.33. Although the simulations slightly over-estimated the excess pore water pressures, they compare quite favourably with the laboratory measurements. Generally, the simple method will result in a lower predicted value of u

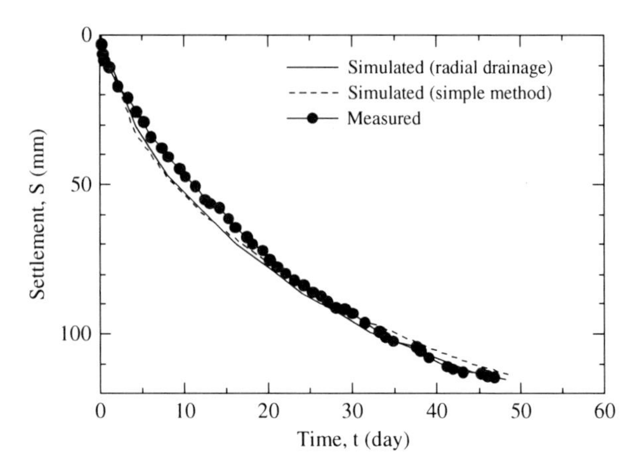

Fig. 3.32 Comparison of the settlement – time curves (measured data from Saowapakpiboon et al. 2010)

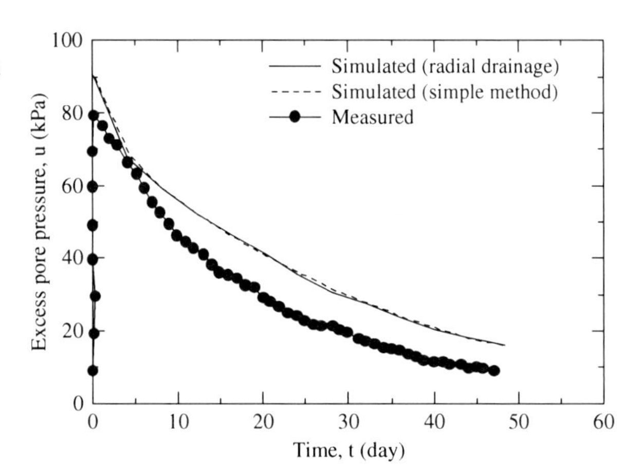

Fig. 3.33 Comparison of excess pore water pressures at the transducer location (measured data from Saowapakpiboon 2010)

near the top drainage surface and a higher value of u close to the bottom of the PVD. At the location of the piezometer (near the middle of the model), the simple method yielded almost identical result with those predicted using individual drainage elements.

3.6 Application to a Case History

The simple modelling method was applied to the analysis of a test embankment at the Hangzhou-Ningbo (HN) Expressway in eastern China. Analyses using drainage elements (Chai et al. 1995) to represent the effect of the PVDs were also conducted. It is worth mentioning that using the concept of equivalent hydraulic conductivity, the FEM analysis can be conducted using standard commercial finite element programs. The results obtained using the simple modelling method are compared with those obtained using drainage elements in terms of settlements, excess pore water pressures, and lateral displacement profiles. The results of both types of FEM analysis are also compared with the measured field data.

3.6.1 Test Embankment on a Soft Clay Deposit in Eastern China

The Hangzhou-Ningbo (HN) expressway is located on the southern coast of Hangzhou Bay, as shown in Fig. 3.34. It starts from Hangzhou, the capital of Zhejiang Province and extends to Ningbo, the biggest harbor city in the same Province. The total length of the HN expressway is 145 km, of which about 92 km passes over a soft clay deposit (in China it is called 'mucky clay'). In order to obtain reliable field data and experience that would guide the design and construction of the entire expressway, 12 full scale field test embankments with a total length of 3.15 km were constructed and investigated (Wang et al. 1998). The PVD-improved section of soft ground analyzed here underlay one of these test embankments.

The generalized soil profile and the properties of the soft soil deposit at the test site are shown in Fig. 3.35. The overall thickness of the soft soil layers is about 23 m. A thin weathered crust (TC) having a thickness of 1 to 1.5 m overlies a silty clay layer (SC1) approximately 4 m thick. The third layer is the very soft mucky clay (MC) with a thickness of approximately 10 m. The fourth layer is also a soft clay, called mucky-silty clay (MSC), and it has a thickness of approximately 4 m. The fifth layer is a medium to stiff silty clay layer (SC2) approximately 3 ~ 5 m thick. Below the fifth layer, it is a clayey sand layer. Prior to embankment construction the subsoils were in lightly overconsolidated to normally consolidated states. The overconsolidation ratio (OCR) of the top crust was about 5. The soft silty clay and

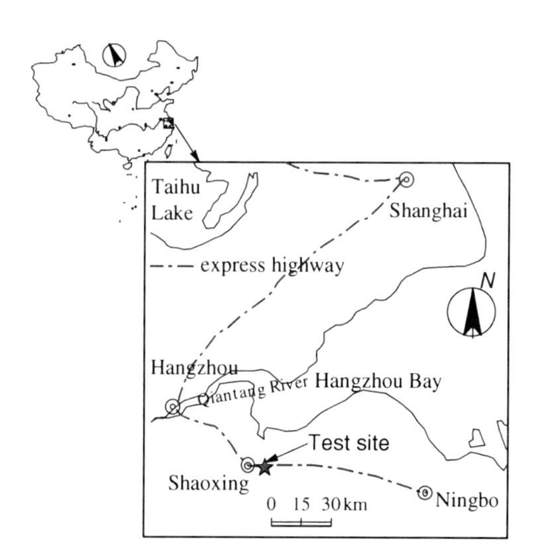

Fig. 3.34 Location of the test site in China (after Chai et al. 2001)

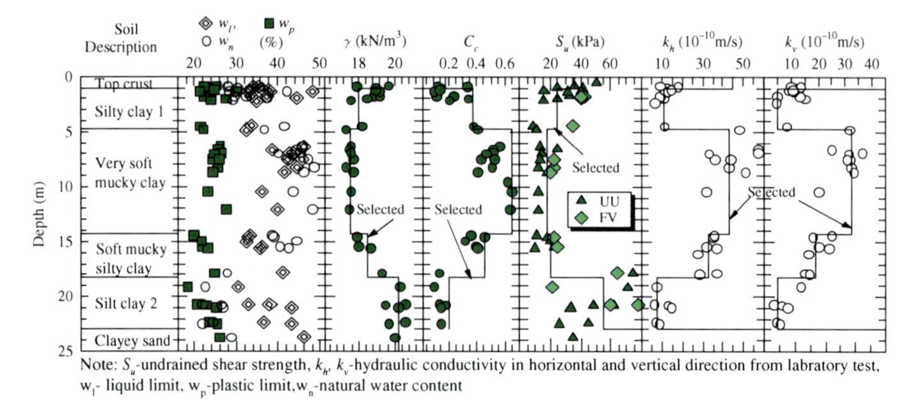

Fig. 3.35 Soil profile and properties at the test site in China (after Chai et al. 2001)

mucky clay have initial water contents greater than their corresponding liquid limits, low hydraulic conductivity and low shear strength. The groundwater level was about 1.5 m below the ground surface.

Figure 3.36 shows the geometry of the embankment on the PVD-improved subsoil and the main instrumentation points. A 0.5 m thick sand mat was placed on the top of the soft ground. Decomposed granite was used as the fill material which was compacted in layers to a unit weight of about 20 kN/m^3. The height of the constructed embankment was 5.88 m. PVDs were installed to 19 m depth in a triangular pattern with a spacing of 1.5 m

3.6.2 FEM Modelling and Model Parameters

The modelled subsoil was 29 m deep from the ground surface, and 120 m horizontally from the embankment centreline. 120 m is about 4 times the embankment half width, which is considered sufficient to substantially reduce the boundary effects in the numerical model. The displacement boundary conditions were as follows: at the bottom, both vertical and horizontal displacements were fixed and at the left hand vertical boundary (under the embankment centreline) and the right hand vertical boundary (away from the centreline), the horizontal displacement was fixed but vertical movement was allowed. The drainage boundary conditions adopted in the model were as follows: the ground surface and the base of the model (a sand layer) were both considered as drainage boundaries. The left and right hand vertical boundaries were assumed to impermeable, except for selected locations on the right boundary (away from the centreline) corresponding to the locations of sand layers, which were assumed to be drainage points.

Figure 3.37 shows the finite element mesh adopted for the analysis of this field case. The elements used were 8 noded quadrilaterals. In the analyses which included

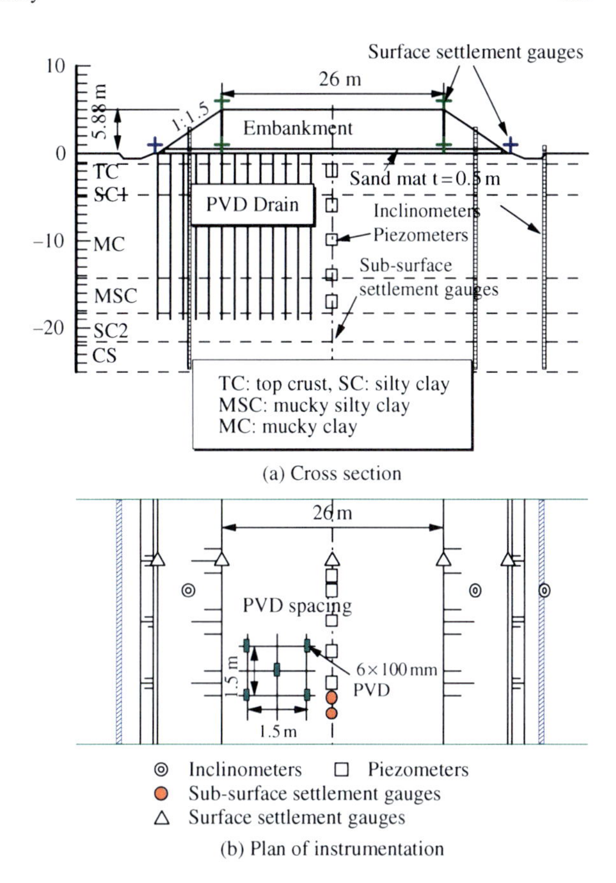

Fig. 3.36 Cross-section of embankment and location of field instrumentation (after Chai et al. 2001)

discrete drainage elements, the 1D drainage elements were set at every other vertical line of the finite element mesh in the PVD-improved zone as, indicated in Fig. 3.37. The behaviour of the clay layers was represented by the Modified Cam-clay model (Roscoe and Burland 1968). All parameters defining the mechanical behaviour of the clay layers were estimated based on laboratory test results, except for the Poisson's ratio and hydraulic conductivity of the subsoil. The hydraulic conductivity was back-calculated from the response of the test embankments constructed on natural subsoil at the same site (Shen et al. 2000). During the consolidation calculations, the hydraulic conductivity was varied with void ratio according to Taylor's equation (Taylor 1948):

$$k = k_0 \cdot 10^{-(e_0 - e)/C_k} \tag{3.51}$$

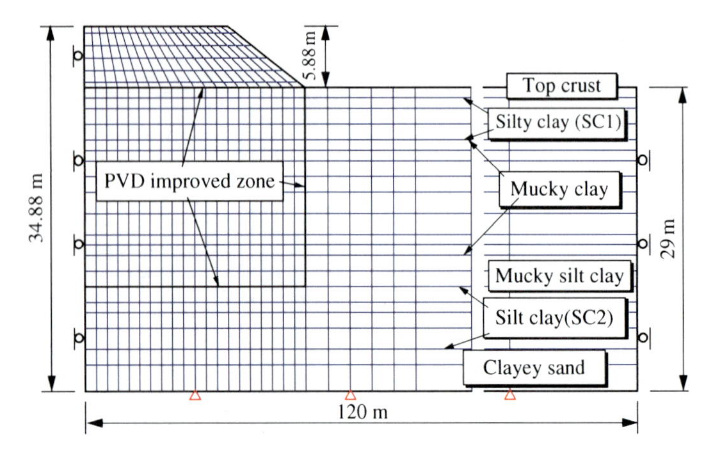

Fig. 3.37 Finite element mesh in the analysis (after Chai et al. 2001)

where k_0 = the initial hydraulic conductivity, e_0 = the initial void ratio, k = the current hydraulic conductivity, e = the current void ratio, and C_k = a constant ($C_k = 0.5e_0$, according to Tavenas *et al.* 1986). The sand layers and the embankment fill materials were treated as elastic materials, and values of the Young's modulus, E, and Poisson's ratio, ν, were assumed empirically. The model parameters so determined for the subsoil are listed in Table 3.8. Based on the back-analysis of an embankment on natural subsoil at the same site, the value of C_f (in Eq. (3.11)) was estimated to be 6 (Shen et al. 2000). The parameters related to the PVD behaviour are listed in Table 3.9, which were determined by using the method proposed by Chai and Miura (1999). The embankment construction history was simulated closely in the numerical model.

Table 3.8 Model parameter for subsoil in the test site in eastern China

Layer	E (kPa)	ν	κ	λ	M	e_0	γ_t (kN/m^3)	k_h (10^{-8} m/s)	k_v (10^{-8} m/s)	k_{ve} (10^{-8} m/s)
TC	–	0.3	0.008	0.08	1.0	0.81	19.3	3.51	3.51	39.8
SC1	–	0.35	0.016	0.16	1.0	1.07	18.5	0.65	0.26	8.45
MC	–	0.35	0.028	0.28	0.8	1.36	17.3	2.94	1.96	33.5
SMC	–	0.35	0.018	0.18	0.8	1.10	17.9	2.39	1.17	27.9
SC2	–	0.3	0.010	0.10	1.0	0.81	19.3	0.45	0.21	6.03
CS	25,000	0.25	–	–	–	–	19.5	29.98	29.98	–
Fill	30,000	0.2	–	–	–	–	20.0	–	–	–

Table 3.9 Parameters related to the behaviour of PVD

Item	Symbol	Unit	Value
Drain diameter	d_w	Mm	53.0
Unit cell diameter	d_e	M	1.58
d_e/d_w	n	–	29.7
Smear zone diameter	d_s	Mm	355
Ratio of k_h over k_s in field	$(k_h/k_s)_f$	–	13.8
d_s/d_w	s	–	6.7
Discharge capacity	q_w	m^3/year	100
Ratio of field to laboratory values of k_h	C_f	–	6

3.6.3 Comparison of Results

Figure 3.38 shows a comparison of the predicted and observed curves of settlement versus time. The simple method has resulted in a slightly slower predicted consolidation rate compared with that predicted using the discrete drainage elements. The maximum difference in the plotted settlements is less than 5% (or 0.1 m). Figure 3.39 compares the predicted excess pore water pressures with the measured values, and it indicates that the simple method (using k_{ve}) predicted slightly higher excess pore water pressures than those predicted using the drainage elements at the piezometer locations, which is consistent with the results for settlement. When compared with the measured data, the FEM analyses predicted a faster excess pore water pressure dissipation rate. For lateral displacement, the two numerical methods yielded almost identical results (Fig. 3.40). From these comparisons, it can be observed that from a practical standpoint the simple method can predict results that are as good as those obtained using discrete drainage elements.

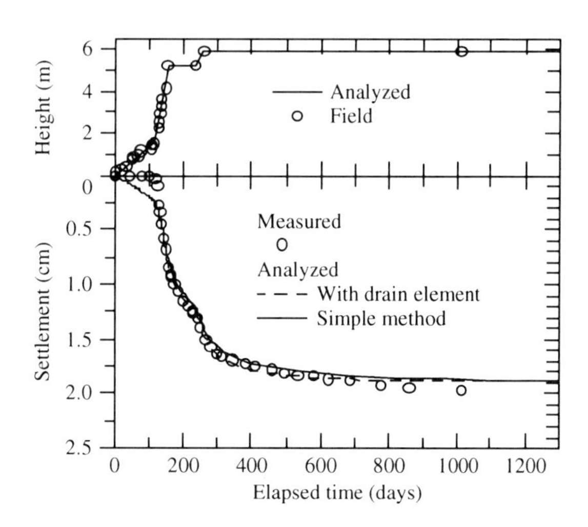

Fig. 3.38 Comparison of the settlement curves (after Chai et al. 2001)

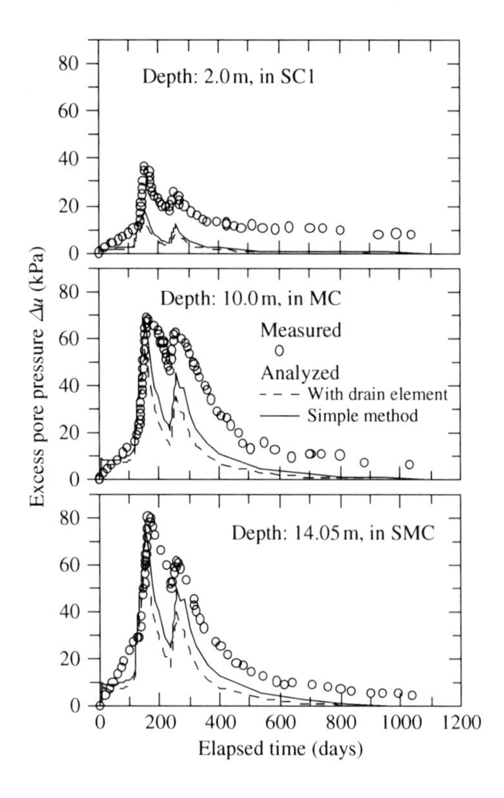

Fig. 3.39 Calculated excess pore water pressure variations (after Chai et al. 2001)

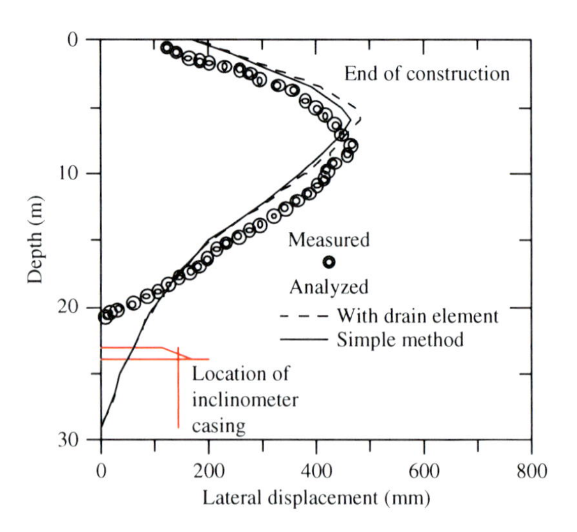

Fig. 3.40 Lateral displacement profiles (after Chai et al. 2001)

3.7 Summary

For calculating the consolidation settlement of subsoil improved by the installation of prefabricated vertical drains (PVDs), analytical solutions are only available for a homogeneous subsoil deforming under one-dimensional (1D) conditions. Normally this type of solution is called a unit cell consolidation theory. The most commonly used unit cell consolidation theories are those presented by Barron (1948) and Hansbo (1981). In addition to the horizontal coefficient of consolidation of a deposit (k_h), the factors which have a significant effect on PVD-induced consolidation are the hydraulic conductivity (k_s) and the diameter (d_s) of the smear zone that forms around the PVD during installation, and the discharge capacity of the PVD (q_w). The available test data as well as the methods for determining these parameters have been described in this chapter.

It has been shown that for a deposit with two-way drainage, full penetration of the PVDs into the deposit may not be an economical choice. An equation for calculating optimum PVD installation depth has also been developed and described in this chapter.

For many practical problems, such as those requiring the simulation of embankment construction, two-dimensional (2D) finite element analysis is often a computational necessity, and this usually requires the assumption of plane strain conditions. Several methods have been developed for numerically modelling PVD-improved subsoil in plane strain finite element analyses. These methods can be classified into 4 groups. In the first group the PVDs are modelled by solid elements after modifying the value of the hydraulic conductivity in the horizontal direction in order to obtain matching solutions for 50% of consolidation in both the axisymmetric unit cell and plane strain models. In the second group, a macro-element is formulated to represent the drainage effects of both the natural soil and the PVDs. The third group employs discrete 1D drainage elements directly in order to model the drainage effects of PVDs. The fourth group combines the drainage effects of the PVDs and the natural soil in the vertical direction into an equivalent vertical hydraulic conductivity. The latter has been referred to as the 'simple method'. Since most commercial software does not include special macro-elements, nor special purpose 1D drainage elements, conducting a finite element analysis of PVD-improved subsoil using these techniques can be a troublesome task. The major advantage of the simple method is that PVD-improved subsoil can be analyzed in the same way, using the same type of solid elements, as for the unimproved soil. The only difference lies in the choice of values for the effective hydraulic conductivity in the vertical direction.

The effectiveness of the simple modelling method has been demonstrated and verified by successfully simulating a large scale laboratory model test and a test embankment constructed in eastern China. These comparisons show that for most practical purposes the simple method can yield a result as good as that obtained using discrete drain elements to represent the behaviour of the PVDs.

References

Abaqus Inc. (2008) Abaqus version 6.7 documentation, Dassault Systémes SIMULIA, Providence, RI. See also: http://www.simulia.com/

Asaoka A (1978) Observational procedure of settlement prediction. Soil Found 18(4):87–101

Barron RA (1948) Consolidation of fine-grained soils by drain wells. Trans Am Soc Civil Eng 113:718–742

Bergado DT, Anderson LR, Miura N, Balasubramaniam AS (1996) Soft ground improvement in lowland and other environments. ASCE Press, New York, NY, p 427

Biot MA (1941) General theory of three dimensional consolidation. J Appl Phys 12:155–164

Carrillo N (1942) Simple two and three dimensional cases in the theory of consolidation of soils. J Math Phys 21:1–5

Chai J-C, Bergado DT (1993) Performance of reinforced embankment on Muar clay deposit. Soil Found 33(4):1–17

Chai J-C, Miura N (1999) Investigation of factors affecting vertical drain behavior. J Geotech Geoenviron Eng ASCE 125(3):216–226

Chai J-C, Miura N, Sakajo S, Bergado DT (1995) Behavior of vertical drain improved subsoil under embankment loading. Soil Found 35(4):49–61

Chai J-C, Bergado DT, Miura N, Sakajo S (1996) Back calculated field effect of vertical drain. Proceedings 2nd international conference on soft soil engineering, vol 1, Nanjing, China, pp 270–275

Chai J-C, Miura N, Sakajo S (1997) A theoretical study on smear effect around vertical drain. Proceedings 14th international conference on soil mechanics and foundation engineering, vol 3, Hamburg, Germany, pp 1581–1584

Chai J-C, Shen S-L, Miura N, Bergado DT (2001) Simple method of modeling PVD improved subsoil. J Geotech Geoenviron Eng ASCE 127(11):965–972

Chai J-C, Miura N, Kirekawa T, Hino T (2009) Optimum PVD installation depth for two-way drainage deposit. Geomechan Eng An Int J 1(3):179–192

Hansbo S (1979) Consolidation of clay by band shaped prefabricated drains. Ground Engineering 12(5):16–25 (Thomas Telford, Ltd., London)

Hansbo S (1981) Consolidation of fine-grained soils by prefabricated drains. Proceedings of 10th international conference on soil mechanics and foundation engineering, vol 3, Stockholm, pp 677–682

Hansbo S (1983) How to evaluate the properties of prefabricated drains. Proceedings of 8th European conference on soil mechanics and foundation engineering, vol 2, Helsinki, Paper 6.13

Hansbo S (1987) Fact and fiction in the field of vertical drainage. In: Joshi RC, Griffiths FJ (eds) Prediction and performance in geotechnical engineering. A.A. Balkema, Rotterdam, The Netherlands, pp 61–72

Hansbo S (1994) Foundation engineering. Elsevier, London, p 519

Hird CC, Pyrah IC, Russell D (1992) Finite element modelling of vertical drains beneath embankments on soft ground. Géotechnique 42(3):499–511

Indraratna B, Redana IW (1997) Plane train modeling of smear effects associated with vertical drains. J Geotech Geoenviron Eng ASCE 123(5):474–478

Indraratna B, Redana IW (2000) Numerical modeling of vertical drains with smear and well resistance installed in soft clay. Can Geotech J 37:132–145

Jamiolkowski M, Lancellotta R (1981) Consolidation by vertical drains: uncertainties involved in prediction of settlement rates. Proceedings of 10th international conference on soil mechanics and foundation engineering, vol 4, A.A. Balkema, Rotterdam, pp 593–595

Jamiolkowski M, Lancellotta R, Wolski W (1983) Pre-compression and speeding up consolidation. General report. Special session 6, Proceedings of 8th Europe conference on soil mechanics and foundation engineering, A.A. Balkema, Rotterdam, pp 1201–1226

Kamon M, Ito Y (1984) Function of band-shaped prefabricated plastic board drains. Proceedings 19th annual meeting of Japanese society on soil mechanics and foundation engineering. Tokyo, Japan, pp 1575–1576 (in Japanese)
Madhav R, Park YM, Miura N (1993) Modelling and study of smear zones around band shaped drains. Soil Found 33(4):135–147
Miura N, Park YM, Madhav MR (1993) Fundamental study on the discharge capacity of plastic board drain. J Geotech Eng JSCE 35(III):31–40 (in Japanese)
Miura N, Chai J-C, Toyota K (1998) Investigation on some factors affecting discharge capacity of prefabricated vertical drain. Proceedings of 6th international conference on geosynthetics, Atlanta, pp 845–850
Olson R (1998) Settlement of embankments on soft clays. J Geotech Geoenviron Eng ASCE 124(4):278–288
Onoue A (1988) Consolidation by vertical drains taking well resistance and smear into consideration. Soil Found 28(4):165–174
Onoue A (1991) Permeability of disturbed zone around vertical drains. 26th annual meeting of the Japanese Society of Soil Mechanics and Foundation Engineering, Japanese Society of Soil Mechanics and Foundation Engineering, Tokyo, pp 2015–2018 (in Japanese)
Rixner JJ, Kraemer SR, Smith AD (1986) Prefabricated vertical drains. Vol. I, Engineering guidelines, FWHA/RD-86/168. Federal Highway Administration, Washington, DC
Roscoe KH, Burland JB (1968) On the generalized stress-strain behavior of 'wet' clay. In: Heyman J, Leckie, FA (eds) Engineering plasticity. Cambridge University Press, Cambridge, pp 535–609
Saowapakpiboon J (2010) Behavior of smeared zone and performance of PVD under surcharge and vacuum preloading with and without heat. Dissertation, Asian Institute of Technology, Bangkok, Thailand
Saowapakpiboon J, Bergado DT, Voottipruex P, Lam LG, Nakakuma K (2010) PVD improvement combined with surcharge and vacuum preloading including simulations. Geotext Geomembr 29:74–82
Sekiguchi H, Shibata T, Fujimoto, A, Yamaguchi H (1986) A macro-element approach to analyzing the plane strain behavior of soft foundation with vertical drains. Proceedings of 31st symposium, Japanese Geotechnical Society, pp 111–116 (in Japanese)
Shen S-L, Yang CW, Miura N, Chai J-C (2000) Field performance of PVD improved soft clay under embankment. In: Nakase and Tsuchida (eds) Coastal geotechnical engineering in practice, vol 1. A. A. Balkema, Rotterdam, pp 515–520
Shinsha H, Hara H, Abe Y, Tanaka A (1982) Consolidation settlement and lateral displacement of soft ground improved by sand drains. Tsuchi-to-Kiso, Japanese Soc Soil Mech Found Eng 30(5):7–12 (in Japanese)
Tavenas F, Tremblay M, Larouche G, Leroueil S (1986) In situ measurement of permeability in soft clays. ASCE specialty conference on use of in-situ tests in geotechnical engineering, Blacksburg, pp 1034–1048
Taylor DW (1948) Fundamentals of soil mechanics. Wiley, New York, NY
Wang ZM, Xu LX, Jiang ZH (1998) Field experimental study on the soft ground treatment of Hongzhou-Niingbo (HN) expressway foundation. In: Cai TL (ed) Soft ground treatment in Hangzhou-Ningbo expressway. Hongzhou Press, Hongzhou, pp 183–222 (in Chinese)
Yoshikuni H, Nakanodo H (1974) Consolidation of soils by vertical drain wells with finite hydraulic conductivity. Soil Found 14(2):35–46
Zeng GX, Xie KH (1989) New development of the vertical drain theories. Proceedings of 12th international conference on soil mechanics, vol 2, Rio de Janeiro, pp 1435–1438

Chapter 4
Vacuum Consolidation

Abstract This chapter describes the methods commonly used for conducting vacuum consolidation of soft soils in the field. It then presents the general theory of vacuum consolidation and describes the major characteristics of vacuum consolidation of soft ground, the methods adopted for calculating the deformations induced by application of vacuum pressure, as well as some example analyses based on case histories and laboratory studies. Distinctions are drawn between the behaviour of soft soil layers with one and two-way drainage and the case of fully and partially penetrating pre-fabricated vertical drains (PVDs) installed in the soft clay. The field case studies considered in this chapter were sourced from Japan and China.

4.1 Introduction

Preloading is a commonly used method for improving the mechanical performance of soft clayey deposits. The effective surcharge responsible for providing the preload to the soil can arise from either the weight of imposed fill material, e.g., due to embankment construction, or it may be due to a vacuum pressure applied directly to the soil, or some combination of both. Using a vacuum pressure has several advantages over embankment loading, e.g., no fill material is required, construction periods are generally shorter and there is no need for heavy machinery. In addition, the vacuum pressure method does not put any chemical admixtures into the ground and consequently it can be regarded as an environmentally friendly ground improvement method. Numerous applications in which the vacuum consolidation method has been used to improve soft clayey deposits have been reported in the literature (e.g., Bergado et al. 1998; Tang and Shang 2000; Chu et al. 2000; Chai et al. 2008; Saowapakpiboon et al. 2010).

The mechanism of vacuum consolidation is different from that which occurs when a conventional surcharge load is applied to the ground surface to induce soil consolidation. With the load originating from the weight of embankment fill, there will be consolidation settlement resulting from volume change in the soil as well as shear stress-induced deformations. The latter will normally be manifested most obviously as outward lateral displacement of the ground beneath the embankment. In contrast, the application of a vacuum pressure causes only isotropic stress increments in the ground. These will generally induce settlement as well as inward lateral displacement of the ground, toward the centre of the vacuum loaded region of soil.

J. Chai, J.P. Carter, *Deformation Analysis in Soft Ground Improvement*, Geotechnical, Geological and Earthquake Engineering 18, DOI 10.1007/978-94-007-1721-3_4,

Furthermore, the effects of the drainage boundary conditions are generally different in both cases (Chai et al. 2005). For a soil deposit with two-way drainage, vacuum pressure is normally applied at only one boundary, i.e., the upper boundary and perhaps also internally via drains installed in the ground. At the other, lower drainage boundary effectively no vacuum pressure can be applied. Hence there will always be a head difference between the top and bottom boundaries of the soil layer, and the final condition (i.e., at 100% consolidation) will correspond to a steady flow of water through the deposit toward the upper drainage boundary at which the vacuum pressure is applied. In contrast, if a conventional surcharge load is applied to the surface of a soil layer with two way drainage, the flow of pore water will tend to occur towards both the top and bottom drainage boundaries, driven by the gradients of excess pore water pressure that vary throughout the layer.

This chapter presents the theory of vacuum consolidation, the characteristics of vacuum consolidation and methods for calculating vacuum pressure-induced ground deformation. Example calculations with case histories are included in order to illustrate the application of this theory in practice.

4.2 Field Methods for Vacuum Consolidation

In practice, there are two main methods used to conduct vacuum consolidation in the field. Each is described in this section.

4.2.1 Air-Tight Sheet Method

The first method for conducting vacuum consolidation in the field was originally proposed by Kjellman (1952). It involves placing an air-tight sheet on the ground surface in the first instance and sealing the periphery of the sheet by embedding it in the ground. Air and water in the soil below the sheet are then sucked from the ground by a vacuum pump (Fig. 4.1). In order to accelerate the consolidation process vacuum consolidation is normally combined with the prior installation of prefabricated vertical drains (PVDs) into the ground. The role of these PVDs is usually to enhance the soil drainage and, unless they are connected and sealed to the vacuum system, they will not normally provide a means of directly applying the vacuum pressure at depth in the soil layer.

It is obvious that the effectiveness of this method of vacuum consolidation depends very much on the air-tightness of the covering sheet. There will be situations under which it is difficult to maintain this air-tightness. Three typical examples of when this kind of difficulty may arise are described as follows.

1. The presence of a high air/water permeability layer at the ground surface. To avoid air leakage, the air-tight sheet must be embedded below this layer, or an air/water cut-off wall penetrating through this layer and into the underlying lower

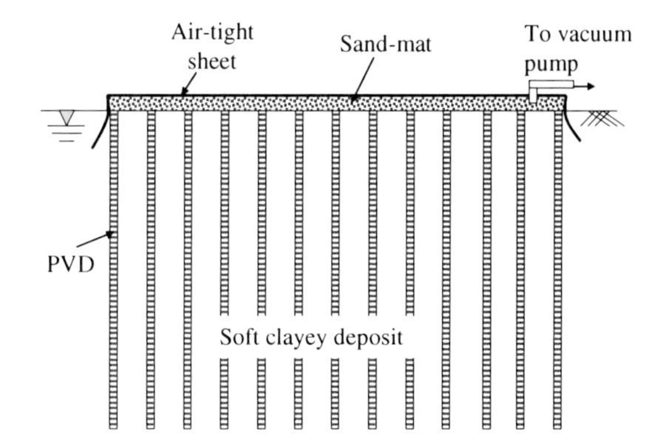

Fig. 4.1 Illustration of original vacuum consolidation technique

permeability layer must be constructed around the perimeter of the preloading area. In practice, these options are costly and sometimes their effectiveness cannot be guaranteed.

2. Combining vacuum pressure with an embankment load. Combining embankment loading with vacuum pressure should increase the overall preloading pressure and reduce the consequent lateral displacement of the ground (Indraratna et al. 2005; Chai et al. 2006). However, after placing embankment fill, any damage to the air-tight cover sheet cannot be identified and repaired. Consequently, the effect of vacuum pressure will be reduced in cases where the sheet has sustained damage during construction (Bergado et al. 1998).
3. Consolidating a soft clayey deposit under water. There are situations where a soft soil layer lying entirely beneath the water table needs to be consolidated by applying a vacuum pressure. Particular examples include the use of vacuum preloading for reclaimed land in coastal areas and consolidating soft sea bed deposit (Takeya et al. 2007). Usually it is not convenient to place an air or water tight sheet below water level.

4.2.2 Vacuum-Drain Method

A relative new technique of applying vacuum pressure to soft clayey deposits involves application of the vacuum pressure directly to the prefabricated vertical drains which are specially capped for this purpose. They are usually referred to as capped PVDs or simply CPVDs (Fujii et al. 2002; Chai et al. 2008). As illustrated schematically in Fig. 4.2, the cap provides an interface between the rectangular shaped PVD and the drainage hose, which is usually circular, allowing suction to be applied to the soil directly via the PVDs. This method is usually implemented in combination with the use of a surface or subsurface soil layer acting as an air sealing layer. In such cases there is often no need to place an air-tight sheet on the ground surface. With this method a vacuum pressure is applied directly to each PVD through the hose, with a value just below the cap of the CPVD which will

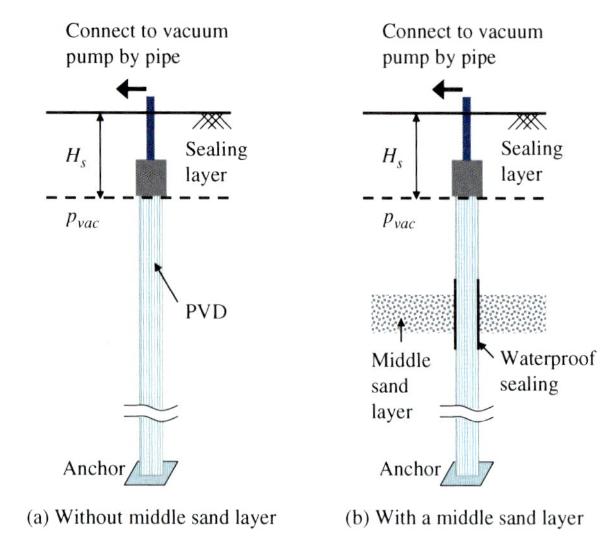

Fig. 4.2 Vacuum consolidation with CPVD (after Chai et al. 2008). (**a**) Without middle sand layer; (**b**) With a middle sand layer

be designated as p_{vac}. Each CPVD is connected to a geosynthetic pipe placed on the ground surface, which in turn is connected to a vacuum pump. This technique is usually denoted as the 'vacuum-drain method'. Consolidation of soft clayey deposits underwater can be conducted quite easily using this method (Takeya et al. 2007).

Within the top sealing layer of soil the vacuum pressure varies from the value of p_{vac} at the bottom of this layer, i.e., at the cap location, to zero at the ground surface. Therefore, any compression of the sealing layer under the action of the vacuum pressure is likely to be less than that which would result by applying a vacuum pressure under an air-tight sheet placed on the ground surface. For the layers of soil below the sealing layer, the vacuum drain method will result in similar compression under vacuum as that which would be experienced if the air tight sheet method was adopted instead.

The required thickness of the surface sealing layer (H_s) can be estimated using a simple model, i.e., if the vacuum pressure at the bottom of the sealing layer is assumed to be p_{vac} and zero at the ground surface it can be demonstrated (Chai et al. 2008) that:

$$H_s = \frac{p_{vac}}{\gamma_w Q_a} k_{air} \cdot A \tag{4.1}$$

where γ_w = the unit weight of water, k_{air} = the permeability to air flow of the sealing layer, A = the area of treatment, and Q_a = the capacity of the vacuum pump. For example, if $k_{air} = 10^{-5}$ m/s, $p_{vac} = 80$ kPa, $Q_a = 0.1$ m^3/s, and $A = 100$ m^2, then $H_s = 0.8$ m can be calculated from Eq. (4.1). k_{air} is generally larger than the coefficient of hydraulic conductivity (k). Although the ratio k_{air}/k depends on the degree of saturation of the soil, Mike (1971) reported that for sandy soils k_{air}/k is roughly the same as the reciprocal of the ratio of the coefficient of viscosity of air and water, i.e., about 55. In cases where the sealing layer is saturated, Eq. (4.1) can still be used to estimate H_s by substituting the hydraulic conductivity k and the pump capacity for water (Q_w) in place of k_{air} and Q_a, respectively. However, if there is no water supply from the ground surface, in the longer term the sealing layer will become partially saturated and air flow will eventually become the dominant factor rather than water flow.

Figure 4.2b shows the situation where there is a sand layer in the middle of a clayey deposit. To avoid vacuum pressure loss through this sand layer, a sealing sheet is required over the filter of that section of PVD passing through the sandy layer. This sealing sheet is usually attached to the filter using an appropriate glue. To use the technique shown in Fig. 4.2b effectively, an intelligent construction system is also needed. One method that has been employed successfully involves placing a piezocone on the tip of the mandrel used to install the CPVDs (Nomura et al. 2007). The piezocone readings obtained during any given CPVD installation can then be used to infer the locations of any sealing sheets that might be required on the adjacent CPVDs.

4.3 Theory of Consolidation Due to Vacuum Pressure

Mathematically, vacuum consolidation is an initial value problem involving a known negative value of the initial excess pore water pressure applied and usually maintained at one boundary of a soil layer, and either zero excess pore water pressure (two-way drainage) or zero hydraulic gradient (one-way drainage) maintained at the other boundary. The solutions presented below have been developed under the assumption that the vacuum pressure propagates into a soil deposit one-dimensionally (1D) for both single layer and two-layer systems. For the case of a soil into which PVDs have been installed, the consolidation theory is normally derived using a unit cell model (a cylinder of soil including the PVD), as described in Chapter 3. In the theory presented here it is assumed that the consolidation of the soil is mainly due to the presence of the PVDs allowing horizontal flow of pore water towards the PVDs, along which it is extracted from the soil, with negligible vertical flow of pore water through the soil itself.

4.3.1 Single Layer System (Without PVDs)

Under the same assumptions as those made in Terzaghi's 1D consolidation theory (Terzaghi 1943), the governing equation and the boundary conditions for the generation and dissipation of excess pore water pressure in a saturated soil layer under a vacuum pressure loading are as follows:

$$c_v \frac{\partial^2 u}{\partial z^2} = \frac{\partial u}{\partial t} \tag{4.2}$$

$$u(0,t) = -p_{vac} \tag{4.3}$$

$$u(H,t) = 0 \text{ for } t > 0 \text{ (two-way drainage)} \tag{4.4}$$

$$\frac{\partial u(H,t)}{\partial z} = 0 \text{ for } t > 0 \text{ (one-way drainage)} \tag{4.4a}$$

where z = the spatial coordinate; t = time; u = the excess pore water pressure; c_v = the coefficient of consolidation of the soil; p_{vac} = the magnitude of the applied vacuum pressure at $z = 0$; and H is the thickness of the deposit.

Conte and Troncone (2008) derived a solution for the 1D consolidation problem for a uniform layer with harmonic pore water pressure variations with time applied at the upper boundary. Using their solutions, various time-dependent pore water pressure boundary conditions can be simulated by developing the appropriate Fourier series approximation of the relevant time function. Following this method, and introducing a constant excess pore water pressure at the boundary where the vacuum pressure is applied, a theory for vacuum consolidation for a single layer system can be easily obtained.

For vacuum consolidation, the final state is not a condition with zero excess pore pressure in the deposit. Therefore, the solution to the governing equation must consist of two parts, namely the steady state solution ($Y(z)$) and the transient solution ($v(z,t)$). With the boundary condition of Eq. (4.3), $u(z, t)$ can be expressed in the following form:

$$u(z,t) = -p_{vac}\left(Y(z) - v(z,t)\right) \tag{4.5}$$

where the term $-p_{vac}Y(z)$ is the final steady state excess pore water pressure distribution and $p_{vac}v(z, t)$ is the time-dependent component of the excess pore water pressure. Closed-form mathematical expressions for both functions $Y(z)$ and $v(z, t)$ have been derived for two-way and one-way drainage conditions, respectively.

4.3.1.1 Two-Way Drainage

In this case the excess pore water pressure distribution in the soil is given by:

$$u(z,t) = -p_{vac}\left[\left(1 - \frac{z}{H}\right) - \frac{2}{\pi}\sum_{n=1}^{\infty}\frac{1}{n}\sin(\lambda_n z)\, e^{\lambda_n^2 \cdot c_v \cdot t}\right] \tag{4.6}$$

where $\lambda_n = n\pi/H$. In the bracket of the right hand side, the first term represents $Y(z)$ and the second term represents $v(z, t)$. The average degree of consolidation is given by:

$$U(t) = 1 - \frac{8}{\pi^2}\sum_{n=1}^{\infty}\frac{1}{(2n-1)^2}e^{-\frac{c_{vs}\cdot t}{H^2}\cdot(2n-1)^2\pi^2} \tag{4.7}$$

4.3.1.2 One-Way Drainage

In this case the excess pore water pressure distribution is given by:

$$u = -p_{vac}\left[1 - \frac{4}{\pi}\sum_{n=1}^{\infty}\frac{1}{2n-1}\sin(a_n z)\,e^{-a_n^2 \cdot c_v \cdot t}\right] \quad (4.8)$$

where $a_n = \frac{(2n-1)\pi}{2H}$. In this case, $Y(z) = 1$, and the second term in the bracket is $v(z,\ t)$. The average degree of consolidation is:

$$U = 1 - \frac{8}{\pi^2}\sum_{n=1}^{\infty}\frac{1}{(2n-1)^2}e^{-\frac{c_v \cdot t}{4H^2}\cdot(2n-1)^2\pi^2} \quad (4.9)$$

From Eqs. (4.7) and (4.9), it can be seen clearly that under vacuum pressure loading, the rate of consolidation for a deposit with two-way drainage is faster than that of a deposit with one-way drainage, even through for both cases the pore water can only be drained from the boundary where the vacuum pressure is applied.

4.3.2 Single Layer with PVD Improvement

The theory of consolidation under vacuum pressure loading for a PVD-improved soil deposit is derived here for the case of one-way drainage conditions using a unit cell model. The latter consists of a PVD surrounded by an annular smear zone generated during PVD installation and the outer annular zone of soil whose overall drainage is to be improved by the presence of the PVD. The unit cell model is shown in Fig. 4.3. In order to avoid vacuum pressure loss at the bottom drainage boundary of a two-way drainage deposit, the PVDs are normally installed so that they only partially penetrate the soft subsoil. In this case, there is an optimum PVD penetration depth, which will be discussed further in Section 4.5. The governing equation for consolidation is as follows:

$$\frac{\partial u}{\partial t} = c_h\left[\left(\frac{\partial^2 u}{\partial r^2}\right) + \frac{1}{r}\left(\frac{\partial u}{\partial r}\right)\right] \quad (3.3\text{ bis})$$

where r = the radial distance and c_h = the coefficient of consolidation in the horizontal direction. The boundary conditions are:

$$\frac{\partial u(r_e, z, t)}{\partial r} = 0 \quad (4.10)$$

$$\frac{\partial u(r, 0, t)}{\partial z} = 0,\quad \frac{\partial u'(r, 0, t)}{\partial z} = 0 \quad (4.11)$$

$$\frac{\partial u(r, l, t)}{\partial z} = 0,\quad \frac{\partial u'(r, l, t)}{\partial z} = 0 \quad (4.12)$$

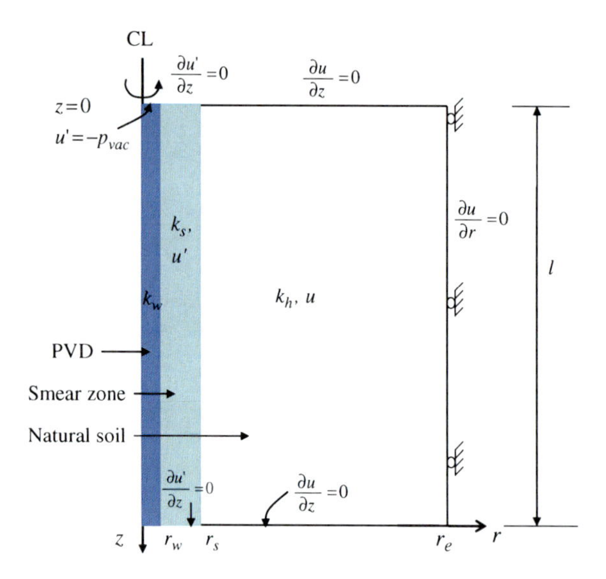

Fig. 4.3 Unit cell model and boundary conditions

$$u' = (r_w, 0, t) = -p_{vac} \tag{4.13}$$

where u and u' = the pore water pressures in the undisturbed zone and the smear zone, respectively (Fig. 4.3), z = depth from the ground surface, r_w = equivalent radius of a PVD, and r_e = radius of the unit cell. The boundary condition at $r = r_w$ is derived from the flow continuity conditions. The first flow continuity condition is that the total inflow of pore water through the boundary of a cylinder with a radius of r has to be equal to the change in volume of the hollow cylinder with outer radius of r_e and inner radius of r. Under the assumption of equal vertical strain this condition can be expressed as follows:

$$\frac{\partial u}{\partial r} = \frac{\gamma_w}{2k_h}\left(\frac{r_e^2}{r} - r\right)\frac{\partial \varepsilon_v}{\partial t} \quad \text{for} \quad (r_s \le r \le r_e) \tag{4.14a}$$

$$\frac{\partial u'}{\partial r} = \frac{\gamma_w}{2k_s}\left(\frac{r_e^2}{r} - r\right)\frac{\partial \varepsilon_v}{\partial t} \quad \text{for} \quad (r_w \le r \le r_s) \tag{4.14b}$$

where ε_v = vertical strain, r_s = radius of the smear zone, γ_w = unit weight of water, and k_h and k_s = hydraulic conductivities of the undisturbed zone and the smear zone in the horizontal direction, respectively. The second continuity condition is that the pore water flow into the PVD from a horizontally cut soil slice is equal to the change of vertical flow rate in the PVD:

$$\frac{\partial u'\,(r_w, z, t)}{\partial r} = \frac{r_w}{2}\frac{k_w}{k_s}\left(\frac{\partial^2 u'\,(r_w, z, t)}{\partial z^2}\right) \tag{4.15}$$

where k_w = the hydraulic conductivity of the PVD (the discharge capacity of a PVD is $q_w = \pi r_w^2 k_w$). Substituting Eq. (4.15) into Eq. (4.14b) provides:

$$\frac{\partial^2 u'}{\partial z^2} = \frac{\gamma_w}{k_w}\left(n^2 - 1\right)\frac{\partial \varepsilon_v}{\partial t} \tag{4.16}$$

where $n = r_e/r_w$. Using the boundary conditions expressed in Eqs. (4.12) and (4.13), and integrating Eq. (4.16) results in the following boundary condition at $r = r_w$:

$$u'(r_w, z, t) = \frac{\gamma_w}{k_w}\left(n^2 - 1\right)\left(lz - \frac{z^2}{2}\right)\frac{\partial \varepsilon_v}{\partial t} \tag{4.17}$$

The interface condition between the smear zone and the undisturbed zone is described by:

$$k_h \frac{\partial u(r_s, z, t)}{\partial r} = k_s \frac{\partial u'(r_s, z, t)}{\partial r} \tag{4.18}$$

For the problem defined above, $Y(z) = 1$ in Eq. (4.5), which is the same as for the case of one-way drainage of a soil layer without PVDs. Assuming there are no vacuum pressure induced cracks around the periphery of the unit cell, the solution for the time dependent part (v in Eq. (4.5)) will be the same as the solution under an applied surcharge load. As discussed in Chapter 3, there are several solutions available. By neglecting vertical flow in the soil, Hansbo's (1981) solution is relatively simple and can take into account both the smear effect and the well resistance, and so it is adopted here. In this case, v is not only a function of z and t, but also a function of radial distance, r. If $v(r, z, t)$ and $v'(r, z, t)$ are used to denote the time dependent excess pore water pressures in the undisturbed zone and the smear zone, respectively, the solution under a vacuum pressure $-p_{vac}$ can be expressed as:

$$u(r, z, t) = -p_{vac}[1 - v(r, z, t)] \text{ for } (r_s < r \le r_e) \tag{4.19}$$

$$u'(r, z, t) = -p_{vac}[1 - v'(r, z, t)] \text{ for } (r_w < r \le r_s) \tag{4.20}$$

For a unit pressure applied at the top of the PVD, Hansbo's (1981) solution with some rearrangement, provides the expressions for v and v' as follows:

$$v'(r, z, t) = \frac{k_h}{k_s \cdot r_e^2 \cdot \mu}\left[r_e^2 \ln\frac{r}{r_w} + \frac{r^2 - r_w^2}{2} + \frac{k_s}{k_w}\left(n^2 - 1\right)\left(2lz - z^2\right)\right]\exp\left(\frac{-8T_h}{\mu}\right) \quad \text{for } (r_w \le r \le r_s) \tag{4.21}$$

$$v(r,z,t) = \frac{1}{r_e^2 \cdot \mu}\left[r_e^2 \ln\frac{r}{r_s} - \frac{r^2 - r_s^2}{2} + \frac{k_h}{k_s}\left(r_e^2 \ln s - \frac{r_s^2 - r_w^2}{2}\right) + \frac{k_h}{k_w}\left(n^2 - 1\right)\left(2lz - z^2\right)\right]$$
$$\exp\left(\frac{-8T_h}{\mu}\right) \text{ for } (r_w \leq r \leq r_s) \tag{4.22}$$

where

$$T_h = \frac{c_h t}{4r_e^2} \tag{3.5 bis}$$

and

$$\mu = \ln(n/s) + (k_h/k_s)\ln(s) - \frac{3}{4} + \pi \cdot z(2l - z)\,k_h/q_w \tag{3.8 bis}$$

where $s = r_s/r_w$, and $l =$ the drainage length of a PVD. The expressions appearing in Eqs. (4.21) and (4.22) are only applicable for $t > 0$. The average degree (U_h) of consolidation of the unit cell is:

$$U_h = 1 - \exp(-8T_h/\mu) \tag{3.7 bis}$$

For average well resistance the expression for μ becomes:

$$\mu = \ln(n/s) + (k_h/k_s)\ln(s) - \frac{3}{4} + \frac{2\pi \cdot l^2 \cdot k_h}{3q_w} \tag{3.8a bis}$$

4.3.3 Two-Layer System

4.3.3.1 Governing Equations

Consider now the situation defined in Fig. 4.4, where a two-layer soil system is subjected to a vacuum pressure applied at the top surface, from which drainage also occurs. The underside of the lower soil layer may be either impermeable or a drainage boundary. In the general case the hydraulic and mechanical properties are not the same in each layer. For 1D deformation conditions where the total stress remains constant for time, $t > 0$, the governing equations for the consolidation of the layered soil system are as follows:

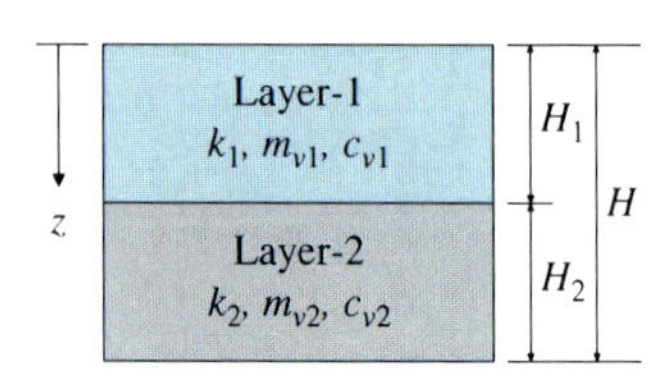

Fig. 4.4 Two-layer system

$$\frac{\partial u_1(z,t)}{\partial t} = c_{v1}\frac{\partial^2 u_1(z,t)}{\partial z^2} \quad 0 \le z \le H_1 \tag{4.23}$$

$$\frac{\partial u_2(z,t)}{\partial t} = c_{v2}\frac{\partial^2 u_2(z,t)}{\partial z^2} \quad H_1 \le z \le H \tag{4.24}$$

where $u_1(z, t)$ and $u_2(z, t)$ = the excess pore water pressures in layer-1 and layer-2, respectively, c_{v1} and c_{v2} = the coefficients of consolidation in the vertical direction of layer-1 and layer-2, respectively; and H_1 = the thickness of layer-1. The boundary conditions are the same as for the single layer system, i.e., Eqs. (4.3) and (4.4). However, for a two-layer system there are interface conditions that also need to be satisfied:

$$u_1(H_1,t) = u_2(H_1,t) \tag{4.25}$$

$$k_1\frac{\partial u_1(H_1,t)}{\partial z} = k_2\frac{\partial u_2(H_1,t)}{\partial z} \tag{4.26}$$

where k_1 and k_2 = the hydraulic conductivities in the vertical direction of layer-1 and layer-2, respectively.

4.3.3.2 Steady State Solution

For a two-layer system with two-way drainage and a vacuum pressure applied at the surface, at the final state (i.e., at large time) there will be a steady flow of pore water towards the boundary where the vacuum pressure is applied. Two possible excess pore water pressure distributions for the long term steady state condition are illustrated in Fig. 4.5. Assuming for convenience that the vacuum pressure at the upper boundary is unity, and denoting the vacuum pressure at the interface between the two layers at large time as u_m, the hydraulic gradients in layer-1 (i_1) and layer-2 (i_2) can be expressed as:

$$i_1 = \frac{1-u_m}{H_1}, \quad i_2 = \frac{u_m}{H_2} \tag{4.27}$$

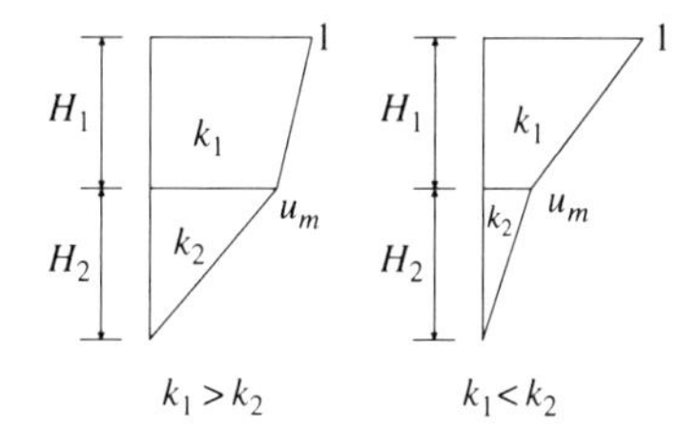

Fig. 4.5 Possible steady state vacuum pressure distributions for a two-layer system

Using the flow continuity condition:

$$i_1 k_1 = i_2 k_2 \tag{4.28}$$

the pore pressure at the interface u_m can be determined as follows:

$$u_m = \frac{k_1 \cdot H_2}{k_1 H_2 + k_2 H_1} \tag{4.29}$$

where H_2 = the thickness of layer-2. The final (steady state) vacuum pressure distribution in the system, $Y(z)$, will then be:

$$Y(z) = \begin{cases} 1 + \left(\frac{u_m - 1}{H_1}\right) z & 0 \le z \le H_1 \\ \left(\frac{u_m}{H - H_1}\right)(H - z) & H_1 < z \le H \end{cases} \tag{4.30}$$

For a two-layer system with one-way drainage, under a unit vacuum pressure applied at $z = 0$, the final vacuum pressure distribution $Y(z)$ will be:

$$Y(z) = 1 \quad 0 \le z \le H \tag{4.31}$$

4.3.3.3 Transient Solution

Again, for convenience, it is assumed that unit excess pore water pressure is maintained at the top boundary, $z = 0$, and the transient solution for this problem is denoted as $v(z, t)$. Consider the case where the two-layer soil system is consolidating under the influence of an initial excess pore pressure distribution. In order to be specific, the case of a spatially bi-linear (two-way drainage) or uniform (one-way drainage) distribution of initial excess pore water pressure throughout the system is considered, and is defined by:

$$v(z, 0) = \begin{cases} 1, & z = 0 \\ u_m, & z = H_1 \quad 2 - \text{way drainage} \\ 0, & z = H \end{cases} \tag{4.32}$$

$$v(z, 0) = \begin{cases} 1, & z = 0 \\ 1, & z = H_1 \quad 1 - \text{way drainage} \\ 1, & z = H \end{cases} \tag{4.33}$$

The solution for $v(z, t)$ is derived from the governing Eqs. (4.23) and (4.24), the interface continuity conditions Eqs. (4.25) and (4.26) (replacing the variable u by v) and the following boundary conditions

$$v(0, t) = 0 \tag{4.34}$$

$$v(H,t) = 0\ (\text{2-way} \quad \text{drainage}) \tag{4.35}$$

$$\left.\frac{\partial v}{\partial z}\right|_{z=H} = 0\ (\text{1-way} \quad \text{drainage}) \tag{4.36}$$

The solution to this problem is a special case of the solution given by Zhu and Yin (1999) for a two-layer system in which the initial excess pore water pressure varies with both time and depth. The time variation assumed by Zhu and Yin was defined as a ramp function while the initial spatial distribution was bi-linear, the same as defined in the case considered here. Following Zhu and Yin the following variables are defined:

$$p = \frac{\sqrt{(k_2 m_{v2})} - \sqrt{(k_1 m_{v1})}}{\sqrt{(k_2 m_{v2})} + \sqrt{(k_1 m_{v1})}}, \quad q = \frac{H_1\sqrt{c_{v2}} - H_2\sqrt{c_{v1}}}{H_1\sqrt{c_{v2}} + H_2\sqrt{c_{v1}}} \tag{4.37}$$

$$\alpha = \frac{1}{2}(1+q), \quad \beta = \frac{1}{2}(1-q) \tag{4.38}$$

and

$$T_v = \frac{c_{v1} c_{v2} t}{\left(H_1\sqrt{c_{v2}} + H_2\sqrt{c_{v1}}\right)^2} \tag{4.39}$$

In this case the solution to the governing equations may then be expressed conveniently as:

$$v(z, T_v) = \sum_{n=1}^{\infty} T_n(T_v) Z_n(z) \tag{4.40}$$

where, for two-way drainage:

$$T_n(T_v) = \frac{b_n}{c_n} \exp\left(-c_n^2 T_v\right) \tag{4.41}$$

and

$$Z_n = \begin{cases} \dfrac{\sin\left(c_n \alpha \left[z/H_1\right]\right)}{\sin(c_n \alpha)} & 0 \le z \le H_1 \\[2ex] \dfrac{\sin\left(c_n \beta \left[(H - z)/H_2\right]\right)}{\sin(c_n \beta)} & H_1 \le z \le H \end{cases} \tag{4.42}$$

The constant b_n appearing in Eq. (4.41) is given by:

$$b_n = \frac{m_{v1} H_1 / [\alpha \sin(c_n \alpha)] + m_{v1} H_1 / \left(\alpha^2 c_n\right)(u_m - 1) + m_{v2} H_2 / \left(\beta^2 c_n\right) u_m}{m_{v1} H_1 / \left[2\sin^2(c_n \alpha)\right] + m_{v2} H_2 / \left[2\sin^2(c_n \beta)\right]} \tag{4.43}$$

and c_n is the nth positive root of the following equation in the variable θ:

$$\sin\theta + p\sin(q\theta) = 0 \tag{4.44}$$

For one-way drainage boundary conditions:

$$T_n(T_v) = \frac{d_n}{e_n}\exp\left(-e_n^2 T_v\right) \tag{4.45}$$

and

$$Z_n = \begin{cases} \dfrac{\sin\left(e_n\alpha\left[z/H_1\right]\right)}{\sin(e_n\alpha)} & 0 \le z \le H_1 \\ \dfrac{\cos\left(e_n\beta\left[(H-z)/H_2\right]\right)}{\cos(e_n\beta)} & H_1 \le z \le H \end{cases} \tag{4.46}$$

In this case the constant d_n appearing in Eq. (4.45) is given by:

$$d_n = \frac{m_{v1}H_1/\left[\alpha\sin(e_n\alpha)\right]}{m_{v1}H_1/\left[2\sin^2(e_n\alpha)\right] + m_{v2}H_2/\left[2\cos^2(e_n\beta)\right]} \tag{4.47}$$

and e_{n} is the nth positive root of the following equation in the variable θ:

$$\cos\theta - p\cos(q\theta) = 0 \tag{4.48}$$

4.4 Characteristics of Vacuum Consolidation

4.4.1 Settlement

4.4.1.1 One-Way Drainage

Vacuum loading involves application of an isotropic incremental consolidation pressure to the soil. It will induce vertical settlement and at the same time it has a tendency to induce inward lateral (horizontal) displacement of the ground. If inward lateral displacement occurs, the settlement induced by application of a vacuum pressure will be less than that induced by a surcharge load of the same magnitude. The necessary condition for a vacuum pressure of magnitude $\Delta\sigma_{vac}$ to induce inward lateral displacement is as follows (Chai et al. 2005):

$$|\Delta\sigma_{vac}| > \frac{K_0 \cdot \sigma'_{v0}}{1 - K_0} \tag{4.49}$$

where K_0 = the at-rest coefficient of earth pressure, and σ'_{v0} = the initial vertical effective stress.

4.4.1.2 Effect of Drainage Boundary Conditions

In the case of two-way drainage, at the bottom boundary the excess pore water pressure is fixed at zero and effectively no vacuum pressure can be applied at this location. It is obvious therefore that vacuum consolidation involving two-way drainage should result in less settlement than in the case of one-way drainage. Under a vacuum pressure, for both one-way and two-way drainage conditions, water is drained out of the soil layer only at the boundary where the vacuum pressure is applied. Therefore, for both cases, the length of the drainage path is the same. However, it should be noted that the final state is different for these two cases, and the rate of consolidation for the two-way drainage case is 4 times faster than that for the case of one-way drainage (compare Eqs. (4.7) and (4.9)).

4.4.1.3 Two-Layer Soil Systems

Generally, the effect of the initial vertical effective stress (σ'_{v0}) in the soil on the consolidation characteristics of a two-layer soil system to which vacuum pressure has been applied should be similar to that for a uniform soil layer. However the effects of the drainage boundary conditions are strongly affected by the order of the soil layers relative to the boundary where the vacuum pressure is applied (Chai et al. 2009a).

(a) One-way drainage. The order of soil layering will not influence the final settlement. However, it will influence the rate of consolidation. Under the condition of no lateral displacement occurring, when a layer with a relatively low value of the hydraulic conductivity (k) is located adjacent to the drainage boundary, the consolidation rate will be slower; and this is the same for the case of consolidation induce by the application of a surface surcharge load (e.g., Pyrah 1996).

(b) Two-way drainage. Under two-way drainage conditions and vacuum pressure loading, the final state involves a steady flow of pore water toward the boundary where the vacuum pressure is applied. In a two-layer soil system at steady state, the flow continuity condition (Eq. 4.28) must hold, which implies that a layer with a lower value of k must have a higher hydraulic gradient (i). In such cases changing the order of the soil layers will not only influence the rate of consolidation, but also the distribution of vacuum pressure in the system, as illustrated in Fig. 4.5. The final settlement will be a function of both the relative values of the hydraulic conductivity (k) and the compression indices of the soil layers (C_c).

4.4.2 Laboratory Oedometer Tests

4.4.2.1 Test Details

The problem of one and two layer consolidation under vacuum loading and also under conventional surcharge loading has also been studied experimentally using laboratory oedometer apparatus. The equipment used in the laboratory study described here (Chai, 2010, private communication) was a Maruto Multiple Oedometer Apparatus manufactured in Tokyo, Japan. It has five consolidation cells, which can be used either as individual consolidation cells or connected together to form a multi-layer system. Each sample was 60 mm in diameter and typically about 20 mm thick. Figure 4.6 shows the set-up of one consolidation cell. A vacuum pressure can be applied to the top of the sample. During testing, the settlement, the excess pore water pressure at the bottom of the sample and the horizontal earth pressure at the mid height of the consolidation ring were measured. In order to test a two-layer system, two cells were connected together, as illustrated schematically in Fig. 4.7. For the two-layer cases involving only surcharge loading, the same surcharge load was applied to the tops of both cells. However, in the case of vacuum pressure loading, the desired vacuum pressure was applied only at the top of layer-1.

Two types of soil were tested by Chai (2010, private communication). One was reconstituted Ariake clay and another was a mixture of reconstituted Ariake clay and sand (passing the 2 mm sieve) with a mix ratio of 1:1 by dry weight. Some of the physical and mechanical properties of the samples used in this study are given in Table 4.1. In this table, W_L and W_P are the liquid limit and the plastic limit of the samples, respectively. Values of the hydraulic conductivity (k) and the coefficient of one-dimensional volumetric compressibility (m_v) were calculated from the consolidation test results corresponding to a consolidation stress increment from 80 to 160 kPa. Values of the friction angle (ϕ') listed in Table 4.1 are from the results of consolidated undrained triaxial compression tests with measurement of pore water pressures.

Ariake clay contains mainly the smectite clay mineral (Ohtsubo et al. 1995) and consequently it generally has a high liquid limit and plasticity index. Geologically, the Ariake clay originated from volcanic ash and there is a considerable amount of diatoms in the upper non-marine deposit (Hasuike formation) from which the soil samples used in this testing were recovered. The presence of diatoms is known to increase the friction angle considerably (Shiwakoti et al. 2002).

The test procedure adopted was as follows. First, soil slurry with a water content higher than its liquid limit was de-aired under a vacuum pressure of approximately 100 kPa. Then the de-aired slurry was put into a mould with a diameter of 60 mm and a height of 60 mm (20 mm high consolidation ring plus a 40 mm collar), and pre-consolidated under a surcharge pressure of 20 kPa. Following initial consolidation, the sample was cut to 20 mm thick for further consolidation testing. For samples prepared by this method the degree of saturation was close to 100%. All B values measured before the consolidation test were larger than 0.9, where B is the ratio

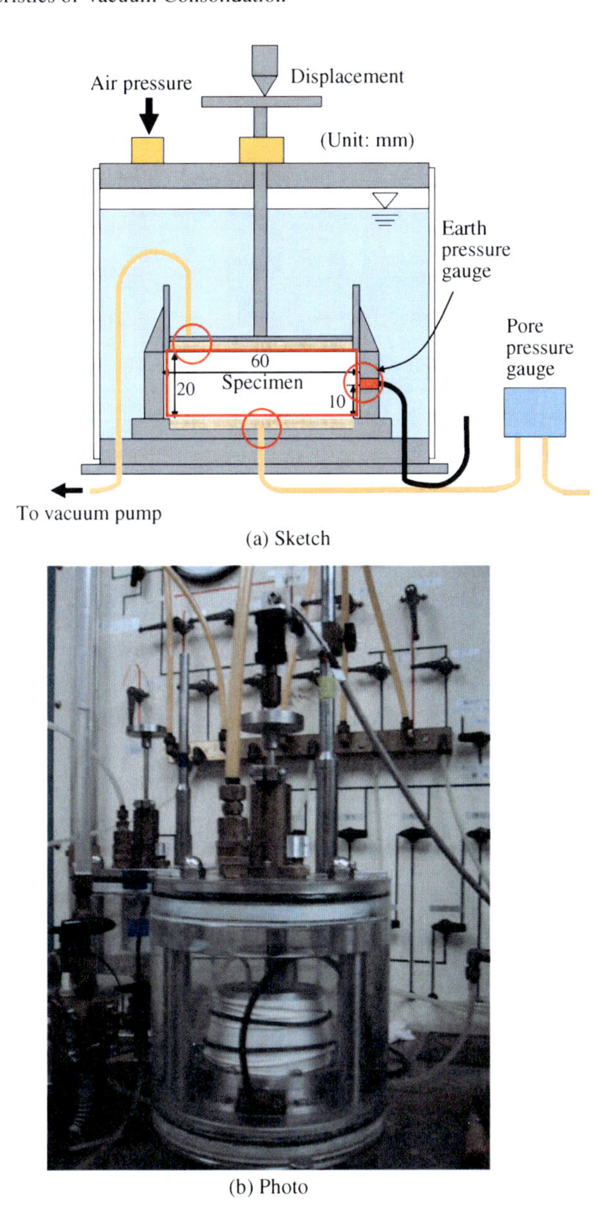

Fig. 4.6 Setup of the consolidation cell. (**a**) Sketch; (**b**) Photo

of the measured excess pore water pressure response to the applied surcharge load under undrained conditions. To investigate the effect of the initial vertical effective stress (σ'_{v0}) in the sample on the behaviour during subsequent vacuum consolidation, samples were first consolidated under a predetermined surcharge load in the range $\sigma'_{v0} = 0 \sim 120$ kPa. Then an incremental consolidation surcharge load ($\Delta\sigma_v$)

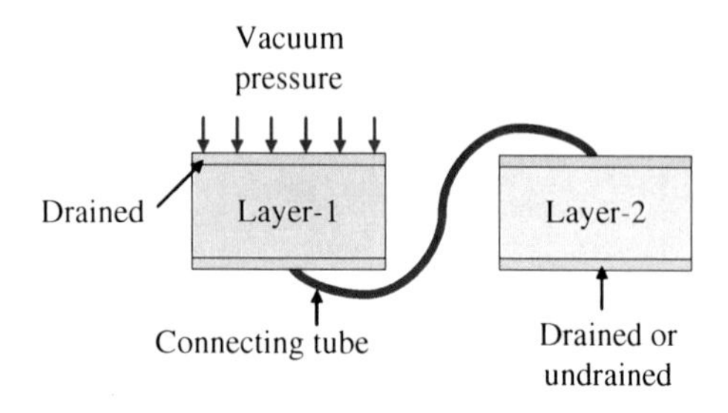

Fig. 4.7 Two-layer configuration

Table 4.1 Basic soil properties

	Grain size distribution (%)							
Soil	Clay (<5 μm)	Silt	Sand	W_L (%)	W_p (%)	m_v (m^2/kN)	k (m/s)	ϕ' (°)
Ariake clay	60.5	38.3	1.2	120.5	60.3	1.47×10^{-3}	2.33×10^{-9}	36.6
Mixed soil	–	–	–	–	–	1.35×10^{-3}	2.98×10^{-9}	35.6

or a vacuum pressure ($\Delta\sigma_{vac}$) or a combination of both was also applied. Samples involving both single layer and two-layer systems with one-way as well as two-way drainage conditions were tested. The test duration was 48 h for a single layer system and 72 h for a two-layer system to ensure the completion of primary consolidation. All cases tested are summarized in Table 4.2.

4.4.2.2 Method for Calculating Earth Pressure Coefficient (K)

The horizontal earth pressure measured during each test is actually a total stress. However, using the consolidation theory presented in Section 4.3, the value of the excess pore water pressure at the location of the earth pressure measurement device was calculated, thus allowing an estimate to be made of the lateral effective stress

Table 4.2 Details of all cases tested

No. of soil layers	Soil type	σ'_{v0} (kPa)	$\Delta\sigma_v$(kPa)	$\Delta\sigma_{vac}$ (kPa)	Drainage condition
1	Ariake clay	0, 40, 80, 120	–	–80	One-way
			80	–	
	Mixed soil		–	–80	
			80	–	
1	Ariake clay	0, 40, 80	0, 80, 120	–80	One-way
			80[a]	–80	
2	A+M[b]	100	–	–50	One-way;
	M+A[b]				Two-way

[a]Surcharge load was applied stepwise; [b]A=Ariake clay; M=mixed soil; A+M=layer-1 A, layer-2 M; M+A=layer-1 M, layer-2 A

in the sample at the location of the earth pressure measurement. In turn this allowed the value of the earth pressure coefficient K to be evaluated (in terms of effective stress components). In the consolidation analysis, a constant coefficient of consolidation (c_v) was used and at certain elapsed times there were discrepancies observed between the calculated and the measured excess pore water pressures at the bottom of the sample. The following method has been adopted to overcome this problem.

(a) At a given time, the ratio of the calculated value of the excess pore water pressure at the bottom of the sample and at the location of the earth pressure measurement was obtained.
(b) Using the measured value of excess pore water pressure at the bottom of the sample and the ratio obtained from (a), the corresponding value at the location of the earth pressure measurement was calculated.

In evaluating the effective stresses in the samples, thus allowing the value of K to be estimated, the vacuum pressure is included in the estimate of the vertical effective stress. However, when estimating the horizontal effective stress, only the effective stress arising from the confinement of the ring is taken as positive. In cases where a gap occurs between the ring and the sample, only the vacuum pressure is assumed to act in the horizontal direction, and for purposes of making a clear distinction this case is arbitrarily treated as a negative effective horizontal earth pressure (the earth pressure gauge used was able to measure the suction) resulting in a negative value of K. This convention was adopted in order to distinguish those cases where a gap forms between the soil sample and the confinement ring from those where it does not.

4.4.3 Test Results

4.4.3.1 Single Layer with Vacuum Pressure or Surcharge Load

A comparison of the settlement-time curves is given in Fig. 4.8a, b for the Ariake clay samples and the mixed soil samples, respectively. In these figures, the curves corresponding to the surcharge load cases generally have larger settlements at early times than those observed for the vacuum pressure cases. When consolidation was essentially completed under a given value of σ'_{v0}, the drainage valve was closed and the predetermined surcharge load was then applied. There was a time period (about 1 h) required for the excess pore water pressure at the bottom of the sample to reach more than 90% of the corresponding applied pressure. During this period some settlement (mainly elastic deformation) occurred and, in order to make comparisons with the vacuum pressure case, this settlement component has been included in the reported total settlement. It can be seen that when $\sigma'_{v0} < 80$ kPa, the settlement induced by vacuum pressure is smaller than that induced by the surcharge load of the same magnitude. It is generally accepted that the direction of the major plastic strain occurring in the soil is usually coaxial with the direction of the principal

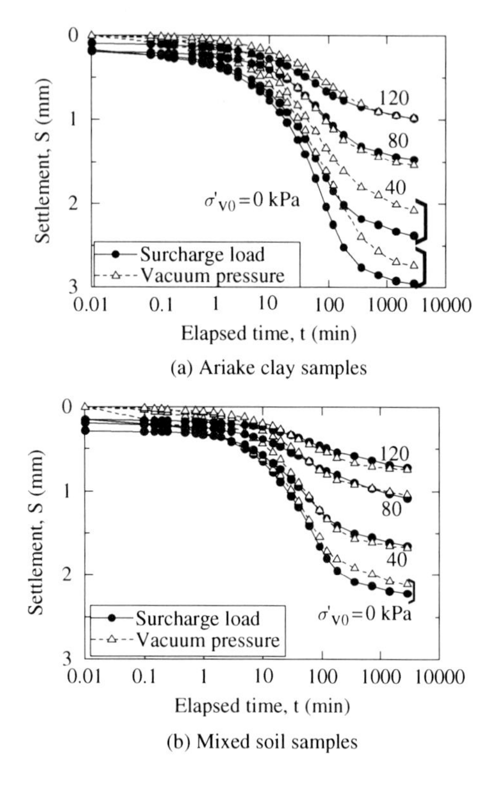

Fig. 4.8 Comparison of settlement-time curves for a single layer system. (**a**) Ariake clay samples; (**b**) Mixed soil samples

effective stress. With the propagation of vacuum pressure into the sample, and when the horizontal effective stress acting on the sample is directed inward, i.e., when the vacuum pressure is dominant, then inward lateral displacement will occur and the settlement will generally be less than that induced by the surcharge load.

For the Ariake clay samples, the calculated values of K are depicted in Fig. 4.9a, b for the surcharge load and the vacuum pressure cases, respectively. For the surcharge loading cases, the initial value of K is influenced by the value of σ'_{v0}, but the final values of K are almost independent of σ'_{v0} (Fig. 4.9a). Using the Jaky (1948) equation, for normally consolidated soils, the at-rest coefficient of earth pressure (K_0) can be approximated as:

$$K_0 = 1 - \sin\phi' \tag{4.50}$$

Adopting the values of ϕ' listed in Table 4.1, K_0 was estimated to be about 0.41. The final value of K measured in these tests was about 0.4, which is very close to the value calculated from the empirical relationship corresponding to at-rest conditions and expressed in terms of ϕ' (Eq. (4.50)). However, for the vacuum pressure cases, the final values deduced for K depend on the initial consolidation pressure, σ'_{v0} (Fig. 4.9b). When inward lateral displacement occurred and a gap between

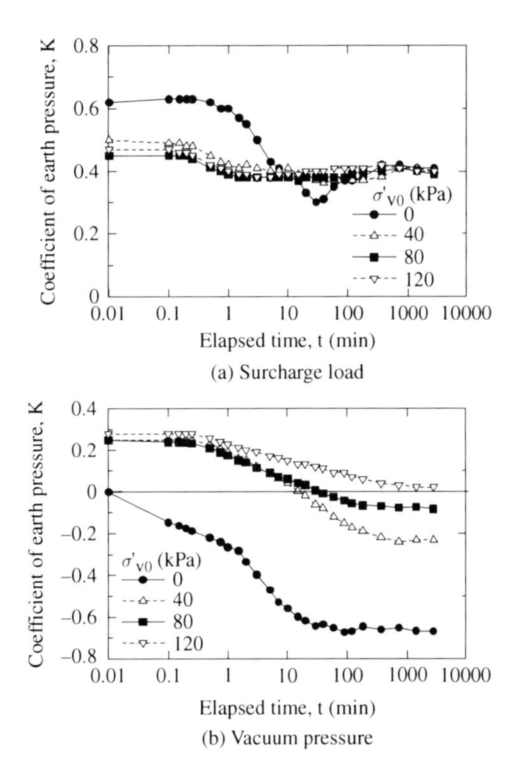

Fig. 4.9 Variations of K for Ariake clay samples. (**a**) Surcharge load; (**b**) Vacuum pressure

the ring and the sample formed, the value of K was arbitrarily assigned a negative value. Under oedometer conditions, $K \geq 0$ indicates that the sample deforms one-dimensionally (1D) and the vacuum pressure-induced settlement is the same as the settlement induced by a surcharge load. The relationship between the final value of K and σ'_{v0} is shown in Fig. 4.10 for both the Ariake clay and the mixed soil samples.

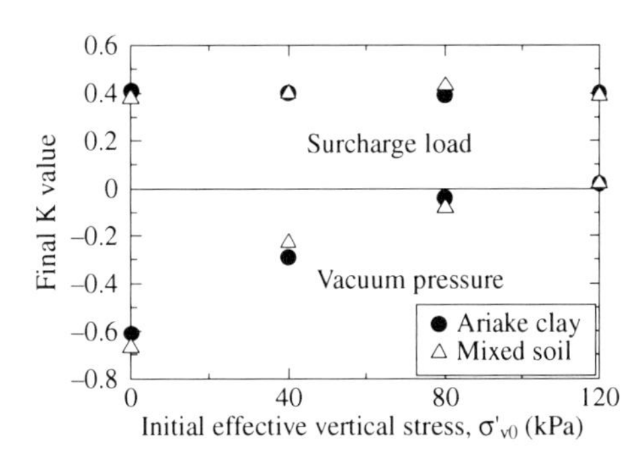

Fig. 4.10 Final K value versus σ'_{v0}

4.4.3.2 Combined Vacuum Pressure and Surcharge Loading

In the field, when $K = K_0$, there should be no lateral displacement associated with the consolidation of the soil deposit. Figure 4.9 shows that during the consolidation process under surcharge loading, $K > K_0$, and for the vacuum pressure cases, $K < K_0$. It may be possible therefore to keep the value of K close to K_0 by controlling the combination of vacuum pressure with surcharge load. This would have important practical implications for cases in which the lateral displacement is a crucial design requirement.

Two types of test were carried out on single layer samples to investigate the effects of combining vacuum pressure and surcharge loading, i.e., Type-1: in which a vacuum pressure and a surcharge load were applied together instantaneously; and Type-2: in which a vacuum pressure was applied instantaneously but the surcharge was applied in discrete steps over time, i.e., 'stepwise' loading. In the latter case, a surcharge increment of 20 kPa was applied every 30 min until the total surcharge pressure was reached.

Figure 4.11 shows the results of Type-1 tests in which $\sigma'_{v0} = 0$. It can be seen that at the early stages of each test, $K > K_o$, which tends to induce outward lateral displacement, and at the later stage $K < K_o$, which tends to induce inward lateral displacement. The reason behind this phenomenon is that lateral displacement induced by the surcharge load occurs mainly due to the shear stresses in the sample which is therefore largely an immediate deformation. Conversely the lateral displacement induced by the vacuum pressure occurs continuously during the consolidation process. In terms of the final value of K, the effect of the surcharge load is the same as the increase in the value of σ'_{v0} (compare Fig. 4.11 with Fig. 4.9b).

In order to reduce the value of K during the early stages of consolidation, Type-2 tests were conducted. The results of Type-1 and Type-2 tests are compared in Fig. 4.12a to 4.12c for cases in which $\Delta\sigma_{vac} = -80$ kPa, $\Delta\sigma_v = 80$ kPa and $\sigma'_{v0} = 0$, 40 and 80 kPa, respectively. It can be seen that generally, Type-2 tests resulted in smaller values of K during the consolidation process, and under field conditions this loading regime can contribute to reducing the immediate outward lateral displacement generated by the surcharge load.

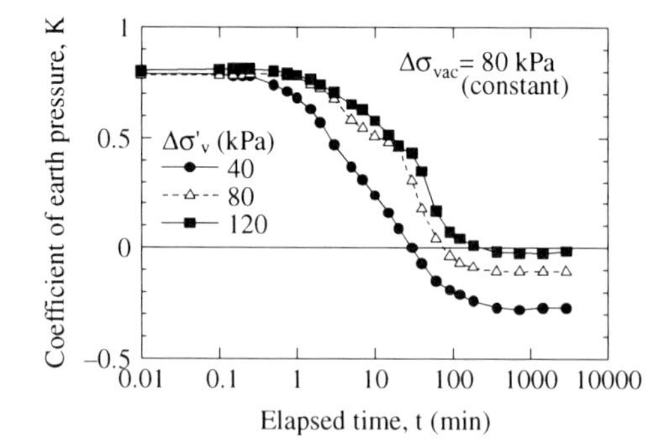

Fig. 4.11 Variation of K under a combination of vacuum pressure and surcharge loading

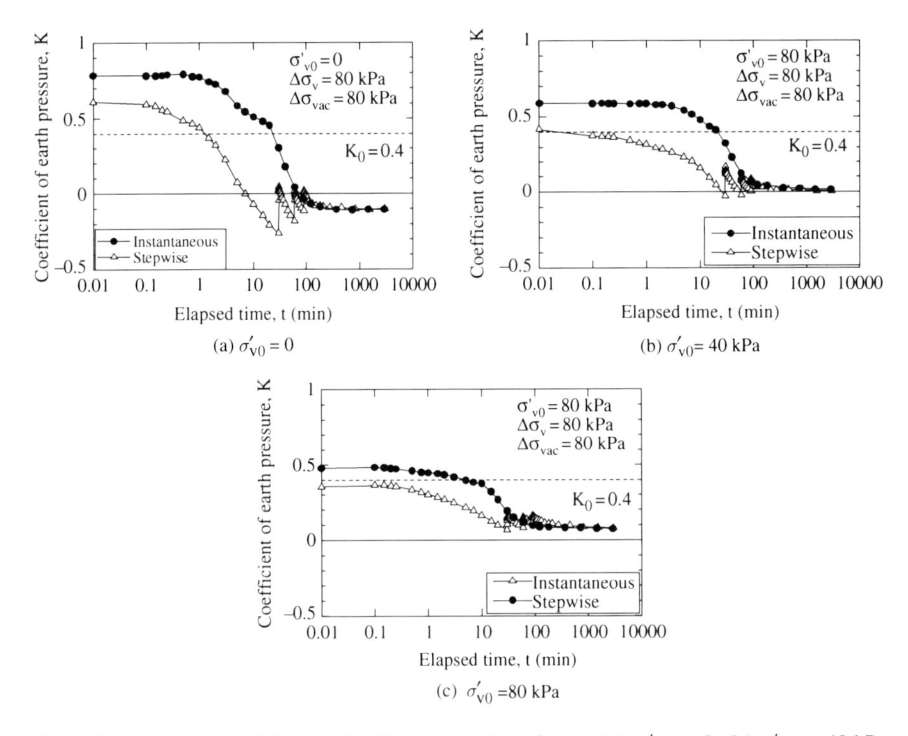

Fig. 4.12 Comparison of K values for Type-1 and Type-2 tests. (**a**) $\sigma'_{v0} = 0$; (**b**) $\sigma'_{v0} = 40$ kPa; (**c**) $\sigma'_{v0} = 80$ kPa

4.4.3.3 Two-Layer System Under Vacuum Loading

The settlement-time curves are shown in Fig. 4.13 for cases where two-layer samples were subjected to vacuum pressure loading only. For the two types of samples tested, the mixed soil had a higher value of k (Table 4.1) and when it was placed at the boundary where the vacuum pressure was applied, the rate of consolidation was faster. This phenomenon can be more clearly observed from the measured excess pore water pressure (u) at the middle of the two layers (Fig. 4.14). Although the final values are almost the same, in the earlier stages of consolidation the case M+A (layer-1: the mixed soil M; and layer-2: the Ariake clay A) exhibited a smaller value of u, i.e., a larger vacuum pressure indicating faster consolidation.

For the case of two-way drainage, the excess pore water pressures measured at the interface of the two layers are depicted in Fig. 4.15. It may be observed that the case M+A not only had a faster rate of excess pore water pressure development, but also a smaller final value, i.e., a larger vacuum pressure. In other words, cases where $k_1 > k_2$ will result in generally larger vacuum pressures being distributed through the system (Fig. 4.5).

Figure 4.16a, b show the variation of the interpreted earth pressure coefficient K during the consolidation process for one-way drainage of the M+A and A+M

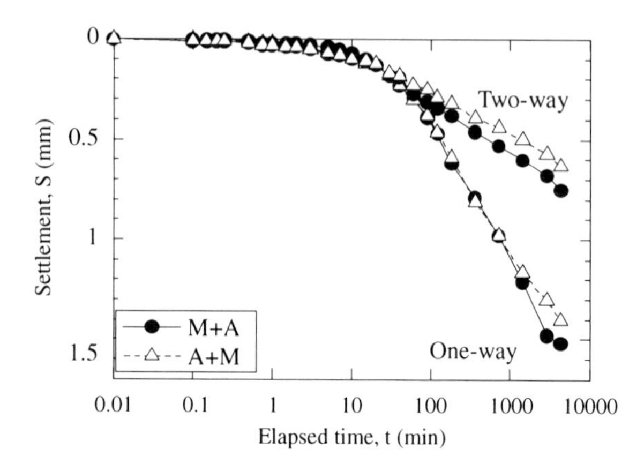

Fig. 4.13 Comparison of settlement curves for two-layer systems under vacuum loading

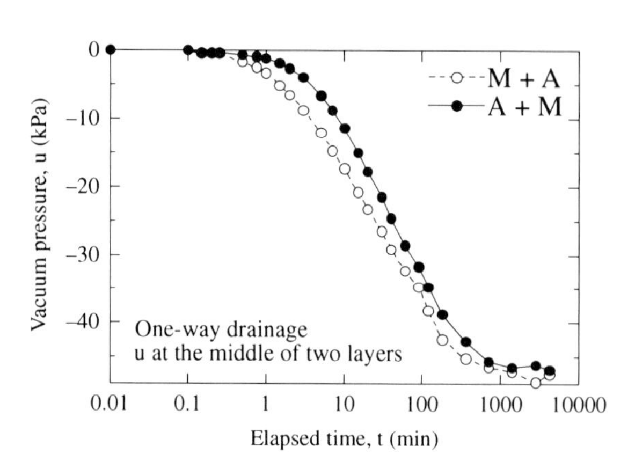

Fig. 4.14 Excess pore water pressure at the middle of two layers under vacuum pressure with one-way drainage

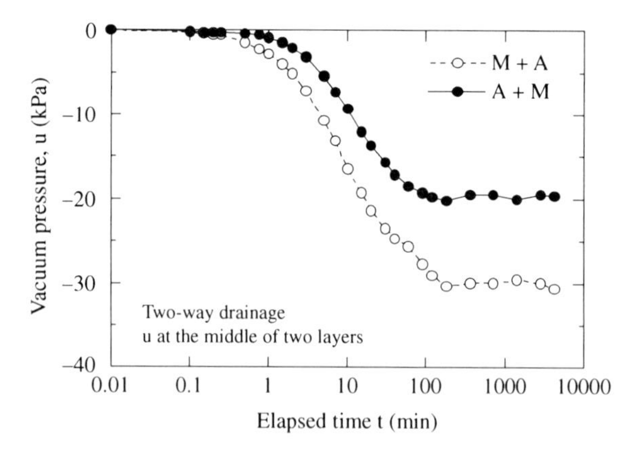

Fig. 4.15 Excess pore water pressure at the middle of two layers under vacuum pressure with two-way drainage

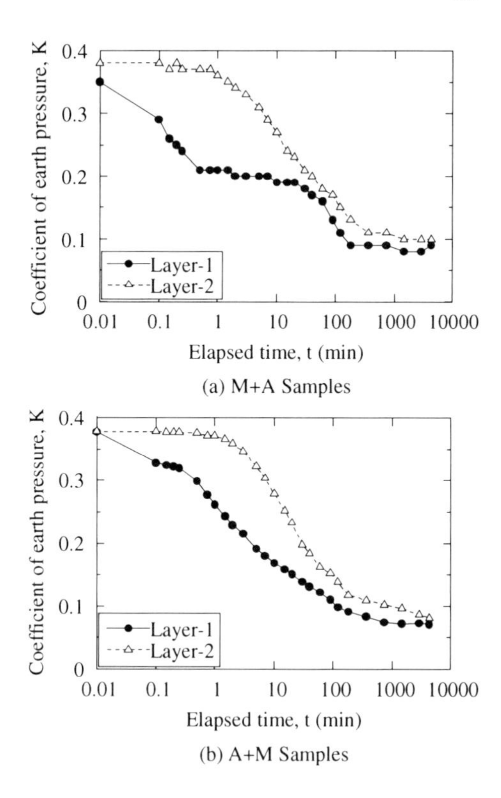

Fig. 4.16 Variation of K for a two-layer system with one-way drainage. (**a**) M+A samples; (**b**) A+M samples

samples, respectively. The final values of K for layer-1 and layer-2 are almost the same, but the value of K for layer-1 reduced faster in the M+A sample than in the A+M case, i.e., the order of the soil layers influenced the rate of consolidation. The results for two-way drainage cases are shown in Fig. 4.17a, b for M+A and A+M samples, respectively. The values of K for the A+M case (Fig. 4.17b) are generally larger than those of the M+A case (Fig. 4.17a) for the corresponding layer-1 and layer-2. The larger K value implies a smaller vacuum pressure at the location of the earth pressure gauge, and this has influenced the consolidation settlement. Further, since the vacuum pressure at the bottom drainage boundary is zero, the vacuum pressure in layer-2 is generally less than that in layer-1, which has resulted in larger deduced values of K.

4.5 Optimum PVD Penetration Depth

4.5.1 Theoretical Prediction

In vacuum consolidation projects conducted in the field, generally the vacuum pressure is applied at the ground surface so that the final vacuum pressure distribution

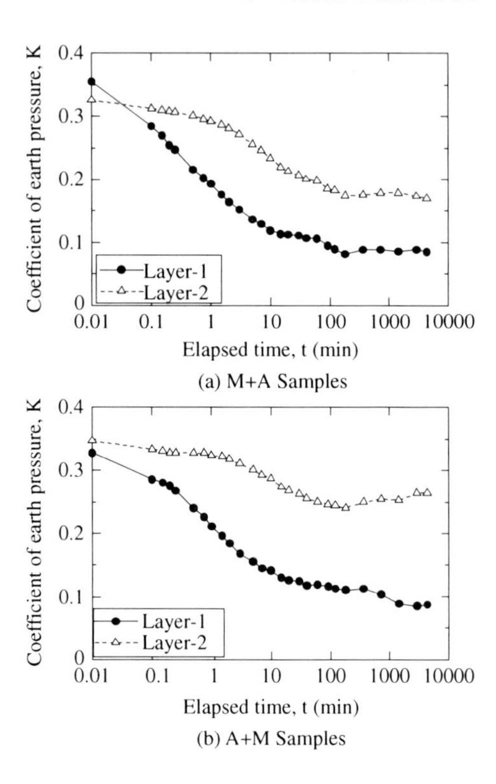

Fig. 4.17 Variation of K for a two-layer system with two-way drainage. (**a**) M + A samples; (**b**) A + M samples

in the ground depends on the properties and drainage boundary conditions of the deposit. Considering a 1D problem and a uniform soil deposit, if the bottom boundary is not drained the final vacuum pressure in the deposit will be uniform. Conversely, if drainage can occur through the bottom boundary such that a vacuum pressure cannot be effectively maintained at this boundary, the vacuum pressure distribution at the end of vacuum consolidation, i.e., when a steady-state distribution of suction (excess) pore pressure is established throughout the soil deposit, will be linear with the maximum value at the ground surface and zero at the bottom of the layer (Chai et al. 2005). This distribution will involve steady upward water flow through the clayey layer. In the case of a non-uniform deposit, the vacuum pressure distribution within the deposit will depend on the relative values of the hydraulic conductivities of the individual layers. For steady upward water flow in a layered deposit the following condition must be satisfied in order to maintain continuity of the flow:

$$i_1 k_{v1} = i_2 k_{v2} = \cdots = i_i k_{vi} \tag{4.51}$$

where i_i and k_{vi} = the hydraulic gradient and hydraulic conductivity of the ith layer, respectively. A layer with a lower value of k must have a higher value of i, and a higher i value implies a larger variation of vacuum pressure within the layer.

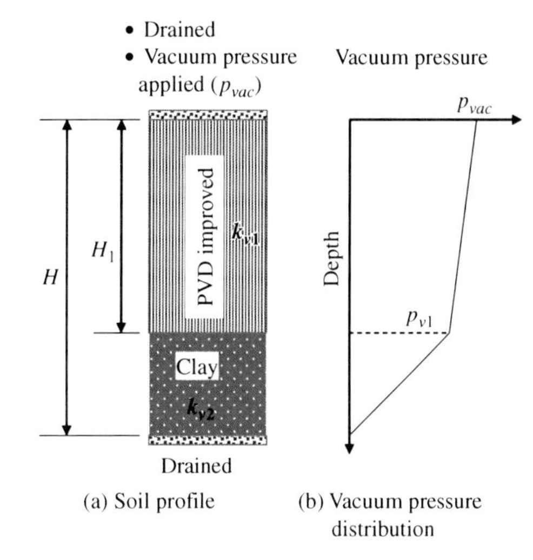

Fig. 4.18 Illustration of vacuum consolidation with PVD improvement

For a clayey deposit with two-way drainage which has been partially penetrated by PVDs, a possible vacuum pressure distribution within the layers at the end of vacuum consolidation is illustrated in Fig. 4.18.

The optimum penetration depth is the depth of installation at which the clay layer will exhibit the largest consolidation settlement under a given vacuum pressure applied at the ground surface. Assuming that the static water table is at the ground surface, the values of k are uniform in both the PVD-improved zone and the unimproved zone, and at the end of vacuum consolidation the vacuum pressure at the ground surface is p_{vac}, and p_{v1} at the base of the PVD-improved zone (Fig. 4.18), the condition of flow continuity requires (Chai et al. 2006):

$$k_{v1} \cdot \frac{p_{vac} - p_{v1}}{H_1} = k_{v2} \cdot \frac{p_{v1}}{H - H_1} \tag{4.52}$$

where k_{v1} and k_{v2} = the vertical hydraulic conductivities of layers 1 and 2, respectively, H_1 = the thickness of the PVD-improved zone, and H = the overall thickness of the soft clayey deposit. If the integration of vacuum pressure with depth is denoted by A, it can be shown that A is expressed as:

$$A = [P_{vac} \cdot H_1] - \left[\frac{1}{2}(p_{vac} - p_{v1})H_1\right] + \left[\frac{1}{2}(H - H_1) \cdot p_{v1}\right] \tag{4.53}$$

The quantity A corresponds to the area under the curve of excess pore suction versus depth. Substituting Eq. (4.52) into Eq. (4.53) to eliminate p_{v1}, an expression for A is obtained as follows:

$$A = \frac{1}{2}p_{vac} \cdot (H_1 + H) - \frac{1}{2}H\frac{p_{vac} \cdot k_{v2} \cdot H_1}{k_{v1} \cdot H - (k_{v1} - k_{v2})H_1} \tag{4.54}$$

The task of finding the maximum value of A corresponds to finding the largest value of the area under the curve of suction versus depth. This is the condition at which in the longer term the application of suction will have the greatest effect on the induced settlement in the clay layer. Hence, differentiating A with respect of H_1 and equating this to zero, provides an expression for the optimum penetration depth H_1 as follows (Chai et al. 2006):

$$H_1 = \left(\frac{k_{v1} - \sqrt{k_{v1} k_{v2}}}{k_{v1} - k_{v2}} \right) H \tag{4.55}$$

Chai et al. (2001) proposed a method to calculate the equivalent vertical hydraulic conductivity of PVD-improved subsoil, which can be used to evaluate the value of k_{v1}, i.e., the mass vertical hydraulic conductivity of the PVD-improved zone:

$$k_{v1} = \left(1 + \frac{2.5 l^2}{\mu d_e^2} \frac{k_h}{k_v} \right) k_v \tag{3.50 bis}$$

where d_e = the diameter of a unit cell (containing a PVD and its improvement area), k_h and k_v = the horizontal and vertical hydraulic conductivities of the natural soil, respectively, $l(= H_1)$ = the drainage length of the PVDs, and q_w = the discharge capacity of the PVDs. The expression for μ is given by Eq. (3.8a bis). Since H_1 in Eq. (4.55) and l in Eqs. (3.50 bis) and Eq. (3.8a bis) have the same meaning, in order to obtain the value of H_1 some iteration is needed. Also note that if the value of k_{v1} in the PVD-improved zone or k_{v2} in the unimproved zone is not uniform, an equivalent hydraulic conductivity (k_{eq}) can be used, but using Eq. (4.55) may not yield an exact optimum penetration depth. For a two-layer case, k_{eq} can be calculated as follows:

$$k_{eq} = \frac{k_{i1} \cdot k_{i2} H_i}{k_{i1} \cdot H_{i2} + k_{i2} \cdot H_{i1}} \tag{4.56}$$

where H_i = the thickness of the PVD-improved or unimproved zone, H_{i1} and H_{i2} = the thicknesses of sub-layer 1 and 2 in H_i, and k_{i1} and k_{i2} = hydraulic conductivities of sub-layer 1 and 2 in H_i, respectively. Furthermore, if the compressibility of the soil around the base of the PVD-improved zone varies significantly, from a settlement point of view, Eq. (4.55) may not guarantee the optimum penetration depth. Note that the method proposed above also has the following limitations.

(a) The effect of horizontal water flow from the surrounding area into the zone of soil improved by vacuum consolidation is not considered.
(b) Only the condition at the end of vacuum consolidation is considered, i.e., when a steady-state distribution of suction (excess) pore pressure is established throughout the soil deposit. The effect of the PVD installation depth on the consolidation process is not considered.
(c) By contrast, in Eq. (4.55), the initial thickness (not the final thickness) of the soft deposit is used to calculate the initial PVD penetration depth. If the whole

deposit compresses uniformly, these contradictory conditions may not cause any significant error, but if the upper soil layer compresses more than the lower layer, some error may be introduced.

4.5.2 Laboratory Model Tests

Chai et al. (2009b) reported the results of a series of laboratory tests aimed at investigating the behaviour of soil samples whose consolidation performance had been enhanced by the introduction of partially penetrating PVDs. These model tests and the important findings of the testing program conducted at Saga University are described as follows.

4.5.2.1 Test Device

A cylindrical soil sample was used in these tests and a sketch of the test set up is shown in Fig. 4.19 (Chai et al. 2009b). The apparatus consisted of a cylinder

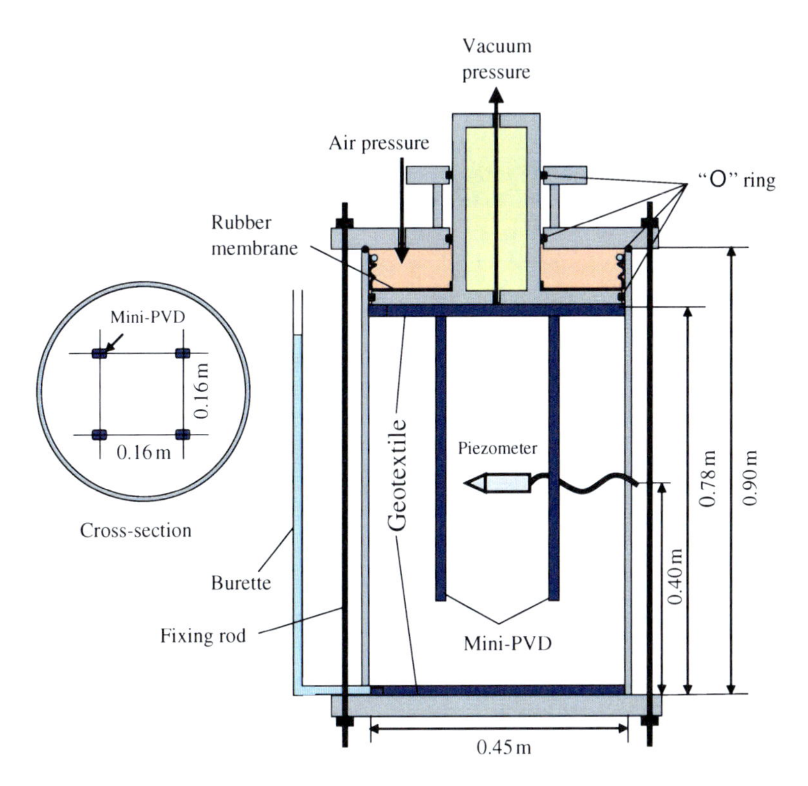

Fig. 4.19 Sketch of the model test device (after Chai et al. 2009b)

containing the soil sample, with 0.45 m inner diameter and 0.9 m height and made of vinyl chloride with a wall thickness of 15 mm; upper and lower pedestals with a thickness of 40 mm; a piston system; and a burette connected to the drainage layer at the bottom of the model. The upper and the lower pedestals are connected by eight 12 mm diameter steel rods. The 40 mm thick piston is made of vinyl chloride and has a hollow shaft with an outside diameter of 100 mm fixed to it. To prevent possible tilting of the piston, a guide was installed on the upper pedestal around the shaft. Sealing between the piston and the cylinder and between the shaft and the upper pedestal was achieved by 'O' rings lubricated with silicon grease. Both air pressure and vacuum pressure can be applied independently as consolidation loads. The air pressure was applied through the upper pedestal to the top of the piston and the vacuum pressure was applied through the hollow shaft of the piston to the bottom of the piston, i.e., to the surface of the soil sample. To prevent leakage of the air pressure and vacuum pressure leakage through the piston, a rubber membrane sleeve with a thickness of 1 mm was installed inside the chamber above the piston. This membrane was initially folded to accommodate vertical displacement of the piston during consolidation of the soil. A KPD – 200 kPa piezometer, manufactured by the Tokyo Sokki Kenkyujo Co. Ltd, Japan, was embedded around the middle height of the soil sample. Settlement of the soil sample was measured at the top of the shaft by a dial gauge.

4.5.2.2 Materials

The soil used in these tests was remoulded Ariake clay with a liquid limit $W_L = 108.8\%$, and plastic limit $W_P = 59.2\%$. The initial water content of the clay was adjusted to about 120% (more than its liquid limit) and the samples were then cured in a plastic container for more than 1 day before putting them into the apparatus. Mini-PVDs were formed by folding non-woven geotextile in 3 layers 30 mm wide and 9 mm in thickness when unconfined. The geotextile used was made of polypropylene and had a mass density of about 131 g/m^2.

4.5.2.3 Test Procedure

To set up the model test, firstly the cylindrical container was installed on the lower pedestal and 3 layers of the geotextile were placed at the bottom of the cylinder as a drainage layer. A thin layer of silicon grease was painted on the wall of the cylinder to reduce the friction between it and the soil sample contained within. The clay was placed inside the cylinder, layer by layer, and when the overall height of the soil sample reached about 0.4 m from the bottom, the piezometer was installed. Further clay was added to the sample, again layer by layer, until the overall height of the soil sample was 0.78 m. After completing installation of the soil in the model ground, 4 mini-PVDs were installed to a pre-determined depth with the plan locations as indicated in Fig. 4.19. The method used to install the mini-PVDs involved inserting a stainless steel rod inside the mini-PVD and pushing it into the soil sample vertically. The rod was then withdrawn leaving the mini-PVD inside the soil sample.

Another 3 layers of the geotextile were placed on the top of the soil sample to act as a drainage layer. Finally, the piston, the rubber membrane sleeve and the upper pedestal were installed and the dial gauge for measuring the vertical displacement was set in place. The pre-determined air-pressure and vacuum pressure were then applied to the composite soil sample and the testing commenced.

4.5.2.4 Test Program

Four tests were conducted using the apparatus described previously with the conditions of each test as listed in Table 4.3. The mini-PVD installation depth varied from 0.48 to 0.73 m leaving an unimproved layer varying in thickness from 0.30 to 0.05 m. Each test lasted for about 15 days. It is well known that vacuum consolidation not only induces settlement, but also inward lateral displacement of the ground (e.g., Chai et al. 2005). If only vacuum pressure were applied in the model tests described above there would have been a distinct possibility of gaps forming between the soil sample and the cylinder wall, due to inward lateral displacement of the soil sample, which may have resulted in leakage of the vacuum pressure. In order to avoid this situation, both surcharge loading derived from air pressure applied to the piston and vacuum pressure loading applied directly to the soil sample were adopted in these tests.

4.5.2.5 Simulation of Model Tests

To obtain detailed information about the likely distribution of excess pore water pressure (vacuum pressure) developed in the model ground and to provide a cross check on the experimental results, the model tests were simulated by finite element analysis (FEA). The mechanical behaviour of the model ground was simulated using the modified Cam-clay model (Roscoe and Burland 1968). The values of the model parameters used in these calculations are listed in Table 4.4. The values of λ, e_0 and k_0 were determined from the results of laboratory consolidation tests involving incremental loading and the values of ν, κ and M were simply assumed based on experience. During the consolidation process, the hydraulic conductivity of the clay soil (k) was varied with void ratio (e) following Taylor's equation (Taylor 1948):

Table 4.3 List of the cases tested

Case	Initial Thickness (m)	Mini-PVD installation depth[a] (m)	Air pressure (kPa)	Vacuum pressure (kPa)
1	0.78	0.48 (0.30)[a]	40	40
2	0.78	0.58 (0.20)	40	40
3	0.78	0.68 (0.10)	40	40
4	0.78	0.73 (0.05)	40	40

[a]The number in parenthesis is the thickness of the layer improved by PVD installation

Table 4.4 Model parameters adopted

λ	κ	e_0	ν	M	k_0 (m/s)
0.265	0.027	3.2	0.3	1.2	2.51×10^{-9}

Note: λ = slope of virgin consolidation line in e-ln p' plot (p' is effective mean stress and e is voids ratio); κ = slope of rebound line in $e - \ln p'$ plot; M = strength parameter for Cam-clay model, i.e., stress ratio q/p' at critical state; ν = Poisson's ratio; e_0 = initial voids ratio; and k_0 = initial hydraulic conductivity

$$k = k_0 \cdot 10^{(e-e_0)/C_k} \tag{4.57}$$

where C_k = a constant (in this study $C_k = 0.4e_0$). Since the entire model ground sample was remoulded, it is considered reasonable to assume that there was no smear zone. When considering the possibility of compression of the mini-PVDs under the confining stresses imposed in these test, back-fitting of some of the observations indicated that the effective diameter of the mini-PVDs should be d_w = 10 mm, and the discharge capacity, q_w = 1 m^3/year (Chai et al. 2007). For the arrangement as shown in Fig. 4.19, the diameter of the unit cell can therefore be evaluated as d_e = 0.225 m.

4.5.2.6 Results of Testing and Analysis

The settlement curves for these four tests are compared in Fig. 4.20, while Fig. 4.21 shows an enlargement of the final part of the consolidation curves. It can be seen that the FEA simulated well the settlement curves for Case-1 and Case-2. For Case-3 and Case-4, the simulated settlement rate is faster than the test data for elapsed times less than 10 days. Theoretically, increasing the mini-PVD installation depth increases the initial settlement rate and it is not clear why the test data did not show this same tendency. Nevertheless, both test data and analysis predictions indicate that Case-2 and Case-3 result in larger settlement than Case-1 and Case-4 (Fig. 4.21). Figure 4.22 shows plots of final settlement at the corresponding mini-PVD installation depth. It can be observed that with respect to this settlement, an optimum mini-PVD installation depth exists between Case-2 and Case-3, although it should be acknowledged that experimental error may account for some of the observed differences in settlement.

As indicated in Fig. 4.19, a piezometer was installed at about the middle of the sample (i.e., about 0.4 m from the bottom). Variations with time of the measured and simulated excess pore water pressures (u) at this piezometer point are compared in Fig. 4.23. Note that these variations include both positive and negative (suction) excess pore pressures. The simulated maximum values of u and the simulated dissipation rates are higher than those measured. Despite taking care to maintain a saturated piezometer filter during the installation process, the results in Fig. 4.23

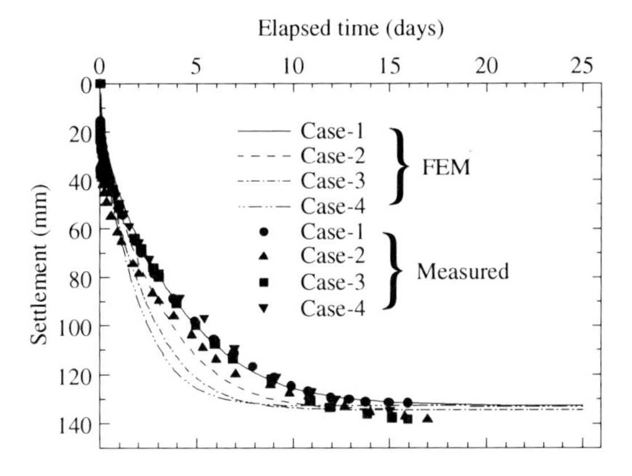

Fig. 4.20 Comparison of settlement time curves

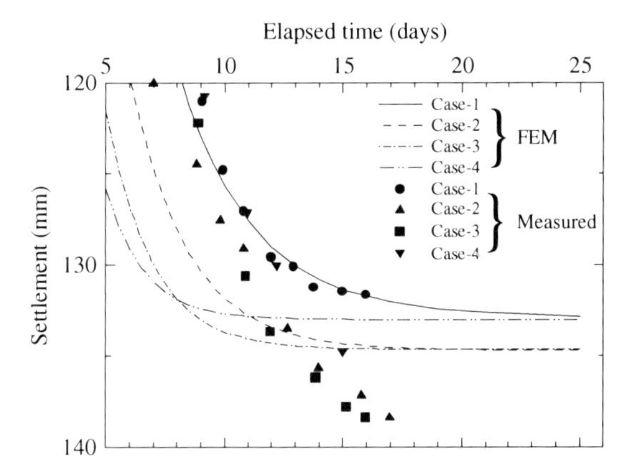

Fig. 4.21 Enlarged comparison of settlements at close to the end of consolidation

show that the piezometer may not have been 100% saturated. At the time of terminating the tests the relative ordering of the values of u for the 4 cases, both measured and analyzed, is different. However, if the maximum simulation time in the FEA is extended to 25 days, the predicted ordering of the values of u is the same as that observed from the measured pore pressures, i.e., the largest is Case-4 (with the smaller vacuum pressure) and the smallest is Case-1. The final vacuum pressure distributions in the model ground are given in Fig. 2.24. This figure shows that the area enclosed by the line corresponding to the pore pressure distribution u, the left or y-axis, and the top or x-axis, i.e., the area A under the curve of suction versus depth is larger for Cases 2 and 3 than it is for Cases 1 and 4. This supports the contention that the PVD penetration depths adopted in Cases 2 and 3 are more effective in terms of the final induced settlement than are Cases 1 and 4.

Using the adopted model parameters, Eq. (4.55) gives an optimum mini-PVD installation depth of about 0.575 m, which is close to the installation depth adopted

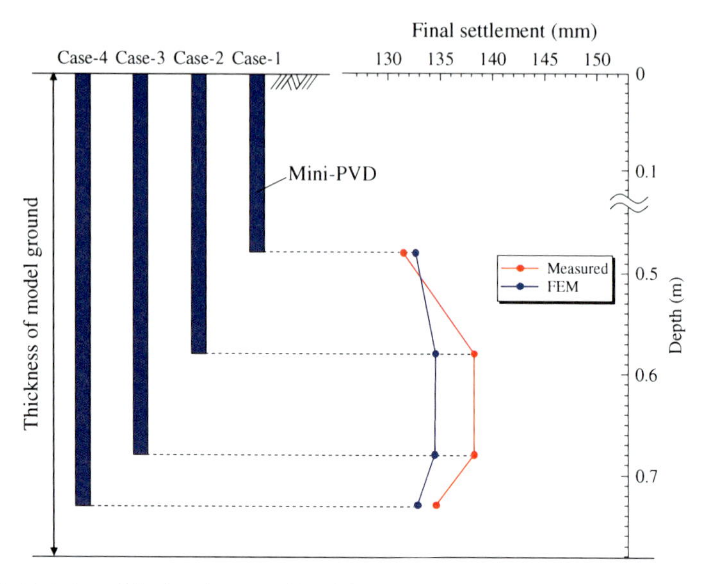

Fig. 4.22 Variation of final settlement with mini-PVD installation depth

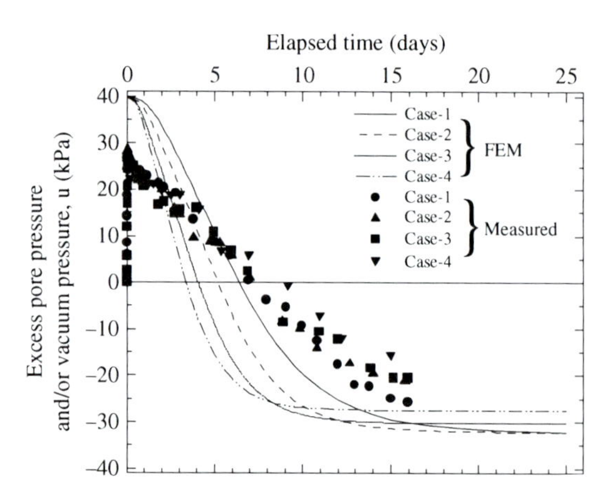

Fig. 4.23 Excess pore water pressure variation at piezometer point

for Case-2 of 0.58 m. From Fig. 4.22, it can be seen that an increase in the mini-PVD installation depth from 0.58 to 0.68 m had almost no effect on the final settlement. From an economic point of view, the optimum installation depth should therefore be close to 0.58 m in this model problem.

From this experimental study it can be concluded that Eq. (4.55) is useful for determining the optimum installation depth of PVDs for a soil deposit with two-way drainage subjected to vacuum pressure applied at the ground surface.

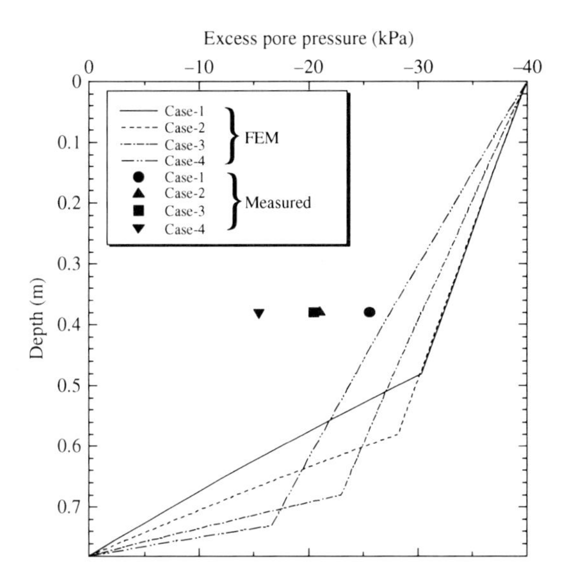

Fig. 4.24 Final vacuum pressure distribution in the model ground

4.6 Estimating Deformations Induced by Vacuum Pressure

The pattern of deformation induced by vacuum pressure loading applied in the field is different in character from that induced by embankment loading. Hence the deformation characteristics of vacuum pressure loading need to be carefully considered when performing calculations to predict the effects of applying a vacuum pressure to soil. Two semi-theoretical methods to predict such deformations have been suggested in the literature (Imai 2005; Chai et al. 2005). These are now considered in the following sections.

4.6.1 Imai's Method

Imai (2005) proposed a method for calculating the final settlement and inward lateral displacement occurring during vacuum consolidation of a soil layer. This method assumes that the imposed vacuum pressure ($-p_{vac}$) is constant throughout the whole deposit. Further, it also assumes active earth pressure conditions in the soil surrounding the area being subjected to the vacuum treatment, as shown in Fig. 4.25. Consequently the effective stress conditions in the area being treated will be as follows.

Initially, the vertical effective stress (σ'_{v0}) in the ground is the effective overburden pressure, and the horizontal effective stress (σ'_{h0}) can be calculated from knowledge of the at-rest earth pressure coefficient (K_0):

$$\sigma'_{h0} = K_0 \sigma'_{v0} \tag{4.58}$$

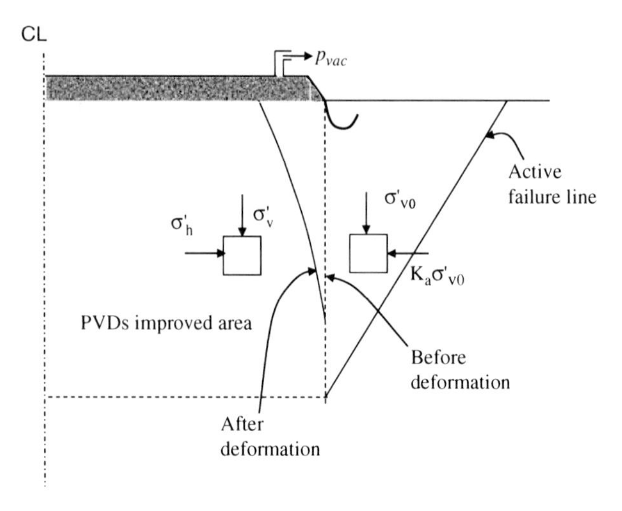

Fig. 4.25 Stress state and deformation pattern assumed in Imai's method

In the longer term, due to sustained application of a vacuum pressure to the soil, incremental stresses must be added to both the initial vertical and horizontal effective stress components. Clearly a value equivalent to the magnitude of the applied vacuum pressure, p_{vac}, should be added to the initial vertical and horizontal effective stresses, but it is also assumed that the initial horizontal effective stress will tend to reduce from the at-rest condition. i.e., from $K_0\sigma'_{v0}$ to $K_a\sigma'_{v0}$ where K_a is the active earth pressure coefficient of the soil, so that finally:

$$\sigma'_v = \sigma'_{v0} + p_{vac} \tag{4.59}$$

$$\sigma'_h = K_a\sigma'_{v0} + p_{vac} \tag{4.60}$$

Hence, the incremental effective stresses are:

$$\Delta\sigma'_v = p_{vac} \tag{4.61}$$

$$\Delta\sigma'_h = p_{vac} - (K_0 - K_a)\,\sigma'_{v0} \tag{4.62}$$

A parameter called the Isotropic Index (I) can be defined as:

$$I = \frac{\Delta\sigma'_h}{\Delta\sigma'_v} = 1 - (K_0 - K_a)\,\frac{\sigma'_{v0}}{p_{vac}} \tag{4.63}$$

Assuming axisymmetric conditions and using Hooke's law, the vertical and horizontal strains (ε_v, ε_h) induced in the improved zone can be expressed as:

$$\varepsilon_v = -a_v\, m_v\, p_{vac} \tag{4.64}$$

$$\varepsilon_h = -a_h m_v p_{vac} \tag{4.65}$$

where m_v = the 1D coefficient of volumetric compressibility and a_v and a_h = strain correction factors given by:

$$a_v = \frac{(1 + K_0) - 2K_0 I}{(1 + 2K_0)(1 - K_0)} \tag{4.66}$$

$$a_h = \frac{-K_0 + I}{(1 + 2K_0)(1 - K_0)} \tag{4.67}$$

For one-dimensional deformation conditions, m_v can be expressed in terms of the effective stress Young's modulus (E') and Poisson's ratio (ν') of the soil skeleton as follows:

$$m_v = \frac{(1 - 2\nu')(1 + \nu')}{(1 - \nu')} \cdot \frac{1}{E'} \tag{4.68}$$

and for elastic soil response the relationship between ν' and K_0 is:

$$K_0 = \frac{\nu'}{1 - \nu'} \tag{4.69}$$

Imai (2005) further argued that when $I = K_0$, there will be no inward lateral displacement induced by the vacuum pressure and at that depth where this condition is satisfied the following equation holds:

$$\frac{\sigma'_{vo}}{|p_{vac}|} = \frac{1 - K_0}{K_0 - K_a} \tag{4.70}$$

Assuming $K_0 = 0.5$, $K_a = 0.3$, $p_{vac} = -70$ kPa, the groundwater level at the ground surface and a submerged unit weight of the soil of 6 kN/m^3, from Eq. (4.70) a depth of about 30 m can be calculated as the zone of influence of the vacuum pressure on lateral displacement, i.e., the depth at which the inward lateral displacement vanishes.

Imai (2005) suggested that if the depth of improvement by the PVDs is less than the depth determined by Eq. (4.70), I can be expressed approximately as:

$$I = 1 - (1 - K_0)\frac{z}{H} \tag{4.71}$$

where z = depth from the ground surface, and H = the depth improved by PVDs. The variations of I, a_v and a_h with depth suggested by Imai are shown in Fig. 4.26.

Imai's (2005) method captures some important aspects of vacuum consolidation such as: the fact that vacuum consolidation induces settlement under one-dimensional conditions that is equal to or less than that which would be induced by a surcharge load of the same magnitude; and the fact that the lateral displacement induced by a vacuum pressure reduces with depth. However, the method also

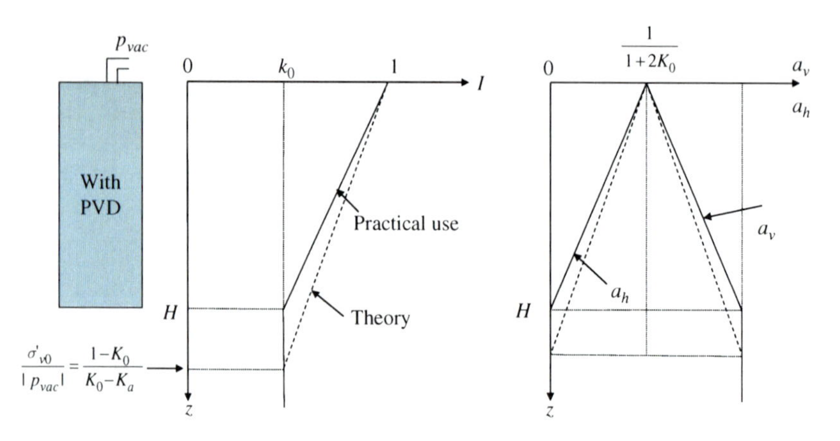

Fig. 4.26 Variation of I, a_v and a_h assumed in the Imai method

has some shortcomings. Assuming active earth pressure at the edge of the vacuum consolidation area is possibly a crude approximation and it ignores the existence of tension cracks which might occur near the ground surface. It is also unclear whether equilibrium is assured by making this assumption. Secondly, for soft clayey soils the deformation induced by application of the vacuum pressure is mainly plastic so that adopting the values of a_v and a_h deduced from elasticity theory may involve significant errors. Further, there is a contradiction involved in using $I = K_0$ to determine the depth where there is no inward lateral displacement, because I is defined by assuming an active stress state in the ground.

4.6.2 *Method of Chai et al.*

4.6.2.1 When Does Lateral Deformation Occur?

Under oedometer conditions, the criterion determining whether vacuum pressure will induce inward lateral displacement is given by Eq. (4.49). However, in the field, conditions are often somewhat different from those in an oedometer test. As illustrated in Fig. 4.27a, at the ground surface, inward lateral displacement induced by vacuum pressure may cause tension cracks with a depth of z_c. For a soil element located at a depth less than z_c, the stress state can be approximated by that shown in Fig. 4.27b, which is the same as in an oedometer test with a vacuum pressure large enough to cause inward lateral displacement. However, below z_c and above z_l, the depth at which no lateral displacement occurs (the value of z_l will be discussed later), the lateral effective stress consists of two parts: one is the vacuum pressure and the other is the earth pressure exerted by the adjacent soil mass. In this zone, the horizontal earth pressure will be between the values corresponding to the at-rest and active states. Denoting the coefficient of earth pressure in this zone as K_{a0}, the stress state of a typical soil element will be like that shown in Fig. 4.27c.

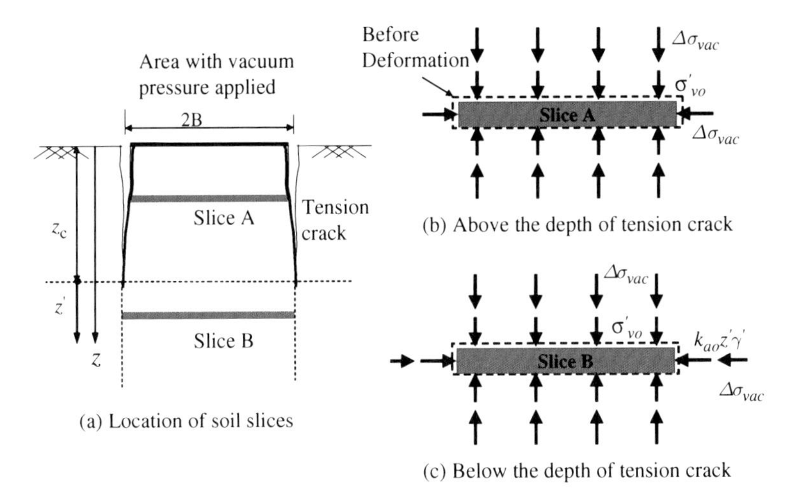

Fig. 4.27 Illustration of stress state and deformation pattern of soil slices in the ground under vacuum loading (after Chai et al. 2005)

In Fig. 4.27b and c, the same symbols, σ'_{v0} and $\Delta\sigma_{vac}$, are used for the initial vertical effective stress and the applied vacuum pressure, but their numerical values can be different in each case. Intuitively, the value of K_{a0} should be close to the active coefficient of earth pressure (K_a) at a depth just below z_c, and close to the at-rest coefficient of earth pressure (K_0) at the depth just above z_l. Another factor which needs to be considered when determining a reasonable value of K_{a0} is that under field conditions, the deeper layers which do not undergo lateral displacement will tend to restrict the inward lateral displacement of the layers above. Considering these factors, an expression is proposed for K_{a0}, as follows:

$$K_{a0} = \beta K_a + (1 - \beta)K_0 \tag{4.72}$$

where β = an empirical factor. Chai et al. (2005) proposed that β should normally be assigned a value in the range 0.67–1.0.

From Rankine earth pressure theory, assuming that the groundwater level is located at a distance z_w below the ground surface, the depth of tension cracking z_c can be expressed as follows:

$$z_c = \frac{2c'}{\gamma_t\sqrt{K_a}}, \quad \text{for } z_c < z_w \tag{4.73a}$$

$$z_c = \frac{1}{(\gamma_t - \gamma_w)}\left(\frac{2c'}{\sqrt{K_a}} - \gamma_w z_w\right), \quad \text{for } z_c > z_w \tag{4.73b}$$

where γ_t = the total unit weight of soil, γ_w = the unit weight of pore water, c' = the effective stress cohesion, $K_a = \tan^2(45 - \phi'/2)$ is the coefficient of active earth pressure, and ϕ' is the friction angle of the soil. Assuming $c' = 5-10$ kPa, $\phi' = 30°$, $\gamma_t = 15$ kN/m^3 and $z_w = 1.0$ m, Eq. (4.73) predicts $z_c = 1.47$ to 4.93 m.

In Fig. 4.27c, if the horizontal stress ($|\Delta\sigma_{vac}| + K_{a0}z'\gamma'$) is larger than the effective stress required to maintain a K_0 stress state, then inward lateral displacement will occur. Equating the effective stress required to maintain a K_0 stress state, $K_0(|\Delta\sigma_{vac}| + \sigma'_{v0})$, and the lateral effective stress, $|\Delta\sigma_{vac}| + \langle z'y'K_{a0}\rangle$, a condition is obtained for determining the depth below which no lateral displacement occurs in the soil, $z_l = z_c + z'(z' = z - z_c)$, i.e.,

$$|\Delta\sigma_{vac}| = \frac{K_0.\sigma'_{v0} - \sigma'_{a0}}{1 - K_0} \tag{4.74}$$

where

$$\sigma'_{av} = \langle z'\gamma'K_{a0}\rangle = \begin{cases} 0 & \text{for } z < z_c \\ K_a z'\gamma' & \text{for } z_l > z > z_c \end{cases} \tag{4.75}$$

and where γ' = the effective unit weight of soil, equal to γ_t above the groundwater level and $(\gamma_t - \gamma_w)$ below the groundwater level. For multilayer soils, σ'_{av} should be calculated using an appropriate summation procedure. With a known value of $\Delta\sigma_{vac}$, the value of z_l can be determined from Eqs. (4.73), (4.74), and (4.75).

The vacuum consolidation method is generally applied to deposits of soft, normal to lightly overconsolidated clayey soils, so that at the end of vacuum consolidation the soil should be in a normally consolidated state. Therefore, the value of K_0 corresponding to the normally consolidated state should be substituted in Eq. (4.74), and this value should typically be less than 1.0. In multilayer subsoils, the value of K_0 for the layer closest to the depth z_l should be substituted in Eq. (4.74).

4.6.2.2 Vertical Deformation

The vertical strain caused by vacuum loading can be expressed as a portion of the vertical strain that would occur under conditions of 1D consolidation with an increment of vertical effective stress given by $\Delta\sigma_{vac}$, i.e.,

$$\varepsilon_v = \alpha\frac{\lambda}{1+e}\ln\left(1 + \frac{\Delta\sigma_{vac}}{\sigma'_{v0}}\right) \tag{4.76}$$

where e = the voids ratio, λ = the virgin compression index in an e-lnp' plot (where p' is effective mean stress) and α = a factor with a value less than or equal to unity. It is proposed that α will have a minimum value (α_{min}) at the ground surface and it will be unity when $Z \geq z_l \left(or\, |\Delta\sigma_{vac}| \leq \frac{K_o \cdot \sigma'_{vo} - \sigma'_{av}}{1 - K_o}\right)$. Chai et al. (2005) proposed the following expression for α:

$$\alpha = \alpha_{\min} + \frac{1 - \alpha_{\min}}{|\Delta\sigma_{vac}|}\left(\frac{K_0\sigma'_{v0} - \sigma'_{av}}{1 - K_0}\right)$$

for

$$|\Delta\sigma_{vac}| \geq \frac{K_0 \cdot \sigma'_{v0} - \sigma'_{av}}{1 - K_0} \tag{4.77}$$

In the field, vacuum consolidation can be conducted under stress conditions that are either close to triaxial (usually in cases where the improved area is close to a circle or a square) as well as plane strain (where the improved area is a long strip). It is convenient to denote the values of $\alpha_{\min}$ for triaxial stress conditions and plane strain conditions as $\alpha_{\min-T}$ and $\alpha_{\min-P}$, respectively. Based on laboratory test results, Chai et al. (2005) proposed that $\alpha_{min-T} = 0.80$ and that α_{min-P} should be larger than $\alpha_{\min-T}$. If it is assumed that both the horizontal and volumetric strains are the same for triaxial and plane strain conditions (the vertical strain is usually larger for plane strain conditions), it can be shown that $\alpha_{\min-P} = (1 + \alpha_{\min-T})/2$, and if $\alpha_{\min-T} = 0.8$ then $\alpha_{\min-P} = 0.9$. If the volumetric strain is the same in each case, it is more logical that both the vertical and horizontal strains in plane strain should be larger than for triaxial stress conditions. Hence, it is assumed that $\alpha_{min-P} = 0.85$.

4.6.2.3 Volumetric Strain

The effective stress path experienced by a soil element undergoing vacuum consolidation varies with depth. For an element at or near the ground surface, the effective stress path is close to isotropic consolidation, while at deeper locations it is closer to 1D consolidation. Strictly speaking, to calculate the volumetric strain accurately, it is necessary to follow the effective stress path experienced by all soil elements with an appropriate elasto-plastic constitutive model. However, for simplicity we consider a semi-empirical equation for calculating the vertical strain (Eq. (4.76)) and also assume that under vacuum loading (for both triaxial and plane strain conditions), the volumetric strain (ε_{vol}) in the ground is the same as that occurring during 1D consolidation, so that:

$$\varepsilon_{vol} = \frac{\lambda}{1+e}\ln\left(1 + \frac{\Delta\sigma_{vac}}{\sigma'_{v0}}\right) \tag{4.78}$$

4.6.2.4 Horizontal Strain

With known values of the vertical (Eq. (4.76)) and volumetric strains (Eq. (4.78)), the average inward (compressive) horizontal strain (ε_h) can be expressed as follows:

$$\varepsilon_h = \frac{1}{2}(\varepsilon_{vol} - \varepsilon_{vv}), \quad \text{for triaxial stress conditions} \tag{4.79}$$

$$\varepsilon_h = (\varepsilon_{vol} - \varepsilon_{vv}), \quad \text{for plane strain conditions} \tag{4.80}$$

Once ε_h is known, the lateral displacement (δ_h) can be approximated quite simply as follows:

$$\delta_h = B \cdot \varepsilon_h \tag{4.81}$$

where B = the half width of the area treated by vacuum consolidation.

For clarity, it should be emphasized that the analysis presented above is for normally consolidated soils only. For a lightly overconsolidated deposit, the same type of calculation can be made in two steps. Step 1 is from the in situ stress state to the maximum pre-consolidation stress previously experienced by soil, using the unloading-reloading compression index (κ) instead of λ in Eqs. (4.76) and (4.78). Step 2 is for stress states from the maximum pre-consolidation stress experienced by the soil to the final stress state and it will be the same as the procedure suggested previously for a normally consolidated soil. The total horizontal strain will be the summation of the values obtained from Steps 1 and 2.

The information needed for calculating the deformations induced by vacuum consolidation according to the method of Chai et al. is as follows.

(a) Vacuum consolidation conditions. Two parameters are required: the magnitude of the vacuum pressure ($\Delta\sigma_{vac}$) at each depth and the half width of the improvement area (B).
(b) Groundwater level and soil parameters. The groundwater level (z_w) and seven soil parameters are required for each soil layer: the total unit weight (γ_t), the initial voids ratio (e), the over consolidation ratio (OCR), the virgin and unloading-reloading compression indices in e-lnp' space (κ and λ respectively), the effective stress friction angle (ϕ') and cohesion (c').
(c) Model parameter α_{min}. It is suggested that for triaxial stress conditions, $\alpha_{\min-T} = 0.80$ and for plane strain conditions, $\alpha_{\min-P} = 0.85$.

It should be noted that the method proposed here does not consider any interaction between the soil strata. In particular, the restraining effect of a deeper layer on an overlying layer of soil is not considered. This may result in a predicted zig zag profile of lateral displacement if the compressibility of adjacent soil layers varies significantly.

4.6.3 Analysis of Field Tests

Several examples of the application of the vacuum consolidation technique in the field have been well documented in the literature. Some of these case study examples are considered here, since they allow an evaluation to be made of the techniques described previously for estimating the ground deformations induced by applying vacuum pressure over a portion of the ground surface.

4.6.3.1 Vacuum Preloading for an Oil Storage Facility

This case history from Tianjin in China was first reported by Chu et al. (2000). The site was a reclaimed section of land and the thickness of the reclaimed soft clay layer was about 4–5 m. Below it was a marine clay layer with an overall thickness of about 10–16 m. This layer can be further divided into a silty clay layer 2–4 m thick, a clay layer 7–8 m thick and a silty clay layer. The marine clay layer was underlain by a stiff sandy silt layer. The treated zone covers an area of approximately 50,000 m^2 and for the purpose of ground improvement it was divided into two sections: Section I (about 30,000 m^2) and Section II (about 20,000 m^2) (see Fig. 4.28).

Before the vacuum loading commenced, partially dried clay fill about 2 m thick (on average) was placed on the site, on top of which was placed a 0.3 m sand mat. PVDs were then installed on a square grid at a spacing of 1.0 m to a depth of 20 m (elevation of about –13.5 m). There is no information about the duration of the fill loading before the vacuum loading commenced. Chu et al. (2000) reported that the ground settled about 0.15 and 0.27 m during installation of the PVDs for Sections I and II, respectively. The vacuum pressure applied at the surface was about –80 kPa and the duration was 4 months. During application of this vacuum loading, the settlement and pore water pressure at different depths and the lateral displacement at the edge of the treated area were monitored. From the measured pore water pressures, it was interpreted that the groundwater level was about 1.0 m below the top surface (elevation about 5.0 m) after the 2.0 m thick clay fill and 0.3 m thick sand mat were constructed, i.e., the groundwater level was almost at the ground surface before application of the clay fill. Pore water pressure measurements at the centre of Section I indicate that before vacuum consolidation commenced, the consolidation induced by the surcharge load, i.e., the clay and sand fill, was not finished. At about –6 m elevation (approximately the middle of the soft clay layer), the measured initial excess pore water pressure was about 20 kPa. However, it is considered that the outward lateral displacement induced by the fill loading was almost finished when vacuum consolidation commenced, i.e., most of the measured lateral displacement was therefore due to vacuum consolidation.

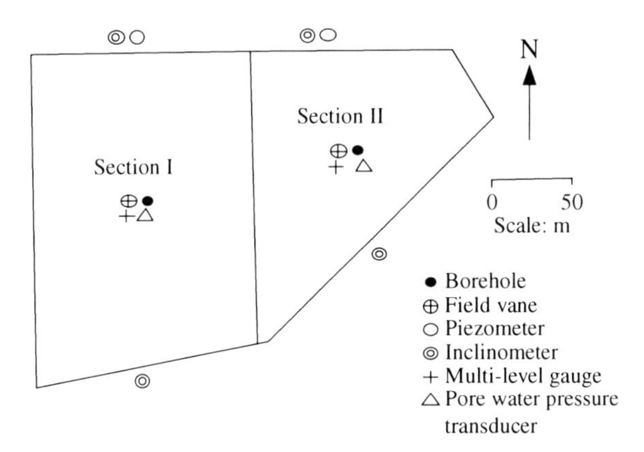

Fig. 4.28 Plan view of treated area and the location of instrumentation at the Tianjin site (after Chu et al. 2000)

Table 4.5 Soil strata and parameters at the Tianjin oil storage site

No.	Description	Elevation (m)	Unit weight, γ_t (kN/m^3)	Voids Ratio, *e*	Compression index (e-lnp'), λ	*OCR*	ϕ' (°)	c' (kPa)
1	Fill + sand mat	6.50 to 4.69	19.0	0.85	0.056	1	31	12
2	Soft clay (reclamation)	4.69 to –0.37	17.9	1.15	0.068	1	30	12
3	Silty clay	–0.37 to –4.28	18.3	1.05	0.092	1	31	–
4	Soft clay	–4.28 to –12.57	17.5	1.30	0.252	1	30	–
5	Stiff silty clay	–12.57 to –16.0	19.0	0.85	0.114	1	31	–
6	Sandy silt	–16.0 to –21.0	19.0	0.85	0.114	1	31	–

The soil parameters required to calculate the final deformations induced by vacuum consolidation are given in Table 4.5 for Section I. Elevations, unit weights and voids ratios were read from the soil profile figures given by Chu et al. (2000). The values of λ were back calculated from the measured compression of each soil layer assuming an applied vacuum pressure of –80 kPa and OCR of 1.0. The back-calculation was made assuming 1D deformation conditions. Considering that field vacuum consolidation might yield less settlement than 1D consolidation, the λ values listed in Table 4.5 are 1.05 times the back-calculated values. Also, it was assumed that the settlement below elevation –12.57 m was due to the compression of the 3.43 m thick stiff silty clay layer and the 5.0 m thick sandy silt underlying it. The measured compression of the clay fill and the sand mat at Section I seems problematic – the amount of compression reduced with elapsed time and was almost zero at the end of vacuum consolidation.

Referring to Section II, 100 mm compression of the clay fill and the sand mat was assumed in calculations. For this site, although the consolidation induced by the surcharge fill may not have finished by the time vacuum pressure was applied, assuming OCR = 1.0 at the commencement of the vacuum treatment should not introduce significant error. Values of the effective stress friction angle (ϕ') and cohesion (c') were assumed. With the parameters listed in Table 4.5, the depth of tension cracks was calculated as about 2.8 m.

Taking into account the shape of the treated area, triaxial stress conditions were assumed in this case with $\alpha_{\min-T} = 0.8$ and a half width of the vacuum consolidation area of 110 m being adopted in calculations of the settlements and lateral displacements. The effects on the calculated deformation of the ground of the value assumed for β (and therefore the value of K_{a0}) in Eq. (4.72) and of the variation of vacuum pressure with depth were investigated by comparing predictions with the field measurements.

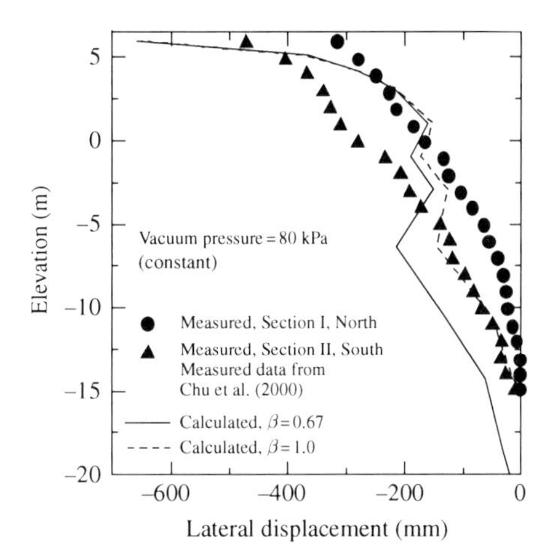

Fig. 4.29 Comparison of measured and calculated lateral displacements at Tianjin, constant vacuum pressure with different β values (after Chai et al. 2005)

Figure 4.29 shows the effect on the predicted lateral displacement of varying β within the range 0.67–1.0, assuming a constant value of vacuum pressure with depth of –80 kPa. The measured field data are from Chu et al. (2000) and correspond to the end of the vacuum treatment and soil consolidation. The zig zag shape of the calculated lateral displacement profile is probably due to a shortcoming of the method which does not consider interaction between soil layers. It can be seen that the value of β mainly influences the lateral displacement at deeper locations and the calculated depth at which the lateral displacement becomes insignificant (z_l). The smaller the value of β (and hence the larger the value of K_{a0}), the larger the calculated lateral displacement and the larger the predicted value of z_l. When $\beta = 1.0$, $z_l \cong 21.5$ m (at an elevation of –15.0 m), and when $\beta = 0.67$, $z_l \cong 29.5$ m (elevation of –23.0 m). It seems that overall $\beta = 1.0$ provides a better simulation of the field data in cases where the vacuum pressure does not reduce with depth. Of course $\beta = 1.0$ corresponds to the active earth pressure state and it will obviously under-estimate the earth pressure for soil at depths near z_l. However, as mentioned previously, the proposed method does not consider the interactions among the different soil strata. In the field, the deeper layers which do not undergo lateral displacement will tend to restrict the inward lateral displacement of the layers above. Using $\beta = 1.0$ indirectly takes into account this effect.

Close to the ground surface the calculation method over-estimated the lateral displacement. This is simply because near the ground surface the initial effective stress due to gravity forces alone is assumed in the analysis to be quite low. In reality, prior to application of a vacuum pressure there may have been some suction above the groundwater level which would tend to increase the initial effective stress. In addition, weathering can increase the stiffness of soil at shallow depths. None of these effects was considered in these calculations.

From the measured pore water pressures at locations near the centre of Section I, it was calculated that there was no significant vacuum pressure reduction with depth,

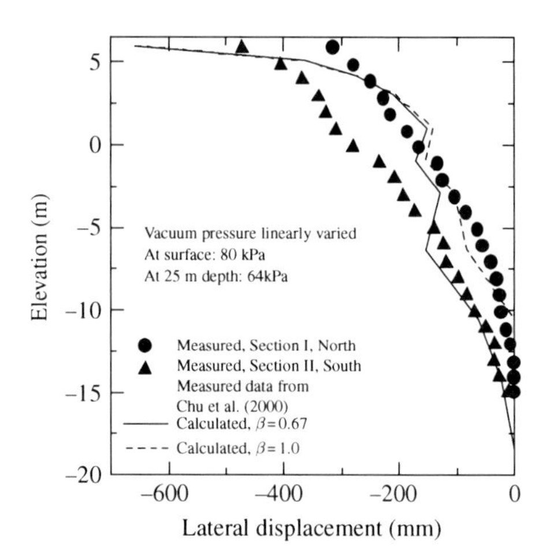

Fig. 4.30 Comparison of measured and calculated lateral displacements at Tianjin, vacuum pressure varied with depth and different β values (after Chai et al. 2005)

down to about 20.5 m (elevation −14.0 m). However, under the edge of the area improved by vacuum pressure there could be significant vacuum pressure reduction with depth. The lateral displacements were also calculated assuming that the vacuum pressure varies linearly from −80 kPa at the ground surface to −64 kPa (80% of the value at the ground surface) at a depth of 25 m (elevation −18.5 m). The predicted lateral displacements are compared in Fig. 4.30. In this case, for both $\beta = 1.0$ and 0.67 the calculated results are comparable with the field data. The corresponding values of z_l are calculated as 17.0 m (elevation −10.5 m) for $\beta = 1.0$ and 22.0 m (elevation −15.5 m) for $\beta = 0.67$.

The calculated values of vertical compression of each soil layer are compared with the field measurements in Table 4.6. Because the values of λ were based on the back calculated values from the measured compression of each layer, comparisons between measured and calculated values are good. The calculated settlements also varied with the assumed value of β, especially for the deeper soil layers. The smaller the value of β, i.e., the larger the value of K_{a0}, the larger is the predicted lateral displacement (Figs. 4.29 and 4.30) and the smaller the predicted compression of the soil layer. Reducing the vacuum pressure with depth reduced the predicted vertical compression of the deeper layers.

4.6.3.2 Field Trial at Yaoqiang Airport

A field trial of vacuum consolidation at the site of Yaoqiang airport in China was reported by Tang and Shang (2000). The test area was 60 m × 40 m. The soil at the site consists of alternate layers of silty sand (2.5 m), silty clay (2.5 m), silt (2.5 m) and soft clay (2.5 m). The groundwater level was about 2.5 m below the ground surface. Before the vacuum consolidation treatment was applied, PVDs were installed to a depth of 12 m over a square pattern at 1.3 m spacing, in an attempt to accelerate the consolidation process. At the ground surface there is a 2.5 m thick silty sand

Table 4.6 Comparison of measured and calculated vertical compression of soil strata at the Tianjin oil storage site

		Vertical compression (mm)				
			Calculated			
			$\Delta\sigma_{vac} = 80$ kPa		$\Delta\sigma_{vac} = 80 - 64$ kPa[a]	
Elevation (m)	Soil layer	Measured	$\beta = 1.0$	$\beta = 0.67$	$\beta = 1.0$	$\beta = 0.67$
6.50 to 4.69	Fill + sand mat	100[b]	87	87	86	86
4.69 to –0.37	Soft clay (reclamation)	154	148	147	145	145
–0.37 to –4.28	Silty clay	113	110	109	105	104
–4.28 to –12.57	Soft clay	417	425	414	392	382
–12.57 to –21.0	Stiff silty clay and sandy silt	167	176	170	148	147

[a] Vacuum pressure linearly varied from 80 kPa at ground surface to 64 kPa at 25 m depth; [b] Assumed by referring to the measured value for Section II

layer. To minimize vacuum pressure loss through this layer, an in situ deep mixing slurry cut-off wall, 1.2 m thick and 4.5 m deep was constructed around the perimeter of the treated area. The vacuum pressure monitored at the ground surface was –70 to –80 kPa, and the reduction in groundwater pressure was measured as being about 65 kPa at depths of 2 and 14 m. However, at 8 to 10 m depth, the measured pore water pressure reduction was only 40–50 kPa. There was some scatter in the measurements of pore water pressure. The duration of vacuum consolidation was recorded as 83 days. The thickness and soil parameters for each significant layer at this site are given in Table 4.7. Unit weights and voids ratios are average values of the ranges reported by Tang and Shang (2000). There are no reported values of the virgin compression index λ. The values listed in Table 4.7 were back calculated

Table 4.7 Soil strata and parameters at the Yaoqiang airport site

No.	Description	Depth (m)	Unit weight, γ_t (kN/m^3)	Voids Ratio, e	Compression index (e-lnp'), λ	OCR	ϕ' (°)	c' (kPa)
1	Silty sand	0–2.5	17.3	0.91	0.046	1	33	12
2	Silty clay	2.5–5.0	19.3	0.87	0.046	1	33	–
3	Silt	5.0–7.5	19.1	0.77	0.036	1	33	–
4	Soft clay	7.5–10.0	18.2	1.35	0.25	1	30	–
5	Silty clay	10.0–16.0	19.8	0.65	0.034	1	33	–

from the reported 1D compression of each layer, assuming that the vacuum pressure in the ground was –65 kPa, and OCR was 1.0. Because the PVDs only penetrated to 12 m depth, Tang and Shang (2000) attributed the settlement below 10 m to a 2 m thick layer (from 10 to 12 m depth) of silty clay. However, because the vacuum pressure measured at 14 m depth was still about –65 kPa, significant compression in the sub-layer from 10 to 16 m was assumed in the calculations reported here. The listed values of the friction angle (ϕ') and cohesion (c') were assumed. The value of c' may seem high for a silty sand layer but, as described previously, an in situ deep mixing slurry cut-off wall was constructed around the vacuum consolidation area and the value of c' was therefore selected to take its effect into account. From Eq. (4.73a), the depth of tension crack was calculated as 2.1 m.

A comparison of the measured and calculated lateral displacements at the end of the field test, assuming a constant vacuum pressure (–65 kPa) with depth, is shown in Fig. 4.31. The measurements were made by an inclinometer located at the middle of one of the longer sides of the treated area and 2.5 m outside the edge of this zone. Plane strain conditions were therefore assumed in the calculations, with a half width of the treated area of 20 m and $\alpha_{\min-P} = 0.85$. The same as for the Tianjin site, it seems that assuming $\beta = 1.0$ yielded a better simulation of the field data. However, it should be noted that the predictions of lateral displacements were made at the edge of the treated zone, while the measurements were recorded 2.5 m away from this edge. The calculated depth (z_l) for no lateral displacement is 14.5 m for $\beta = 1.0$, and this is larger than observed in the field. This is possibly because the method does not consider the restraining effect of a deeper stiff layer undergoing no lateral displacement on an overlaying layer. For the silty sand layer, the proposed method predicted larger lateral displacements near the ground surface and smaller values near the bottom of the layer, relative to the corresponding field measurements. This is probably because assuming an OCR of 1.0 and using a back-calculated compression index for the whole layer will over-estimate the compressibility of the soil near the ground surface (lower initial effective stress), and under-estimate the compressibility of the soil near the bottom of the layer.

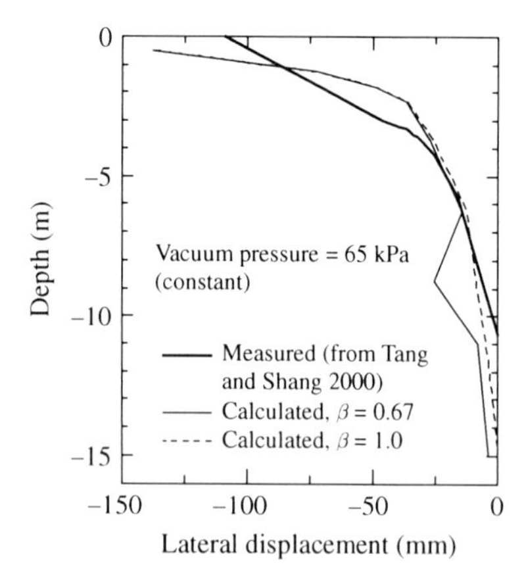

Fig. 4.31 Comparison of measured and calculated lateral displacements at Yaoqiang airport, constant vacuum pressure and different β values (after Chai et al. 2005)

The calculations for the case in which it was assumed that the vacuum pressure reduced with depth, i.e., –65 kPa at the ground surface and linearly reduced to –52 kPa (80% of the value at the ground surface) at 16 m depth, were also conducted. The results are compared in Fig. 4.32. It can be seen that the same as for the Tianjin site, the results for both $\beta = 1.0$ and 0.67 are comparable with the field data.

Comparison of the calculated and measured vertical compression of each soil layer is given in Table 4.8. There is a good agreement for the constant vacuum pressure (–65 kPa) cases because the λ values were back calculated under the same assumption. For cases of reducing vacuum pressure with depth, the calculated compressions in the deeper layers are smaller than the measured data.

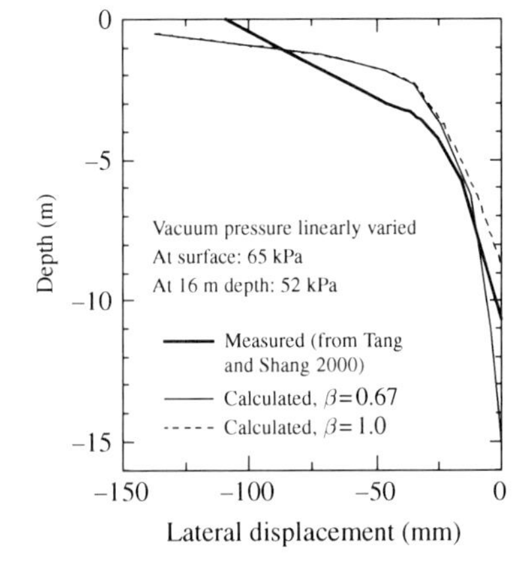

Fig. 4.32 Comparison of measured and calculated lateral displacement at Yaoqiang airport, vacuum pressure varied with depth and different β values (after Chai et al. 2005)

Table 4.8 Comparison of measured and calculated compression of soil strata at the Yao-qiang airport site

		Vertical compression (mm)				
			Calculated			
			$\Delta\sigma_{vac} = 65$ kPa		$\Delta\sigma_{vac} = 65 - 52$ kPa[a]	
Depth (m)	Soil layer	Measured	$\beta = 1.0$	$\beta = 0.67$	$\beta = 1.0$	$\beta = 0.67$
0–2.5	Silty sand	75	83	83	82	82
2.5–5.0	Silty clay	43	44	44	43	42
5.0–7.5	Clayey silt	27	28	28	27	27
7.5–10.0	Soft clay	133	133	131	122	121
10.0–16.0	Silty clay	48	47	46	41	41

[a] Vacuum pressure linearly varied from 65 kPa at ground surface to 52 kPa at 16 m depth

4.7 Deformations Associated with the Vacuum-Drain Method

In the preceding section 4.6, theoretical methods to predict ground deformations due to the direct application of a vacuum pressure to the ground surface, i.e., for the air-tight sheet method, were described. In this section, attention is also focused on predicting the ground deformations that are induced by applying suction to the ground, but in this case the application is via the so called vacuum-drain method. It will be recalled that in the vacuum-drain method (Fig. 4.2), a vacuum pressure is not applied directly to the ground surface. Instead, it is applied at the location of the cap of each capped prefabricated vertical drain (CPVD) and transmitted into the ground directly via the PVDs themselves. In order to make such predictions of deformation, consideration also needs to be given to how the applied suction propagates through the ground.

4.7.1 Distribution of Final Vacuum Pressure

In designing a program of vacuum-drain improvement, there are two assumptions that can be made regarding the final distribution of vacuum pressure in the ground, i.e., at 100% consolidation, as shown in Figs. 4.33 and 4.34 for one-way and two-way drainage deposits, respectively (Chai et al. 2008; Miyakoshi et al. 2007b; Chai et al. 2010). Distribution-1 (Figs. 4.33b and 4.34b) is quite simple but it may not represent the real situation. Since the vacuum pressure is applied at the location of the cap of each CPVD, the vacuum pressure around the ends of a CPVD will propagate approximately radially, as illustrated in Figs. 4.33a and 4.34a, i.e., it is as if the end of the CPVD acts like a point sink in an infinite space,

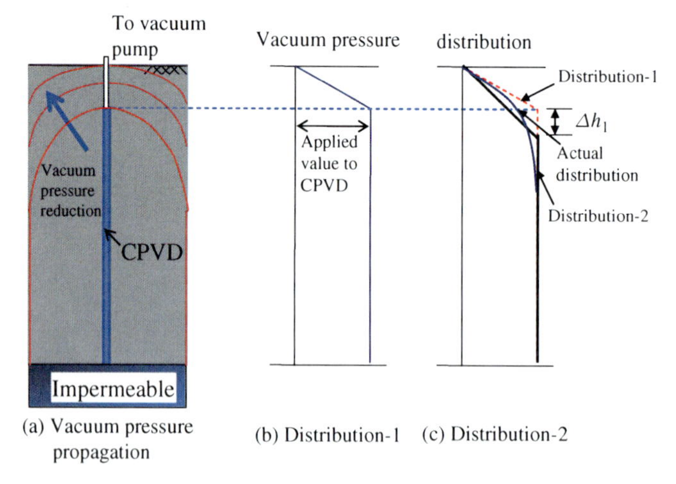

Fig. 4.33 Vacuum pressure distributions for the vacuum-drain method – one-way drainage (modified from Chai et al. 2010)

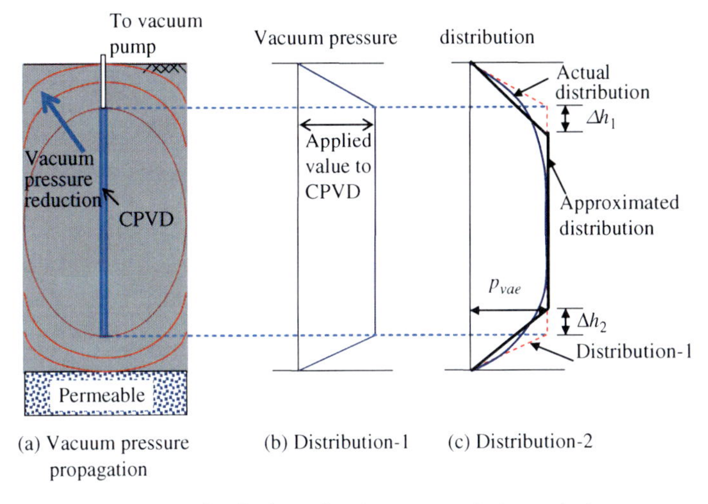

Fig. 4.34 Vacuum pressure distributions for the vacuum-drain method – two-way drainage (modified from Chai et al. 2010)

and so Distribution-1 may over-estimate the final vacuum pressure in the ground. Therefore, Distribution-2, shown in Figs. 4.33c and 4.34c, may be closer to the actual situation. For Distribution-2, Chai et al. (2010) have proposed an empirical equation to determine the value of Δh (either Δh_1 or Δh_2) as shown in Figs. 4.33 and 4.34.

Chai et al. (2010) conducted finite element analyses (FEA) assuming an axisymmetric CPVD unit cell model and investigated the distribution of vacuum pressure in the ground. The results indicated that Δh depends on the relative contributions of radial drainage due to the CPVD and the vertical drainage of the natural soil. In the case of only cylindrical radial drainage with no flow across the external boundary, the steady state is predicted to be a uniform vacuum pressure within the unit cell, as might be expected. However, in the case of vertical and radial drainage around the two ends of the CPVD, the final state is predicted to be steady flow toward the cap of the CPVD at which the vacuum pressure is applied.

Therefore, for a given deposit, the main factors influencing the value of Δh are: (a) factors influencing CPVD consolidation, which include the unit cell diameter (d_e); smear zone parameters, especially the ratio of k_h/k_s (where k_h is the hydraulic conductivity of the natural deposit in the horizontal direction and k_s is the hydraulic conductivity in the smear zone); and the ratio of k_h/k_v (where k_v is the hydraulic conductivity of the natural deposit in the vertical direction); and (b) the thickness of the sealing layer, H_s (Fig. 4.2) and the thickness of the soft clayey layer without the CPVDs at the bottom of the deposit, H_L (in the case of two-way drainage). As for the diameter of the smear zone (d_s), it is generally agreed that $d_s = (2-3)d_m$, where d_m is the equivalent diameter of the mandrel used to install the CPVDs (Chai and Miura 1999). In engineering practice, d_m is about 0.1 m, and so $d_s = 0.3$ m has been adopted. Also, in practice H_s is normally within the range 1–2 m (Fujii et al. 2002; Chai et al. 2008), and FEA results indicate that within this range the influence of H_s and/or H_L on Δh is not significant. An empirical equation for determining Δh has been proposed as follows (Chai et al. 2010):

$$\Delta h = 1.0\left(\frac{d_e}{1.36}\right)^{1.7}\left(\frac{k_h/k_v}{1.5}\right)^{-0.65}\left(\frac{k_h}{k_s}\right)^{0.45} \quad (d_e \text{ and } \Delta h \text{ in metres}) \qquad (4.82)$$

In deriving Eq. (4.82) the ranges of parameters investigated were: $0.9 \le d_e \le 2.26$ m (spacing of 0.8–2.0 m arranged in a square pattern); $1 \le k_h/k_s \le 10$; and $1 \le k_h/k_v \le 10$. Chai et al. (2010) suggested that for $k_h/k_s \le 5$, and $d_e \le 1.70$ m, Eq. (4.82) can be used for all combinations of these parameters. For $d_e = 2.26$ m, Eq. (4.82) can be used for $k_h/k_s \le 2$; and for $k_h/k_s = 10$, it can be used for $d_e \le 1.5$ m.

For a deposit with two-way drainage, at the macro level the layer with the CPVDs installed (i.e., excluding the sealing layer H_s) and the bottom layer without the CPVDs form a two-layer system. The average vacuum pressure at the bottom of the CPVDs (p_{vae} in Fig. 4.34c) can be evaluated approximately by Eq. (4.29), i.e., $p_{vae} = u_m \times p_{vac}$, where p_{vac} is the applied vacuum pressure at the cap location. For the layer with the CPVDs, an equivalent vertical hydraulic conductivity (k_{v1}) is used and it can be calculated by Eqs. (3.50 bis) (Chai et al. 2001).

4.7.2 Deformation Predictions

4.7.2.1 Drainage Conditions

Figure 4.35 shows a schematic illustration of a soil deposit with two-way drainage improved by the vacuum-drain method. The sealing layer has a thickness H_s and the bottom layer without the CPVDs has a thickness H_L. Both are consolidated mainly due to vertical drainage. The middle layer, of thickness H_m, with the installed

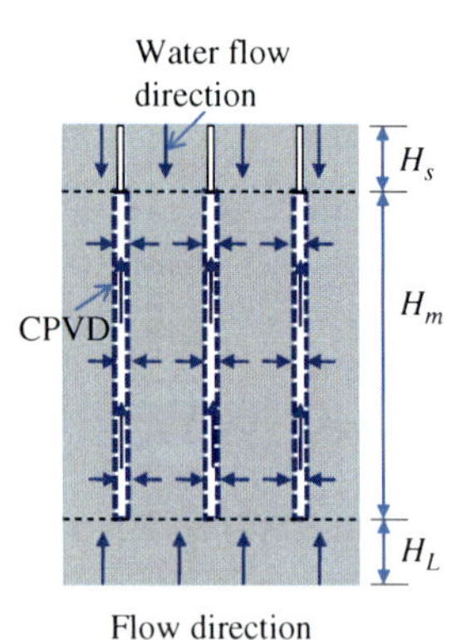

Fig. 4.35 Drainage conditions in the vacuum-drain method (after Chai et al. 2010)

CPVDs, will be consolidated mainly due to cylindrical radial drainage of the soil. For the case of a soil deposit with one-way drainage there is no lower sub-layer. By ignoring the possible effects of shallow cracks around the perimeter of the zone of improvement, the degree of consolidation can be calculated by the consolidation theory given in Section 4.3. For the surface and lower layers, of thickness H_s and H_L respectively, it is proposed that the assumption of one-way drainage conditions is probably reasonable.

Chai et al. (2010) argued that the drainage conditions in the upper and lower layers should be considered to be one-way, as illustrated in Fig. 4.35. However, there will be interactions between all layers, and the average excess pore water pressures at the level of the ends of the CPVDs will change with time. In addition, the actual drainage path will be curved toward each CPVD and not purely vertical. Considering all these factors, it was proposed by Chai et al. that one-way drainage conditions should be assumed for simplicity in the analysis of this problem.

4.7.2.2 Deformation

The deformations occurring at the end of vacuum consolidation can be calculated by the methods given in Section 4.6. If the degree of consolidation and the overall deformation at the end of a project are known, the settlement curve can be predicted. However, in Section 4.3 the degree of consolidation was defined in terms of the excess pore water pressure and not settlement. For a clayey deposit, the relationship between settlement and the effective stress increment is normally logarithmic. When the average degree of consolidation at time t (U) and therefore the average effective stress increment throughout the depth of treatment ($\Delta\sigma'_v$) are known, the corresponding settlement at time t (S) can be calculated from the following equation (Chai et al. 2010):

$$S = S_e \frac{\ln\left[1 + U \cdot \Delta\sigma'_{ve}/(U_e \sigma'_{vo})\right]}{\ln\left(1 + \Delta\sigma'_{ve}/\sigma'_{vo}\right)} \tag{4.83}$$

where S_e = the settlement, U_e = the average degree of consolidation, and $\Delta\sigma'_{ve}$ = the average effective stress increment at the end of the field consolidation, i.e., when the vacuum pressure is removed, respectively.

The procedure for calculating the ground deformations resulting from the vacuum-drain method can be summarized as follows:

(a) Determine the final vacuum pressure distribution in the deposit, i.e., at 100% consolidation, assuming either a bi-linear or tri-linear pattern (Figs. 4.33c or 4.34c).
(b) For a given time (t), calculate corresponding degrees of consolidation, $U(t)$, for each layer using the consolidation theory described in Section 4.3.
(c) Determine the actual vacuum pressure distribution at the end of the field consolidation using the corresponding degree of consolidation (U_e) and the final vacuum pressure distribution determined at step (a).

(d) Calculate the settlement (S_e) at the centre of the improved area and the lateral displacement profile at the edge of the vacuum consolidation area corresponding to the end of the field consolidation.
(e) Calculate the settlements, $S(t)$, for the given time t, using Eq. (4.83).

4.7.3 Analysis of a Case History in Tokyo

4.7.3.1 Site History

Two field test sections were constructed with the vacuum-drain method in Tokyo Bay, Japan (Miyakoshi et al. 2007a; 2007b). Section-A had an area of 60 m × 60 m and Section-B an area of 61.2 m× 61.2 m, and the two sections were almost adjoining, as shown in Fig. 4.36. The soil profile consists of a reclaimed clayey silt layer about 12 m in thickness. Below it is a clayey deposit about 29 m thick which is in turn underlain by a sand layer. The majority of the reclamation was carried out between 2003 and 2005 with a rate of filling of about 3.5 m/year. The total unit weight (γ_t), compression index (C_c) and maximum consolidation pressure (p_c) of the deposits before vacuum consolidation are shown in Fig. 4.37. It can be seen that the reclaimed layer was close to being normally consolidated but the original clayey

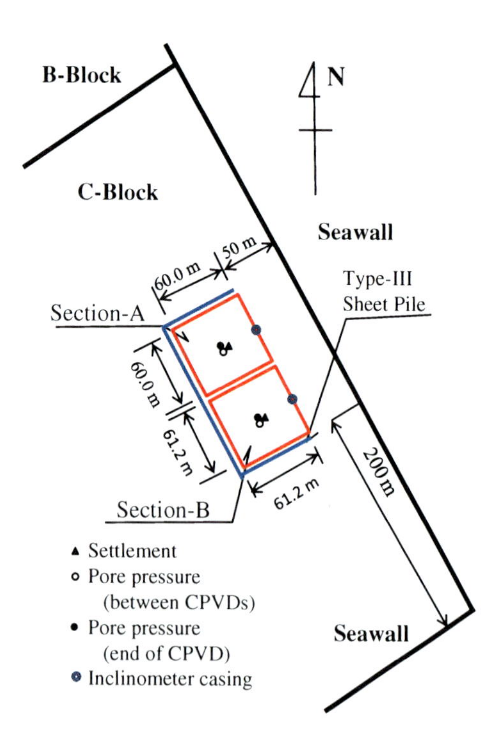

Fig. 4.36 Plan view of the test sections and the key instrumentation points (after Chai et al. 2010)

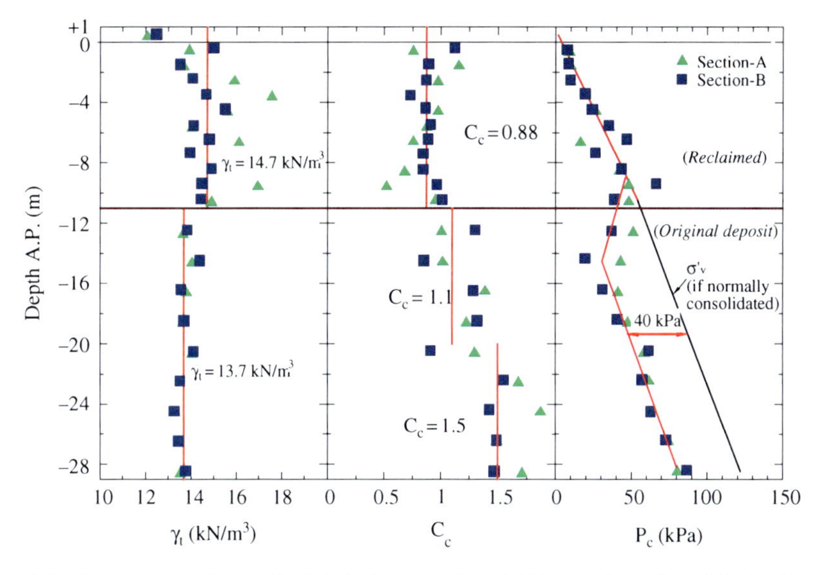

Fig. 4.37 Properties γ_t, C_c and P_c of the Tokyo Bay deposit (Measured data from Miyakoshi et al. 2007a – after Chai et al. 2010)

deposit was in an under-consolidated state. In the original deposit the clay content (with grain sizes less than 5 μm) was more than 50% and for the reclaimed layer the sum of the sand and silt contents was more than 50% (Fig. 4.38).

Takeya et al. (2007) reported that for both the original clayey deposit and the reclaimed layer, the coefficient of consolidation (c_v) was about 0.012 m^2/day. However, from the grain size distribution and the pre-consolidation pressure (p_c)

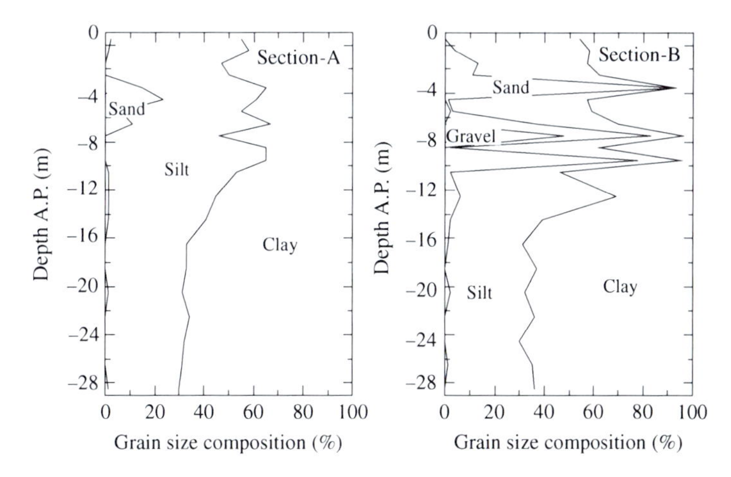

Fig. 4.38 Composition of the subsoil (data from Miyakoshi et al. 2007a – after Chai et al. 2010)

information shown in Figs. 4.37 and 4.38, respectively, it can be argued that the reclaimed layer may have had a higher value of c_v. In the analysis, it was assumed that c_v for the reclaimed layer was 0.024 m^2/day, i.e., twice the value of the original clay deposit. Taking into consideration possible stratification of the soil layers, for all layers the coefficient of consolidation in the horizontal direction was assumed to be twice the corresponding vertical value.

The CPVDs used at this site had a cross-sectional area of 150 mm × 3 mm. At Section-A, the CPVDs had a spacing of 2.0 m, while at Section-B it was 1.8 m on a square pattern. For both sections, CPVDs were installed to a depth of 30 m from the ground surface, and the sealing surface layer had a thickness of about 1.0 m. The field installation of CPVDs started at the beginning of January 2006 for Section-A, and at the beginning of February 2006 for Section-B, and for the both sections the duration of the installation process was about one month. The durations between the end of CPVD installation and just prior to application of the vacuum pressure were 5 and 4 months for Sections-A and B, respectively. Considering half of the period of the CPVDs installation as consolidation time, the partial self-weight consolidation periods before application of the vacuum pressure were estimated to be about 165 and 135 days for Sections-A and B, respectively. From June 30, 2006, a vacuum pressure of –80 to –90 kPa, as measured at the vacuum pump location, was applied and maintained for 204 days. Surface and subsurface settlement gauges, excess pore water pressure (vacuum) gauges, as well as inclinometer casings were installed to monitor the ground response. The key instrumentation points are given in Fig. 4.36.

4.7.3.2 Analytical Model and Soil Parameters

The model adopted in the analysis and the values of the soil parameters assumed for each layer are shown in Fig. 4.39. The values of the initial void ratio (e_0) shown in Fig. 4.39 were calculated from the total unit weights (γ_t) by assuming the soil was 100% saturated and the unit weight of the soil particles (γ_s) was 26.5 kN/m^3. The parameters relating to CPVD consolidation are: the diameter of CPVDs, $d_w =$ 75 mm; the diameter of smear zone, $d_s = 0.3$ m; the discharge capacity, $q_w =$ 1.37 m^3/day (500 m^3/year); and the hydraulic conductivity ratio, $k_h/k_s = 2$ The value adopted for q_w in the analysis was simply an assumed value. The value of k_h/k_s was back fitted to the observed results for the settlements for Section-A under the partial self-weight consolidation, before application of the vacuum pressure.

4.7.3.3 Degrees of Consolidation

The consolidation of the soil layer can be divided into two periods, i.e., Period-1 – after CPVD installation but before vacuum pressure application; and Period-2 – during vacuum consolidation. The analyses were also divided accordingly into these two-parts.

The CPVD-improved zone contains three soil layers (Fig. 4.39). The unit cell consolidation theory (such as Hansbo 1981) was proposed for a uniform soil deposit and cannot be directly applied to a multilayer condition. It is assumed that the degree

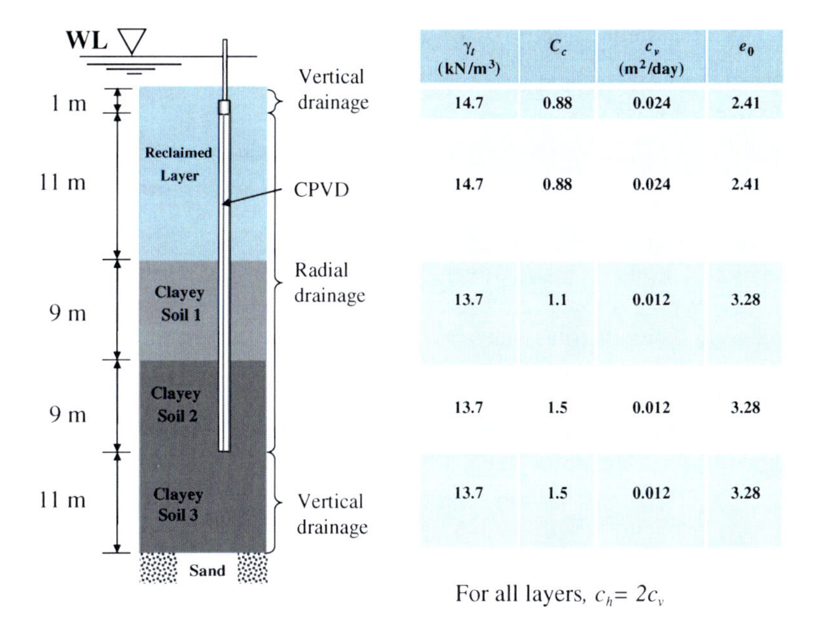

Fig. 4.39 Consolidation analysis model (after Chai et al. 2010)

of consolidation for each layer can be calculated separately using the appropriate coefficient of horizontal consolidation (c_h) for each layer. However, the drainage path length of the CPVDs was assumed to be 29 m. This means that, for example, when calculating the degree of consolidation for the reclaimed layer, it is assumed that there exists a layer with the properties of the reclaimed layer which has a thickness of 29 m. The degrees of consolidation calculated under these assumptions are listed in Table 4.9.

Table 4.9 Calculated degrees of consolidation

		Degree of consolidation, U (%)			
		Part of self-weight induced (before vacuum application)		Vacuum Consolidation	
Soil layer	Thickness (m)	A	B	A	B
Sealing layer	1	–	–	100	100
Reclaimed	11	84.8	85.6	90.4	94.7
Clayey soil 1	9	68.9	70	76.6	83.9
Clayey soil 2	9	68.9	70	76.6	83.9
Clayey soil 3	11	17.8	15.7	20.3	20.3

4.7.3.4 Settlement and Lateral Displacement

Based on the field measured data reported by Miyakoshi et al. (2007a), it was assumed that the final vacuum pressure at the location of the caps of the CPVDs, i.e., 1.0 m below the ground surface, was –70 kPa. The value of vacuum pressure at the bottom end of the CPVDs was calculated to be about –69.3 kPa for both sections. With the values of the initial effective vertical stress (σ'_{v0}), assumed to be equal to the corresponding values of p_c, as shown in Fig. 4.37, and the degrees of consolidation as listed in Tables 3.9, the effective vertical stresses in the ground at the different stages of the consolidation process can be easily calculated.

Figure 4.40 shows distributions of the vertical effective stress at Section-A, initially, before application of vacuum pressure and under self-weight consolidation. The end of the vacuum consolidation and the assumed final vertical effective stress distributions, i.e., at 100% consolidation, are depicted in Fig. 4.41.

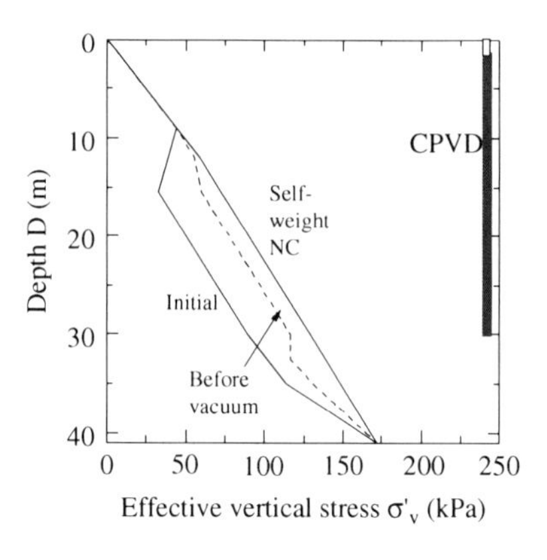

Fig. 4.40 Vertical effective stress distribution at Section-A – initially and before vacuum pressure application (after Chai et al. 2010)

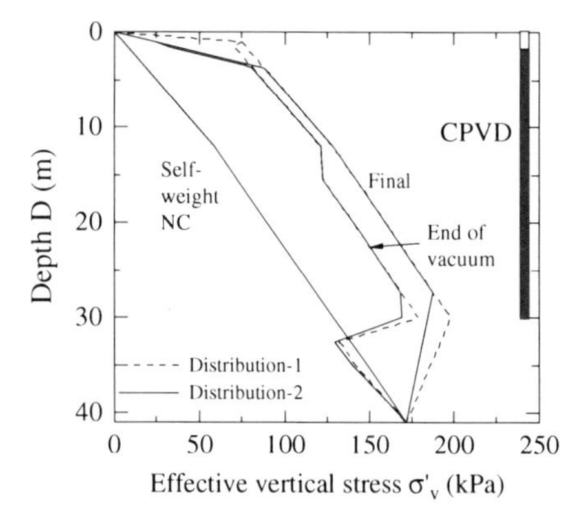

Fig. 4.41 Vertical effective stress distribution at Section-A – end of consolidation (modified from Chai et al. 2010)

With $d_e = 2.26$ and 2.03 m for Sections-A and B, respectively, assuming $k_h/k_s = 2$, and $k_h/k_v = 2$, Eq. (4.82) results in a value of Δh of 2.69 and 2.25 m for Sections-A and B, respectively. Figure 4.41 shows that Distribution-2 results in smaller predicted values of vertical effective stress around the ends of the CPVDs than for Distribution-1 (Fig. 4.34). It is worth mentioning that Eq. (4.82) has been developed assuming that the thickness of the layers above and below the CPVDs was about 1–2 m. For the case considered here, the thickness of the bottom layer without CPVDs was about 11 m. As a general tendency, the thicker the layer without CPVDs, the smaller is the value of Δh. Therefore, there may be an over-estimation of Δh at the bottom of the CPVDs. However, for both Sections A and B, deformation analysis indicates that at the bottom of the CPVDs, varying Δh from 0 to the value given by Eq. (4.82) has no noticeable difference on the resulting settlement curves due to the higher initial effective stress near that location.

Although there was lateral displacement observed after CPVD installation and before vacuum pressure application, it was assumed that the consolidation induced by self-weight was one-dimensional and so only the settlement was calculated. To calculate the settlement-time curves before application of the vacuum loading, using Eq. (4.83), for Section-A the initial effective stress (σ'_{v0}) is the line marked as 'initial', and the stress increment at the end of this period ($\Delta\sigma'_{ve}$) is the difference between the lines labelled 'initial' and 'before vacuum' in Fig. 4.40.

Both the settlement and the lateral displacement induced by vacuum consolidation were calculated. In this case, for Section-A, the initial effective stress (σ'_{v0}) is the line marked with 'before vacuum', and the stress increment at the end of the period ($\Delta\sigma'_{ve}$) is the difference between the lines of 'before vacuum' in Fig. 4.40 and 'end of vacuum' in Fig. 4.41. As noted previously, the deformation calculations depend on whether triaxial or plane strain deformation patterns are assumed in the ground. For the case considered, the improved area of each section was a square and so it is likely that, when considered individually, conditions would have been closer to the triaxial state. But since the two sections were almost joined together, it is also possible that taken together the conditions may have actually been closer to plane strain. In the analysis, it was assumed that the deformations are the average values obtained after making both the triaxial and plane strain assumptions. The additional parameters adopted for calculating the ground deformation induced by vacuum pressure, using the method proposed by Chai et al. (2005), are: the effective stress friction angle, $\phi' = 30°$, cohesion, $c' = 5$ kPa, $\beta = 1.0$ m (Eq. 4.72), the half width of the improved area, $B = 30$ m for Section-A and 30.6 m for Section-B, respectively, and $\alpha_{\min} = 0.8$ for triaxial conditions and 0.85 for the plane strain condition (Eq. 4.77).

The results of the settlement calculations are compared with measured values in Fig. 4.42a and b for Sections-A and B respectively. For the vacuum pressure Distribution-2, the calculated compression of the surface soil layer is obviously smaller than that of Distribution-1 (Fig. 4.34). This is because the initial effective stress in the surface layer was small and the difference in the estimated vacuum pressure can induce considerable differences in the calculated compression.

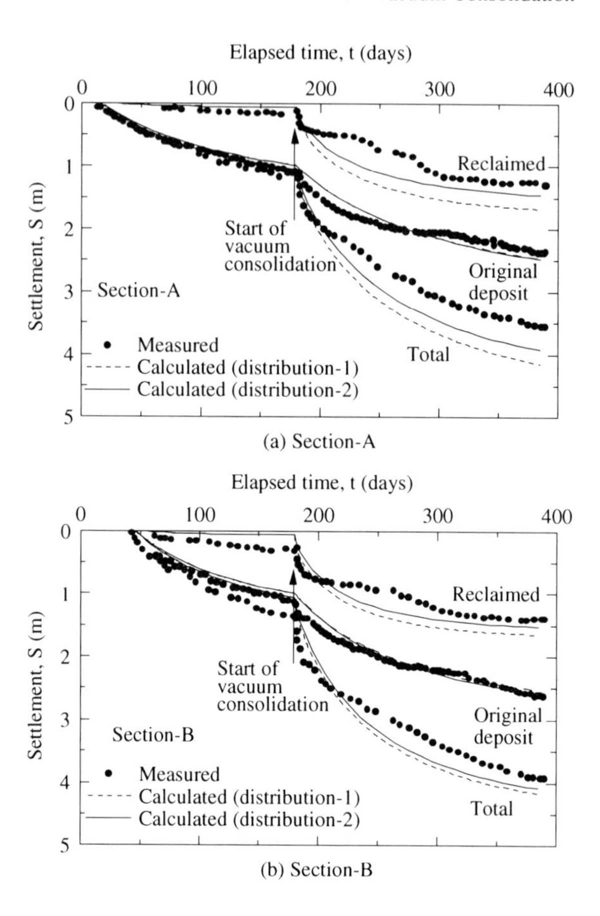

Fig. 4.42 Settlement versus elapsed time curves (Measured data from Miyakoshi et al. 2007a – modified from Chai et al. 2010); (**a**) Section-A; (**b**) Section-B;

Section-A resulted in a larger value of Δh, and hence a larger difference in the calculated settlements. There was no noticeable difference in the calculated vertical compression of the original deposit for both distributions. Generally, using Distribution-2, the predicted values are in better agreement with the measured field data.

Comparison of the lateral displacement profiles at the edge of the improved area is depicted in Fig. 4.43a and b for Sections-A and B, respectively. Firstly, it can be seen that for the vacuum pressure Distribution-2, the calculation yielded a much better prediction for the values of lateral displacement near the ground surface. Secondly, for both vacuum pressure distributions the calculation predicted a smaller lateral displacement for the original soil deposit, which was in an under-consolidated state before application of the vacuum loading. Chai et al. (2008) reported a similar tendency for a site in Yamaguchi, Japan, and suggested that the values of α_{min} proposed by Chai et al. (2005) may only be applicable for a normally consolidated deposit. For an under-consolidated deposit a smaller value of α_{min} should probably be used.

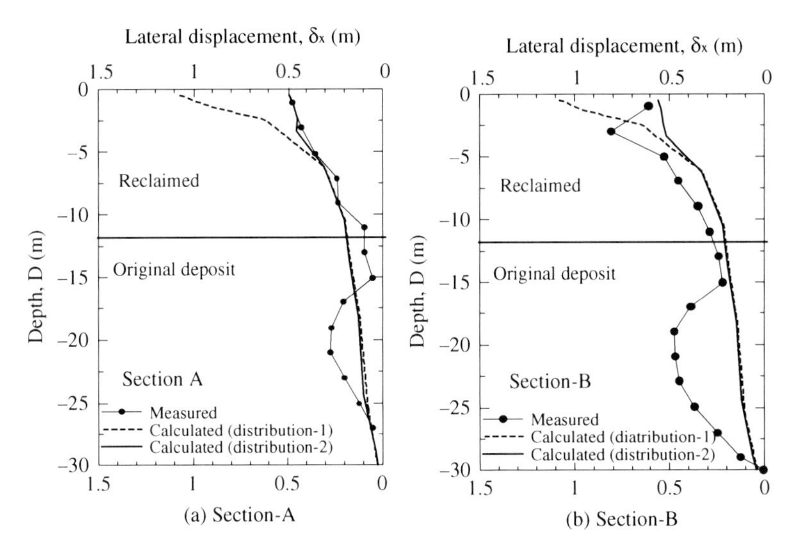

Fig. 4.43 Comparison of lateral displacement (Measured data from Miyakoshi et al. 2007b – modified from Chai et al. 2010). (**a**) Section-A; (**b**) Section-B

4.8 Summary

This chapter described two methods that are used for conducting field vacuum consolidation, viz., the method involving the use of an air-tight sheet placed over the surface of the soil in combination with vacuum loading and the vacuum-drain method. The latter uses a surface soil layer as the air and water-tight layer, rather than an air-tight sheet. Theories of consolidation under vacuum pressure loading have been presented for both single and double layer deposits under one-way as well as two-way drainage boundary conditions. This theory included the case of a soft clay deposit improved by the installation of prefabricated vertical drains (PVDs).

The characteristics of vacuum consolidation have been discussed mainly with reference to laboratory oedometer test results and with emphasis on (a) the amount of settlement induced by the vacuum pressure, and (b) the effects of the drainage boundary conditions. It was noted that application of a vacuum pressure can induce the same or less settlement than that induced by a surcharge load of the same magnitude. When the initial vertical effective stress in a soil sample is small, the vacuum pressure will induce inward lateral displacement and result in a smaller settlement. The effect of the drainage boundary conditions for the case of vacuum consolidation is different from that observed during consolidation under an imposed surcharge load. For a soil layer with two-way drainage, vacuum pressure cannot be applied at the bottom drainage boundary and the final vacuum pressure distribution in the deposit will be triangular, with a maximum value at the location where the vacuum pressure is applied and zero at the bottom drainage boundary. Consequently the consolidation settlement under two-way drainage conditions will be less than that for the case where one-way drainage conditions apply.

For a two-layer system and two-way drainage conditions, the order of the soil layers will influence both the rate of consolidation as well as the final consolidation settlement. Further, for a layer with two-way drainage subjected to vacuum pressure loading there is an optimum PVD installation depth, and a method for determining this depth has also been presented.

Two semi-theoretical methods for calculating the ground deformation (settlement and lateral displacement) induced in the soil by application of a vacuum pressure have been described. The usefulness of one of these methods has been demonstrated by analyzing two case histories previously reported in the literature.

Finally, in the vacuum-drain method the vacuum pressure is applied to each individual capped PVD (CPVD). A method for estimating the final vacuum pressure distribution generated in the deposit when this technique is used has been established and applied successfully to a case history in Tokyo Bay, Japan.

References

Bergado DT, Chai J-C, Miura N, Balasubramaniam AS (1998) PVD improvement of soft Bangkok clay with combined vacuum and reduced sand embankment preloading. Geotech Eng, Southeast Asian Geotech Soc 29(1):95–121

Chai J-C, Miura N (1999) Investigation on some factors affecting vertical drain behavior. J Geotech Geoenviron Eng ASCE 125(3):216–226

Chai J-C, Shen S-L, Miura N, Bergado DT (2001) Simple method of modeling PVD improved subsoil. J Geotech and Geoenviron Eng ASCE 127(11):965–972

Chai J-C, Carter JP, Hayahsi S (2005). Ground deformation induced by vacuum consolidation. Geotech Geoenviron Eng ASCE 131(12):1552–1561

Chai J-C, Carter JP, Hayashi S (2006) Vacuum consolidation and its combination with embankment loading. Can Geotech J 43(1):985–996

Chai J-C, Miura N, Nomura T (2007) Experimental investigation on optimum installation depth of PVD under vacuum consolidation. Proceedings of 3rd Sino-Japan geotechnical symposium, Chongqing, China, China Communications Press, Beijing, pp 87–94

Chai J-C, Miura N, Bergado, DT (2008) Preloading clayey deposit by vacuum pressure with cap-drain: analyses versus performance. Geotext Geomembr 26(3):220–230

Chai J-C, Matsunaga K, Sakai A, Hayashi S (2009a) Comparison of vacuum consolidation with surcharge load induced consolidation of a two-layer system. Géotechnique 59(7):637–642

Chai J-C, Miura N, Kirekawa T, Hino, T (2009b) Optimum PVD installation depth for two-way drainage deposit. Geomechan Eng Int J 1(3):179–192

Chai J-C, Hong Z-S, Shen S-L (2010) Vacuum-drain consolidation induced pressure distribution and ground deformation. Geotext Geomembr 28:525–535

Chu J, Yan SW, Yang H (2000) Soil improvement by the vacuum preloading method for an oil storage station. Géotechnique 50(6):625–632

Conte E, Troncone A (2008) Soil layer response to pore pressure variations at the boundary. Géotechnique 58(1):37–44

Fujii A, Tanaka H, Tsuruya H, Shinsha H (2002) Field test on vacuum consolidation method by expecting upper clay layer as sealing-up material. Proceedings of symposium on recent development about clayey deposit – from microstructure to soft ground improvement, Japan Geotechnical Society, pp 269–274 (in Japanese)

Hansbo S (1981) Consolidation of fine-grained soils by prefabricated drains. Proceedings of 10th international conference on soil mechanics and foundation engineering, vol 3, Stockholm, pp 677–682

Imai G (2005) For the further development of "vacuum-induced consolidation method" – Present understandings of its principles and their applications. Proceedings of Japanese Society of Civil Engineering, No. 798/VI-68, pp 1–16 (in Japanese)

Indraratna B, Rujikiatkamjorn C, Sathananthan I (2005) Analytical modeling and field assessment of embankment stabilized with vertical drains and vacuum preloading. Proceedings of 16th international conference on soil mechanics and geotechnical engineering, Osaka, Japan, Millpress, Rotterdam, pp 1049–1052

Jaky J (1948) Pressure in soils. Proceedings of 2nd international conference on soil mechanics and foundation engineering, vol 1, London, pp 103–107

Kjellman W (1952) Consolidation of clayey soils by atmospheric pressure. Proceedings conference on soil stabilization, Massachusetts Institute of Technology, Massachusetts, USA, pp 258–263

Mike K (1971) Permeability to water and permeability to air. Tsuchi-To-Kiso 19(6):3–4 (in Japanese)

Miyakoshi K, Shinsha H, Nakagawa D (2007a) Vacuum consolidation field test for volume reduction scheme of soft clayey ground (Part-2) – ground characteristic and measured results. Proceedings of 42th annual meeting, Japanese Geotechnical Society, pp 919–920 (in Japanese)

Miyakoshi K, Takeya K, Otsuki Y, NozueY, Kosaka H, Kumagai M, Oowada T, Yamashita T (2007b) The application of the vacuum compaction drain method to prolong the life of an offshore disposal field. Nippon Koei Technical Forum 16:9–19 (in Japanese)

Nomura T, Yoshikawa M, Matuta Y, Miura N (2007) Case histories on using oil pressure as an index to judge the existence of sand layer. J Found Constr September 2007:88–92 (in Japanese)

Ohtsubo K, Egashira K, Kashima K (1995) Depositional and post-depositional geochemistry, and its correlation with the geotechnical properties of marine clays in Ariake Bay, Japan. Géotechnique 45(3):509–523

Pyrah IC (1996) One-dimensional consolidation of layered soils. Géotechnique 46(3):555–560

Roscoe KH, Burland JB (1968) On the generalized stress-strain behavior of 'wet' clay. In: Heyman J, Leckie FA (eds) Engineering plasticity. Cambridge University Press, Cambridge, pp 535–609

Saowapakpiboon J, Bergado DT, Youwai S, Chai J-C, Wanthong P, Voottipruex P (2010) Measured and predicted performance of prefabricated vertical drains (PVDs) with and without vacuum preloading. Geotext Geomembr 28:1–11

Shiwakoti DR, Tanaka H, Tanaka M, Locat J (2002) Influences of diatom microfossils on engineering properties of soils. Soil Found 42(3):1–17

Takeya K, Nagatsu T, Yamashita T (2007) Vacuum consolidation field test for volume reduction scheme of soft clayey ground (Part-1) – field test condition and construction method. Proceedings of 42nd annual meeting, Japanese Geotechnical Society, pp 917–918 (in Japanese)

Tang M, Shang JQ (2000) Vacuum preloading consolidation of Yaogiang Airport runway. Géotechnique 50(6):613–623

Taylor DW (1948) Fundamentals of soil mechanics. Wiley, New York, NY

Terzaghi K (1943) Theoretical soil mechanics. Wiley, New York, NY

Zhu G, Yin J-H (1999) Consolidation of double soil layers under depth-dependent ramp load. Géotechnique 49(3):415–421

Chapter 5
Soil-Cement Columns

Abstract This chapter describes a consolidation theory for a clayey subsoil whose mechanical performance is improved by the installation of soil-cement columns which are stiffer and stronger than the surrounding soil layer. It also describes methods for estimating ground settlements for the case of ground improved by the prior installation of 'floating' columns, a method for predicting the lateral displacement of the ground induced by column installation, and the analyses of some relevant case histories. The various theories presented in this chapter are evaluated and validated by comparing their predictions with the results of measurements obtained in a number of different field trials. The latter include results from laboratory scale testing and case histories from a number of different test embankments in Japan. The case studies used to illustrate the theory for predicting the lateral soil displacements caused by the introduction of the soil-cement columns include a soft clay site as well as a sandy soil.

5.1 Introduction

Deep mixing of cement with soil and water, normally forming soil-cement columns in situ, is a widely used soft ground improvement method (e.g., Broms and Boman 1979; Bergado et al. 1994). During installation of the soil–cement columns either cement slurry (wet mixing) or cement powder (dry mixing) is injected into the ground under pressure. The installation process is often carried out by employing rotary equipment with high torque capacity. The cement is continuously injected and then mechanically mixed with the soil particles by special tools attached to the rotary device. The cement hydrates and reacts with the soil particles eventually forming a solid material usually significantly stiffer and stronger than the original soil layer.

This technique has been used for improving the foundations underlying embankments, stabilizing the base of excavations, as well as forming retaining walls for excavations in soft clay or clay-like deposits, and to prevent liquefaction of loose sandy ground. To reduce construction costs and minimize the impact on the ground environment, an improvement method which uses 'floating' soil-cement columns partially penetrating the soft deposit has been developed (Shen et al. 2001). It has been recently applied in several field construction projects, as reported by Chai et al. (2010).

J. Chai, J.P. Carter, *Deformation Analysis in Soft Ground Improvement*, Geotechnical, Geological and Earthquake Engineering 18, DOI 10.1007/978-94-007-1721-3_5,

This chapter concentrates on the theories for calculating the consolidation settlement as well as the degree of consolidation of clayey soil improved by the installation of floating soil-cement columns. The problem of estimating the lateral ground deformations induced by column installation is also addressed.

5.2 Settlement Predictions

5.2.1 Definitions

For soft soil deposits improved by the inclusion of soil-cement columns, the area improvement ratio (α) and depth improvement ratio (β) are commonly used to describe the relative improvement of the soft layer by the columns. Their definitions are illustrated in Fig. 5.1 and can be expressed as follows:

$$\alpha = \frac{A_C}{A} \tag{5.1}$$

$$\beta = \frac{H_L}{H} \tag{5.2}$$

where A = the total cross-sectional area of the zone improved by a single column, A_C = the cross-sectional area of a column, H = the thickness of the soft clayey deposit, and H_L = the length of a column.

When the column-improved soft subsoil is loaded vertically, e.g., by application of a surcharge load, concentration of vertical stress will generally occur in the stiffer column accompanied by a reduction in vertical stress occurring in the surrounding less stiff clayey soil (Fig. 5.2). The distribution of vertical stress within a unit cell consisting of a column and surrounding softer soil can be expressed by a stress concentration factor (n_s), defined as:

$$n_s = \frac{\sigma_C}{\sigma_S} \tag{5.3}$$

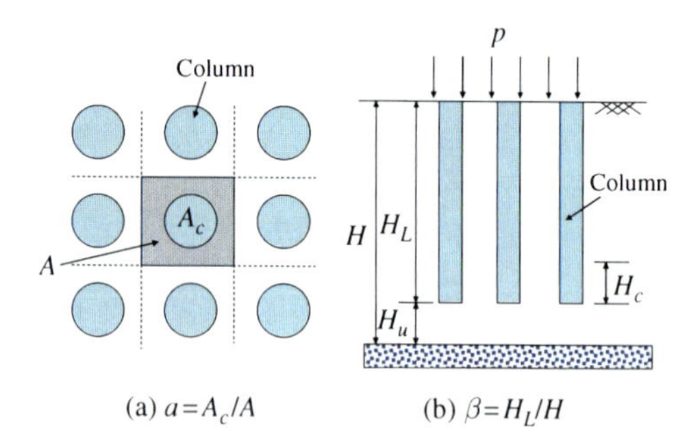

Fig. 5.1 Definition of α and β

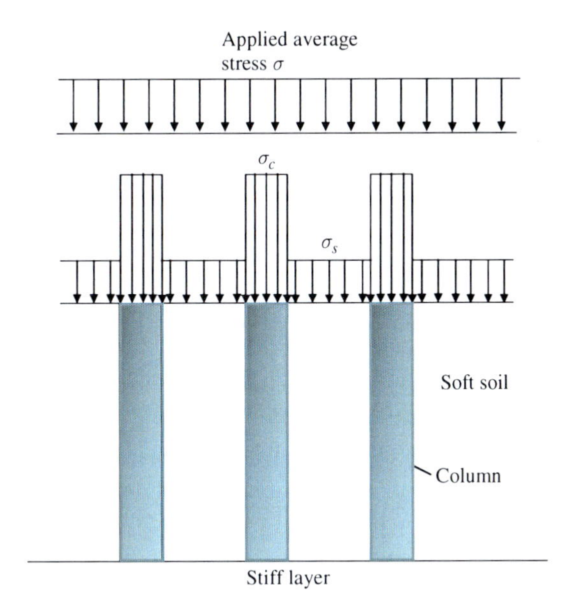

Fig. 5.2 Illustration of stress concentration

where σ_c = the stress in the column, and σ_s = the stress in the surrounding soft soil. The value of n_s depends on the relative stiffness of the column and the surrounding soil and the area improvement ratio (α) for the case where the columns fully penetrate the soft soil layer. The average stress, σ, over the unit cell area corresponding to a given value of α is expressed as:

$$\sigma = \sigma_c \alpha + \sigma_s (1 - \alpha) \tag{5.4}$$

The expressions for the stresses in the column and the soft soil using the stress concentration factor, n_s, are:

$$\sigma_C = \frac{n_s \sigma}{[1 + (n_s - 1)\alpha]} = \mu_c \sigma \tag{5.5}$$

$$\sigma_S = \frac{\sigma}{[1 + (n_s - 1)\alpha]} = \mu_s \sigma \tag{5.6}$$

where μ_C and μ_S = the ratio of stress in the column and the soft soil to the average applied stress over the unit cell area, respectively.

5.2.2 Fully Penetrating Columns

Basically there are two methods for calculating the final settlement of a soil-cement column that fully penetrates a clayey subsoil, namely the equilibrium method

(Aboshi et al. 1979) and the composite modulus method. In the equilibrium method the one-dimensional (1D) settlement of the soft clayey soil is calculated considering the effect of stress concentration in the column. The equation for the final settlement (S_t) is as follows:

$$S_t = m_{vs}\mu_s\sigma H \tag{5.7}$$

where m_{vs} = the coefficient of volumetric compressibility of the soft soil, and H = the thickness of the soft soil layer.

In the composite modulus method it is assumed that the improved zone behaves like a uniform soil mass with a constrained modulus (D) determined as a weighted average of the deformation moduli of the column material and the surrounding soft soil. There are two ways to obtain the value of D. One way is to calculate it from the constrained modulus of the column (D_c) and the soil (D_s) as follows:

$$D = \alpha \cdot D_c + (1 - \alpha)\, D_s \tag{5.8}$$

An alternative is to calculate D from the coefficient of volumetric compressibility (m_v) of the two materials involved (Bergado et al. 1994):

$$D = \frac{1}{m_{vs}\,(1 - \alpha) + m_{vc}\alpha} \tag{5.9}$$

where m_{vs}, m_{vc} = the coefficients of volumetric compressibility of the soft soil and the column, respectively. Normally, the value of D calculated from m_{vs} and m_{vc}, Eq. (5.9), is much smaller than that from D_c and D_s, Eq. (5.8). It is suggested that for cases close to free (equal) stress conditions at the ground surface, Eq. (5.9) should be adopted, and for cases close to an equal strain condition, Eq. (5.8) should be used.

5.2.3 Floating Columns

5.2.3.1 Method of the Japan Institute of Construction Engineering

In order to prevent excessive settlements in road construction on soft ground, bridges and box culverts are normally supported by pile foundations or the soft soil is improved by the installation of soil-cement columns. However, if the soft soil layer is improved by fully penetrating columns, the ground beneath the bridge structure may become so stiff that large differential settlements occur between the structures and the adjacent road or embankment. To overcome this dilemma, 'floating' soil-cement columns are often used to effect ground improvement, as they provide a less stiff foundation option than fully penetrating columns.

The Japan Institute of Construction Engineering (JICE 1999) proposed a method to calculate the settlement of soft subsoil improved by the installation of floating columns. This method, which is widely used in Japan, is described as follows.

For ground improved by floating columns, there will be interactions between the column and the surrounding soft soil. When the value of the improvement area ratio α is small, there will be some contribution to the overall surface settlement derived from the unimproved sub layer beneath the columns as well as from the improved upper sub layer. With an increase in the value of α, the contribution from the improved upper sub layer will be reduced. JICE proposed that when $\alpha < 30\%$, the relative penetration of the columns into the surrounding soft soil is significant, and in settlement calculations the main contribution to the overall settlement will be the unimproved sub layer plus 1/3 of the column-improved sub layer, as shown in Fig. 5.3a. This is effectively the same concept as that proposed by Terzaghi and Peck (1967) for calculating the settlement of a floating pile group. When $\alpha \geq 30\%$, the contribution from the sub layer containing the columns can be ignored and the main contribution to surface settlement will be the unimproved lower sub layer only (Fig. 5.3b). Figure 5.3 also illustrates that when the width of the column-improved area is small, the load spreading effect also needs to be considered and so a spreading angle of 30° has been suggested by JICE.

Obviously, the JICE method only provides a rough estimate of the effect of α on the settlement and it is usually not possible to obtain good agreement between the calculated settlements and field measurements for all possible values of α (Chai et al. 2010). For example, in practical terms the difference between $\alpha = 30\%$ and 29% is not very significant, but with the JICE method the calculated settlement will be quite different in these two cases. Another shortcoming of the method is that it does not consider the effect of the value of β. As a simple illustration, consider a soft clayey deposit with an overall thickness of 10 m, improved by the installation of soil-cement columns 9 m in length with an α value of 20%. Using the JICE method, the thickness of the main compression layer is estimated to be 4 m (1 m of the unimproved sub layer plus 3 m of the column-improved sub layer). In most cases the settlement will be over-estimated assuming an effective thickness of 4 m for the compressible layer.

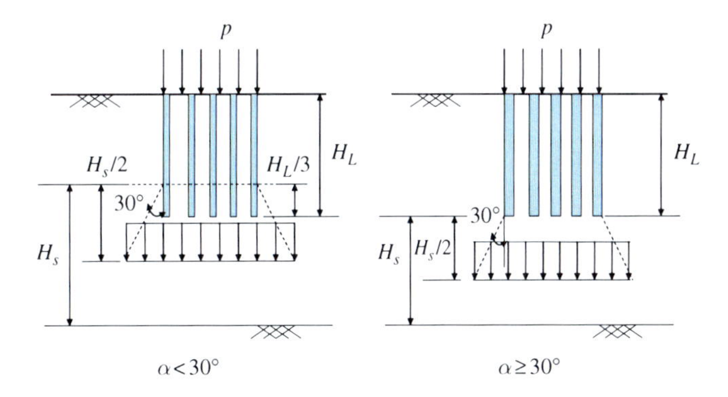

Fig. 5.3 Illustration of JICE method

Furthermore, the JICE method does not clearly specify a way to calculate the compression of the column-improved layer. For consistency with other methods, it is therefore suggested that the area weighted average of the constrained modulus, Eq. (5.8), under one-dimensional (1D) conditions can be used to estimate the contribution of the column-improved sub layer to the overall surface settlement.

5.2.3.2 α–β Method

Although the JICE method has shortcomings, it is simple and easy to use. If, when calculating settlements, the thickness of the part of the column-improved layer to be treated as an unimproved layer can be determined more rationally, the method can be a useful design tool. Chai et al. (2010) conducted a series finite element simulations of the consolidation behaviour of soft clayey soils improved by the installation of floating soil-cement columns. They adopted an axisymmetric unit cell model, i.e., a single column and its surrounding region of soft natural soil. In the analysis, it was assumed that there was also a relatively rigid soil-cement slab constructed on the ground surface, i.e., there was no differential settlement between the column and the soft soil at the ground surface. It has been found that the relative penetration of the column into the surrounding soft soil is mainly influenced by the values of α and β. The loading intensity (p) and the soil type also have a considerable influence on the mechanical interaction between the soil-cement column and the soft soil. For simplicity, and considering that the load intensity p and the parameters used to represent the soil type are directly included in the settlement calculation, Chai et al. (2010) proposed that the thickness (H_c) of the part of the column-improved layer to be treated as an unimproved layer should be a function of α and β as follows:

$$H_C = H_L f(\alpha) g(\beta) \tag{5.10}$$

The physical meaning of H_c is shown in Fig. 5.1b. Bilinear functions have been chosen for $f(\alpha)$ and $g(\beta)$ from the results of the numerical studies, as shown in Fig. 5.4a and b, respectively. These functions can be written mathematically as follows:

$$f(\alpha) = \begin{cases} \frac{8}{15} - \frac{\alpha}{75} & (10\% \le \alpha \le 40\%) \\ 0 & (\alpha > 40\%) \end{cases} \tag{5.11}$$

$$g(\beta) = \begin{cases} 1.62 - 0.016\beta & (20\% \le \beta \le 70\%) \\ 0.5 & (70\% \le \beta \le 90\%) \end{cases} \tag{5.12}$$

The values of α and β appearing in Eqs. (5.11) and (5.12) should be input as percentages. It is considered that the ranges for α and β adopted cover most practical situations. It has also been suggested that the method is suitable for soft clayey deposits with a load intensity from 50 to 160 kPa (Chai et al. 2010).

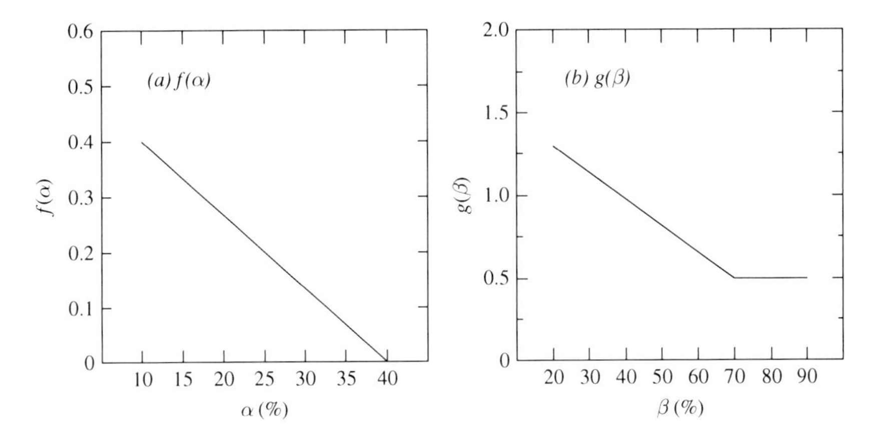

Fig. 5.4 Functions $f(\alpha)$ and $g(\beta)$

Following the same logic, in the settlement calculation the compression of the unimproved layer (of thickness $H_u = H - H_L$, see Fig. 5.1) and part of the column-improved layer (of thickness H_c) should be calculated using the properties of the soft soil. The compression of the improved layer above H_c layer (of thickness $H_L - H_c$) is computed using an area weighted average of the constrained moduli (1D) of the column material and the surrounding soft soil.

In the case of embankment loading, there will be a load spreading effect in the ground. Using the theory of elasticity and assuming plane strain conditions, Gray (1936) derived equations for calculating the stresses in the vertical and the horizontal directions (σ_z and σ_x) in the ground as:

$$\sigma_z = \frac{p}{\pi}\left[\beta + \frac{x\alpha}{a} - \frac{z}{R_2^2}(x - b - a)\right] \tag{5.13}$$

$$\sigma_x = \frac{p}{\pi}\left[\beta + \frac{x\alpha}{a} + \frac{z}{R_2^2}(x - b - a) + \frac{2z}{a}\ln\frac{R_1}{R_0}\right] \tag{5.14}$$

Definitions of essential variables are given in Fig. 5.5. The influence factors (I for $\sigma_z = Ip$, where p is the intensity of surface loading at the embankment centreline) have been published by Osterberg (1957) and are shown here in Fig. 5.6.

5.2.3.3 Application of α–β Method to Three Case Histories

(a) General Description

In 2002, along the national road No. 208 at Shaowa-Biraki, Fukuoka, Japan, three test embankments (referred to here as Cases-1, 2 and 3) were constructed on a soft

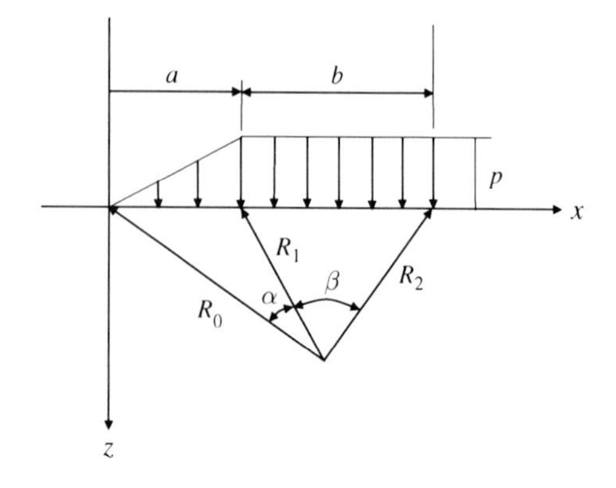

Fig. 5.5 Definitions of variables in Eqs. (5.13) and (5.14)

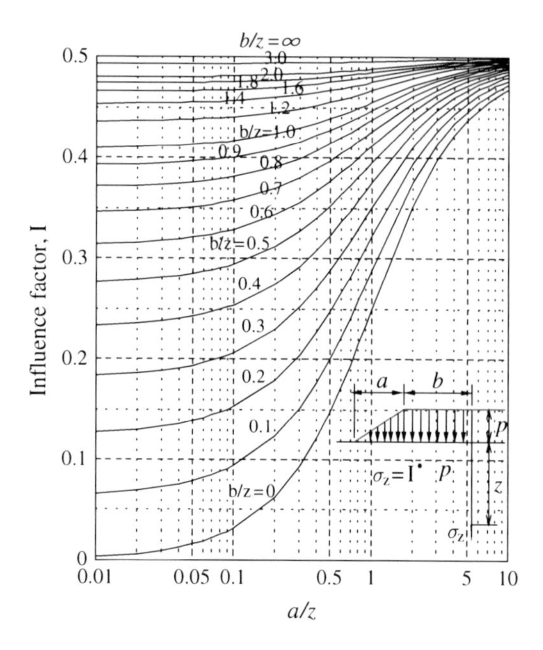

Fig. 5.6 Osterberg's influence factors

clayey silt/silty clay deposit improved by the installation of floating soil-cement columns. The columns were installed either in conjunction with a top slab (Case-1 and Case-2) or without a slab (Case-3) (FNHO 2003; Chai et al. 2010). There was also an embankment constructed on the natural soil deposit which is designated as Case-4. The plan dimensions and the relative locations of the embankments are shown in Fig. 5.7. The embankment for Case-1 had a length of about 64 m, while for Case-2 it was 68.8 m. For Case-3 the embankment was constructed behind a retaining wall and had a length of about 38 m. Case-4 had a base dimension of 40 m by 40 m. For those cases involving the installation of columns, the soil-cement columns were constructed by the dry jet mixing method (Chai et al. 2005). The

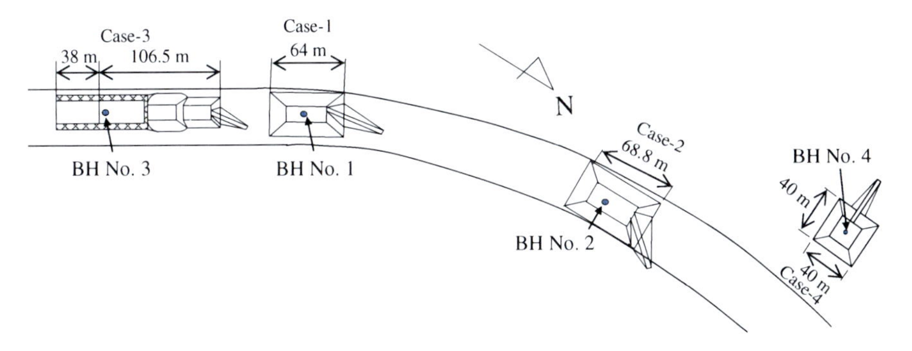

Fig. 5.7 Location and plan dimension of four cases

diameter of each column was about 1.0 m and they were arranged in a square pattern. However, the length of the columns, the improvement area ratio (α) and the design strength of the columns were different in each case. The thickness of the slab and the method for constructing the slab also differed. In all cases the backfill used to construct the embankments was decomposed granite and the compacted backfill had a total unit weight of about 19.0 kN/m^3.

Case-1

The cross-sectional geometry of the embankment and the thickness of the soft layer as well as the thickness of the slab and the length of the soil-cement columns are illustrated in Fig. 5.8 for this case. Based on the results of laboratory tests the amount of cement used in the field for the column construction was 140 kg/m^3 for this case. The value of α value was 21.7% and β was 76%. The slab was constructed by mixing cement with the surface soil using a mixing machine and then compacting the placed mix using a bulldozer. The thickness of the slab for this case was 0.5 m and the amount of cement used in its construction was 80 kg/m^3.

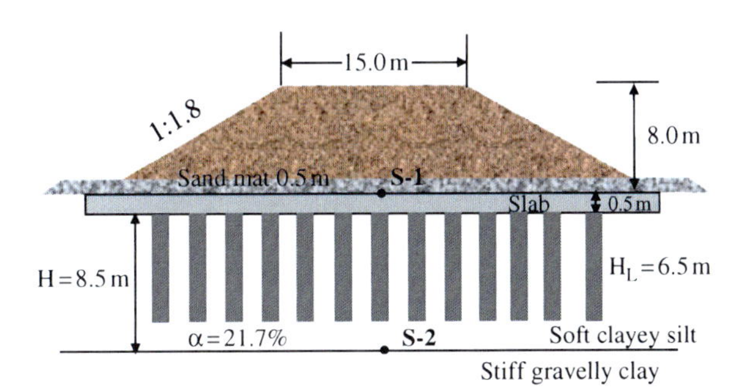

Fig. 5.8 Cross-section of Case-1 (after Chai et al. 2010)

Case-2

The cross-section of Case-2 is shown in Fig. 5.9. The amount of cement used for constructing the columns was 100 kg/m^3 and the value of α was 9% for this case. The length of the columns varied from 5.5 m ($\beta = 85\%$) at the left hand side (about 2/3 of the overall width) to 5.0 m ($\beta = 77\%$) at the right side (about 1/3 of the overall width). This case had a soil-cement slab with a thickness of about 2.5 m which was constructed by overlapping soil-cement columns 3.0 m in diameter. The amount of the cement used for the slab construction was 80 kg/m^3.

Case-3

The longitudinal cross-section of Case-3 is illustrated in Fig. 5.10a and the transverse cross-section at the settlement measuring locations (S-1 and S-2 in Fig. 5.10a) is shown in Fig. 5.10b. The amount of cement used to construct the columns was 130 kg/m^3. The length of the columns was 5.0 m giving an α value of 30% and β value of 47%. For this case there was no soil-cement slab constructed and only a sand mat about 0.5 m thick was placed on the ground surface above the columns. The slope of the embankment was reinforced and it had a steep slope angle (V:H = 1:0.5).

Case-4

In this case the embankment was constructed on natural ground without any ground improvement. The cross-sectional geometry of the embankment is illustrated in Fig. 5.11.

(b) Soil Properties and Calculated Settlement

Case-1

The soil profiles as well as some physical and mechanical properties at borehole No. 1 (Fig. 5.7), which was adjacent to the site of Case-1, are shown in Fig. 5.12 (FNHO 2003). The thickness of the soft soil layer was about 9.0 m and the groundwater level was about 1.1 m below the ground surface. Curves of the measured

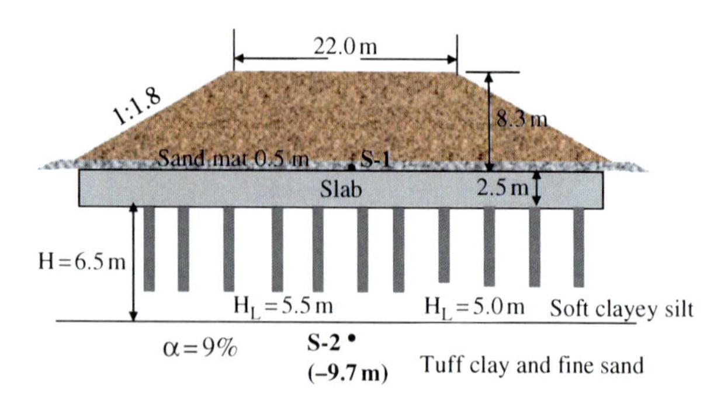

Fig. 5.9 Cross-section of Case-2 (after Chai et al. 2010)

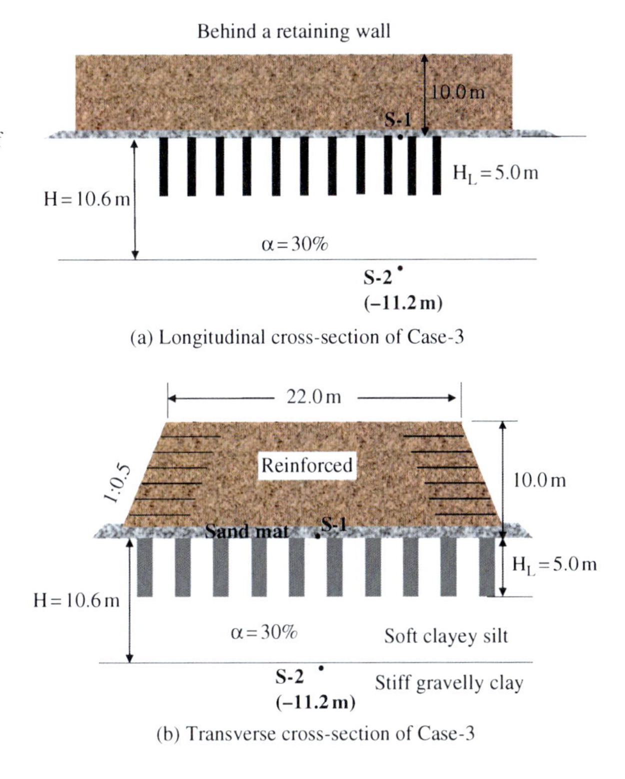

Fig. 5.10 Cross-sections of Case-3 (after Chai et al. 2010). (**a**) Longitudinal cross-section of Case-3; (**b**) transverse cross-section of Case-3

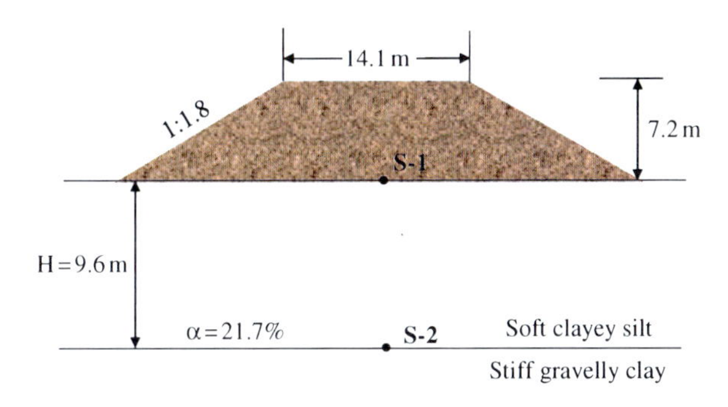

Fig. 5.11 Cross-section of Case-4

settlement (S) versus elapsed time (t) at the measuring points S-1 and S-2 (Fig. 5.8) are depicted in Fig. 5.13. The settlement difference between S-1 and S-2 represents the measured compression (ΔS) of the soft layer improved by the floating column-slab system. Based on the soil properties presented in Fig. 5.12, the parameters adopted for the settlement calculations are listed in Table 5.1. Values of the over consolidation ratio (OCR) were calculated using values of the consolidation yield stress (σ'_p) under 1D conditions and the initial vertical effective stress (σ'_{v0}) in the

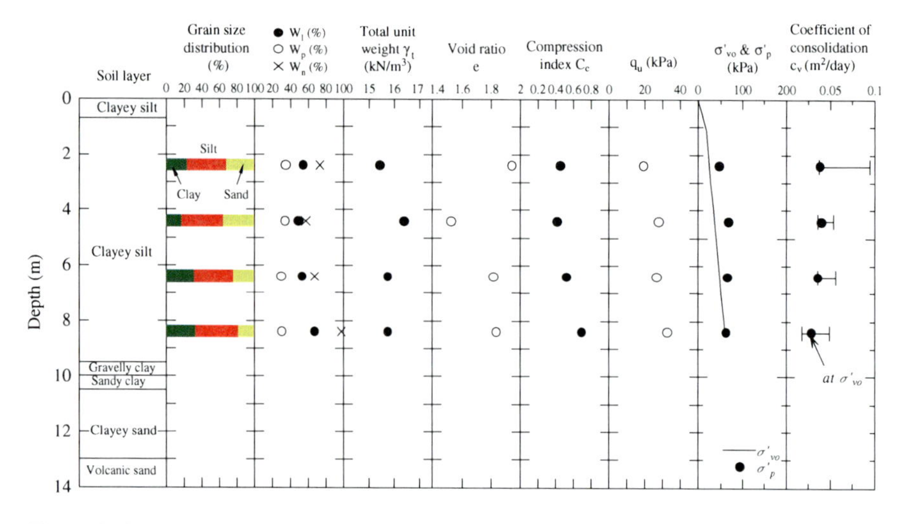

Fig. 5.12 Soil profiles and properties at the site of Case-1

Table 5.1 Soil properties for settlement calculation at sites of Case-1, 2 and 3

		Soil properties				
Case	Depth of groundwater (m)	Depth (m)	λ	e_0	$\gamma_{tт}$ (kN/m³)	OCR
Case-1	1.1	2.4	0.20	1.94	15.46	1.9 (1.94)[a]
		4.4	0.18	1.53	16.39	1.9 (1.89)
		6.4	0.52	1.82	15.75	1.4 (1.38)
		8.4	0.23	1.84	15.75	1.4 (1.06)
Case-2	0.7	1.4	0.21	1.97	15.26	2.5 (2.55)
		3.4	0.19	1.95	15.44	2.5 (2.66)
		5.4	0.18	1.63	16.10	1.5 (1.38)
		7.4	0.26	1.83	15.75	1.5 (1.60)
Case-3	0.9	2.4	0.20	1.20	15.68	2.0 (1.29)
		5.2	0.17	1.53	16.43	2.0 (2.50)
		6.4	0.23	1.80	15.71	1.6 (1.63)
		8.4	0.26	1.62	16.34	1.6 (1.10)
		10.4	0.27	1.52	16.66	1.6 (2.37)
Case-4	0.4	2.4	0.19	2.03	15.50	2.3 (2.33)
		4.4	0.37	1.74	15.68	1.5 (1.0)
		6.4	0.21	1.43	16.47	1.2 (1.18)

[a]Numbers in the parentheses were calculated from the consolidation yield stress (σ'_p) and initial vertical effective stress (σ'_{v0}) in the ground. The numbers not in parentheses were adopted in settlement calculation

ground, providing *OCR* values of 1.9 and 1.4 for the layers above and below 5.0 m depth, respectively. Young's modulus for the columns was assumed as 70,000 kPa, i.e., 100 times the estimated unconfined compression strength (q_u) (Kitazume 1996), and a Poisson's ratio of 0.2 was also adopted for the column material. These values were also used for Cases-2 and 3. The stress increments in the ground under the embankment load were calculated using Osterberg's influence factors. The compressions of the soft layers were calculated by both the method of JICE and the α-β method. Both measured and calculated values are listed in Table 5.2. The measured values correspond to the last measured data points in Fig. 5.13. In this table, the relative error (*RE*) is defined as follows:

$$RE = \frac{(\Delta S)_{cal} - (\Delta S)_{mea}}{(\Delta S)_{mea}} \times 100\,(\%) \tag{5.15}$$

where $(\Delta S)_{cal}$ and $(\Delta S)_{mea}$ = the calculated and measured amount of compression of the soft layer, respectively. It can be seen that both the JICE (1999) method and α–β method over-estimated the amount of the compression, but the α–β method yielded a much better result with a *RE* value of 10.5% compared with 57.0% for the JICE method.

Table 5.2 Measured and calculated compression (ΔS) of the soft layer

		Calculated			
		JICE method		α-β method	
Case	Measured ΔS (m)	ΔS (m)	*RE* (%)	ΔS (m)	*RE* (%)
1	0.277	0.435	57.0	0.306	10.5
2	0.226	0.323	42.9	0.265	17.3
3	0.562	0.536	–4.6	0.575	2.3
4	1.16	–	–	–	–

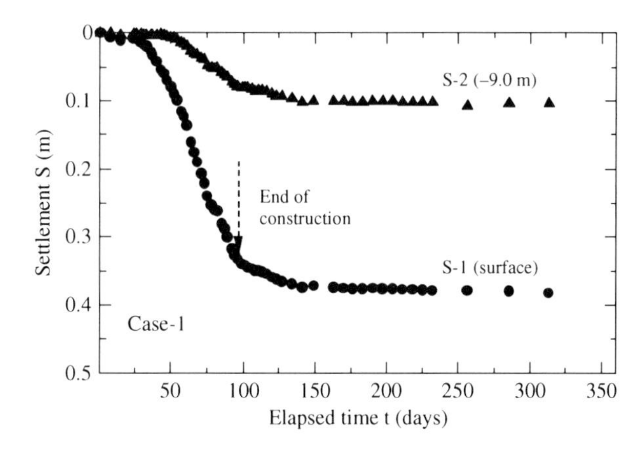

Fig. 5.13 Measured settlement–time curves for Case-1 (after Chai et al. 2010)

Case-2

The soil profile and properties at the site of Case-2 (borehole No. 2 in Fig. 5.7) are shown in Fig. 5.14. The thickness of the soft layer was about 8.0 m, and the groundwater level was about 0.7 m below the ground surface. The measured settlement curves at points S-1 and S-2 (Fig. 5.9) are depicted in Fig. 5.15. The parameters adopted for the settlement (compression) calculation were determined using the same method as for Case-1, as listed in Table 5.1. Both the calculated and measured compressions of the soft layer are given in Table 5.2. Again, the proposed α–β method yielded a better result with a *RE* value of 17.2% compared with 42.9% for the JICE method.

Case-3

The soil profile and the soil properties at the site of Case-3 (borehole No. 3 in Fig. 5.7) are shown in Fig. 5.16. The thickness of the soft layer was about 10.6 m at the settlement measuring point, and the groundwater level was about 0.9 m below the ground surface. The measured settlement curves are given in Fig. 5.17. This case did not have a soil-cement slab constructed at the original ground surface, but considering the improvement area ratio of 30%, and the presence of a sand mat as well as compacted decomposed granite soil backfill above the columns, the relative settlement between the columns and the surrounding soil at the ground surface should be very limited. For this reason the same calculation methods were used to estimate the compression of the soft layer. The computed values are compared with measurements in Table 5.2. For this case, since the α value was 30%, the method

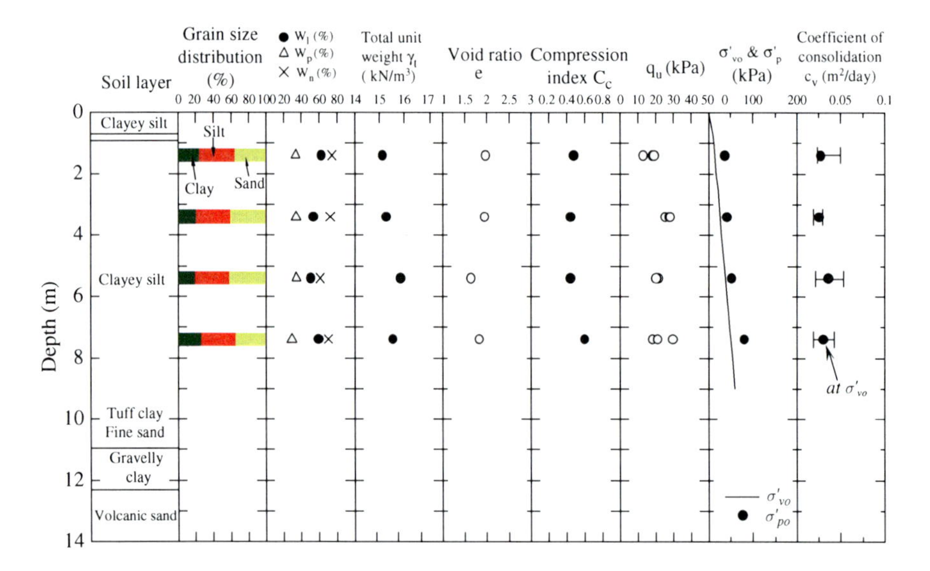

Fig. 5.14 Soil profiles and properties at the site of Case-2

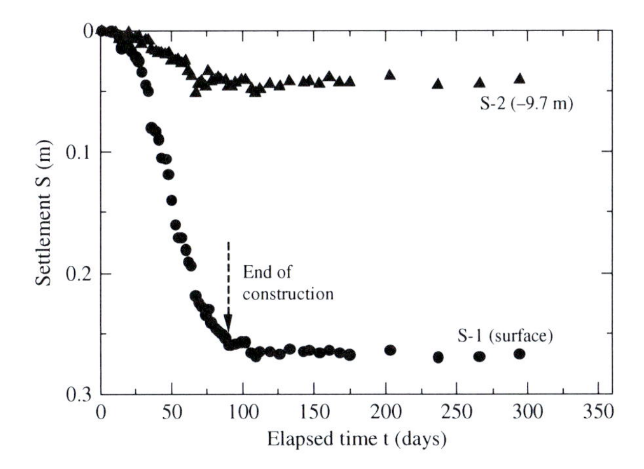

Fig. 5.15 Measured settlement – time curves for Case-2 (after Chai et al. 2010)

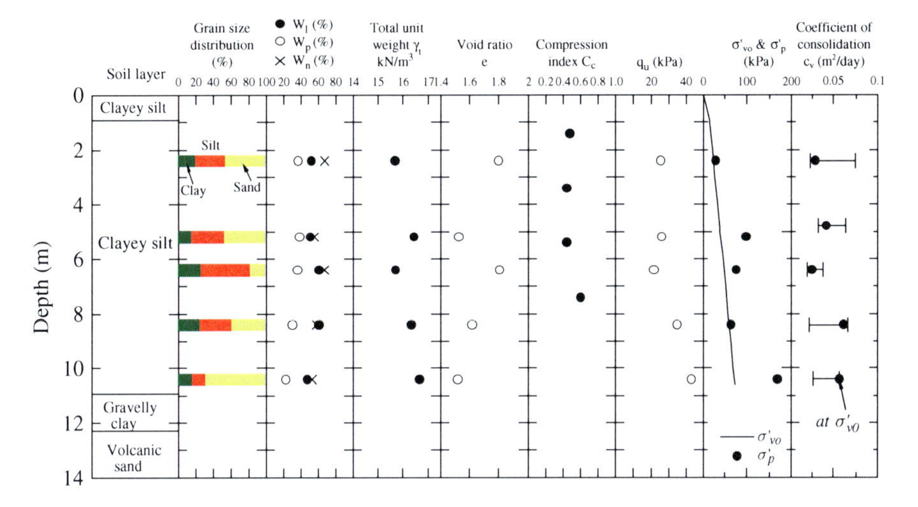

Fig. 5.16 Soil profiles and properties at the site of Case-3

of JICE under-predicted the ΔS value with a *RE* value of -4.8%. The α-β method yielded a slight over-prediction (*RE* = 2.3%). This comparison indicates that the α–β method yielded a reasonable prediction of the consolidation settlement of soft subsoil improved by the installation of floating soil-cement columns.

Case-4

The soil profiles and the soil properties at the site of Case-4 (borehole No. 4 in Fig. 5.7) are shown in Fig. 5.18. The thickness of the soft layer was about 9.6 m at the settlement measuring point, and the groundwater level was about 0.4 m below the ground surface. The measured settlement curves for this case are given

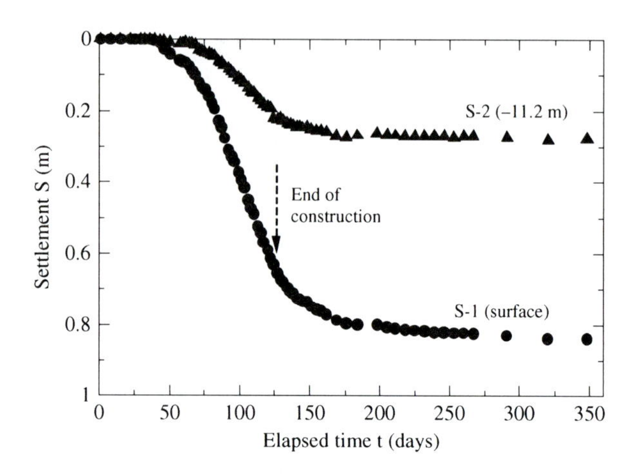

Fig. 5.17 Measured settlement–time curves for Case-3 (after Chai et al. 2010)

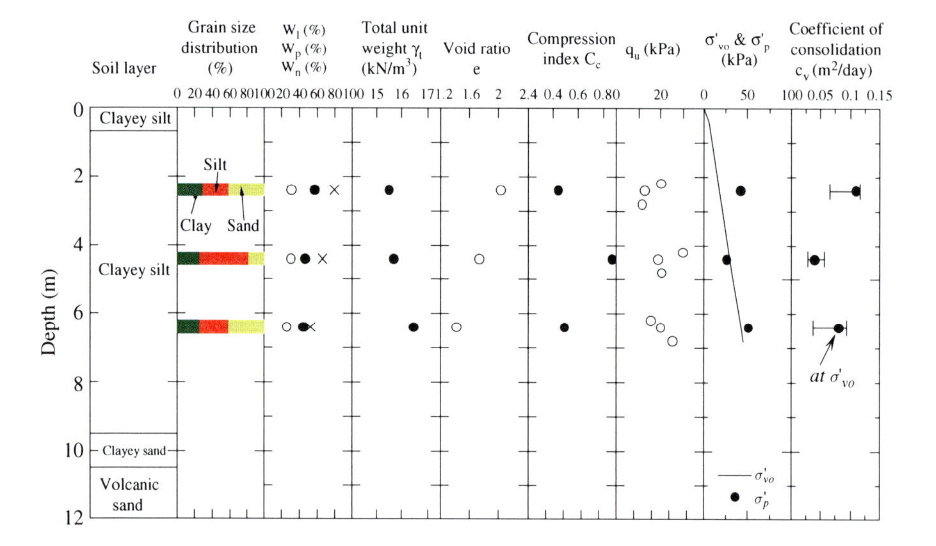

Fig. 5.18 Soil profiles and properties at the site of Case-4

in Fig. 5.19. This case did not involve soil-cement column improvement, and thus it provides a useful reference case for assessing the effects of floating soil-cement columns on reducing consolidation settlement and increasing the rate of consolidation of the soft subsoil. Although the soil profiles are different at each site, making direct comparisons difficult, it is noted that Case-4 has the lowest embankment thickness but the largest surface settlement (Fig. 5.19). The observed compression of the soft layer in Case-4 is about 4 times that measured for Case-1, about 5 times that for Case-2 and about twice that for Case-3 (Table 5.2).

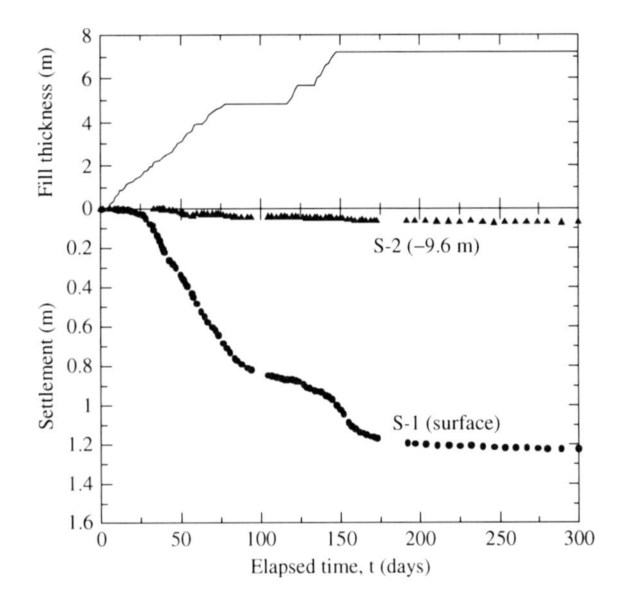

Fig. 5.19 Measured settlement–time curves of Case-4

5.2.4 Summary

Installing soil-cement columns into soft clayey subsoils can reduce the settlement induced by a surcharge load. When the columns fully penetrate the soft layer, the equilibrium method or the composite constrained modulus method can be used to calculate the settlement of the improved subsoil.

In cases where floating soil-cement columns are used to improve the soft subsoil, the stress concentration on the column will vary with the improvement area ratio (α) and the depth improvement ratio (β), as well as vary with depth along the column. To consider these effects when calculating settlements, a method for regarding a lower portion of the sub-layer improved by the installation of soil-cement columns as effectively an unimproved layer of thickness H_C has been described. Two bi-linear functions in the variables α and β have been proposed to determine an appropriate value for H_C. This method, denoted as the α-β method, was adopted to calculate the consolidation settlement (compression) of soft layers for three case histories in Fukuoka, Japan, in which the soft soil layers were improved by floating soil-cement columns and slabs or soil-cement column systems. Comparisons of these settlement predictions with field measured values of settlement showed that the α-β method yielded reasonable predictions.

5.3 Degree of Consolidation

5.3.1 Fully Penetrating Columns

Column inclusions in soil, such as compacted sand piles and stone columns, can accelerate the degree of the consolidation of a soft clayey subsoil due to their higher hydraulic conductivity. The consolidation theory for this kind of improvement is directly analogous to the consolidation theory for vertical drains (Barron 1948; Hansbo 1981). As for a soil-cement column, its hydraulic conductivity is usually in the same order of magnitude as the surrounding soft soil, but due to its higher stiffness its coefficient of consolidation is much higher than that of the soft soil. As a result the soil-cement column may also accelerate the degree of consolidation of the improved ground. Han and Ye (2002) derived a closed form unit cell solution for fully penetrating stone columns which can consider both the effects of the hydraulic conductivity and the stiffness of the column inclusion. It is considered that this solution can also be used for evaluating the effects of fully penetrating soil-cement columns on the degree of consolidation of the improved subsoil. The assumptions used in Han and Ye's solution are as follows:

(a) no vertical flow occurs within the surrounding soil;
(b) the soil is fully saturated and the pore water is incompressible;
(c) the column and the surrounding soil only deform vertically and have equal vertical strain at any depth;
(d) the coefficients of compressibility of the smeared zone and the undisturbed soil are equal;
(e) the load is applied instantaneously and maintained constant during consolidation;
(f) the total vertical stresses within the column and the surrounding soil are generally acknowledged as being different but for the purposes of the modelling are taken as averaged and uniform within each material; and
(g) the excess pore water pressure within the column is averaged and uniform in terms of radius.

In addition, initial and boundary conditions are assumed in terms of the geometry, the compressibility of the column and the soil, and excess pore water pressures defined in Fig. 5.20. In this figure, u_r, u'_r and u_c are excess pore water pressures in the surrounding soft soil, the smear zone and the column, respectively, k_r, and k'_r are horizontal hydraulic conductivities of the surrounding soft soil and the smear zone, respectively, k_c and k_v are the vertical hydraulic conductivity of the column and the surrounding soft soil, respectively, r_c, r_s and r_e are the radii of the column, the smear zone and the unit cell, respectively, H is the drainage path length of the column, z is the depth and r is the radial distance from the centre of the column. The initial and boundary conditions are as follows.

(a) the initial excess pore water pressure within the surrounding soil at the time of instantaneous loading is equal to u_0, i.e., $\bar{u}_r\,|_{t=0} = u_0$;
(b) due to the symmetry of the problem, no flow occurs across the external boundary, i.e., $\left.\frac{\partial u_r}{\partial_r}\right|_{r=r_e} = 0$;
(c) for the column, the ground surface is always free-draining, i.e., $u_c\,|_{z=0} = 0$;

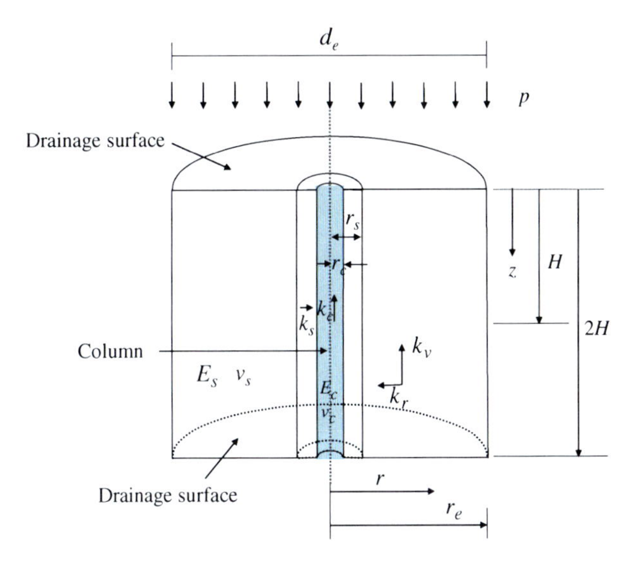

Fig. 5.20 Definition of terms for a unit cell model

(d) within the column, the boundary at $z = H$ is impervious or because of symmetry, no flow occurs across this boundary, i.e., $\frac{\partial u_c}{\partial z}\Big|_{z=H} = 0$;

(e) the quantity of water flowing horizontally through the smeared zone into the column is equal to that flowing vertically out from the column, i.e.,

$$2\pi r_c dz \frac{k_s}{\gamma_\omega} \frac{\partial u_r{}'}{\partial r} \Big|_{r=r_c} = -\pi r_c^2 dz \frac{k_c}{\gamma_\omega} \frac{\partial^2 u_c}{\partial z^2};$$

(f) the excess pore water pressures at the interface between the smeared zone and the undisturbed in situ soil are equal, i.e., $u_r' \big|_{r=r_s} = u_r \big|_{r=r_s}$; and

(g) the excess pore water pressures at the interface between the smeared zone and the column are equal, i.e., $u_r' \big|_{r=r_c} = u_c$.

For this problem the average degree of consolidation due to radial drainage (U_r) can be expressed as:

$$U_r = 1 - e^{-\frac{8}{F_m'} T_{rm}} \tag{5.16}$$

where

$$T_{rm} = \frac{c_{rm}}{d_e^2} t \tag{5.17}$$

$$c_{rm} = c_r \left(H - n_s \frac{1}{n^2 - 1} \right) \tag{5.18}$$

and

$$F_m{}' = \left(\frac{n^2}{n^2-1}\right)\left(\ln\left(\frac{n}{s}\right) + \frac{k_r}{k_s}\ln s - \frac{3}{4}\right) + \left(\frac{s^2}{n^2-1}\right)\left(1 - \frac{k_r}{k_s}\right)\left(1 - \frac{s^2}{4n^2}\right) + \left(\frac{k_r}{k_s}\right)\left(\frac{1}{n^2-1}\right)\left(1 - \frac{1}{4n^2}\right) + \frac{32}{\pi^2}\left(\frac{k_r}{k_c}\right)\left(\frac{H}{d_c}\right)^2 \tag{5.19}$$

and where c_r = coefficient of consolidation of the soft soil due to radial flow, n_s = the stress concentration ratio between the vertical effective stress in the column and in the surrounding soil, t = time, $n = r_e/r_c$, $d_c = 2r_c$ and $s = r_s/r_c$. Regarding the value of n_s, Han and Ye (2001) suggested that n_s can be taken to be the same as the ratio of the coefficient of volumetric compressibility of the surrounding soil to that of the column.

The effect of vertical drainage of the natural soft soil can be evaluated by Terzaghi's 1D consolidation theory. Carrillo's solution (Carrillo 1942) can then be used to combine the effect of the radial drainage due to the presence of the column and the vertical drainage of the natural soil:

$$U_{vr} = 1 - (1 - U_v)(1 - U_r) \tag{5.20}$$

where U_{vr} = the average degree of consolidation, and U_v = the average degree of consolidation due to vertical drainage alone.

5.3.2 Floating Columns

The consolidation behaviour of subsoil improved by the installation of floating soil-cement columns is different from that where the soft layer is improved by fully penetrating columns. For the floating case, during the consolidation process there will be relative penetration of the columns into the surrounding soft soil. As a result, the stress concentration ratio is likely to change with time and depth. Further, this relative penetration is influenced by the values of α and β, the loading intensity and the strength and stiffness of the soft soil (Chai et al. 2010). As yet there is no rigorous theoretical solution for this complicated problem.

For the case of floating column improvement, Lorenzo and Bergado (2003) published a solution to estimate the degree of consolidation, but it ignores the effect of vertical drainage in the natural soft soil. Miao et al. (2008) proposed a method which considers the effect of the higher stiffness of the column on the stress concentration but ignores the drainage effects of the column. Therefore, it should be no great surprise that these methods have had limited success when used to calculate the degree of consolidation of soft subsoils improved by the installation of floating soil-cement columns.

As a pragmatic approach, Chai and Pongsivasathit (2010) proposed that at the macro scale a soft clayey subsoil improved by the installation of floating columns

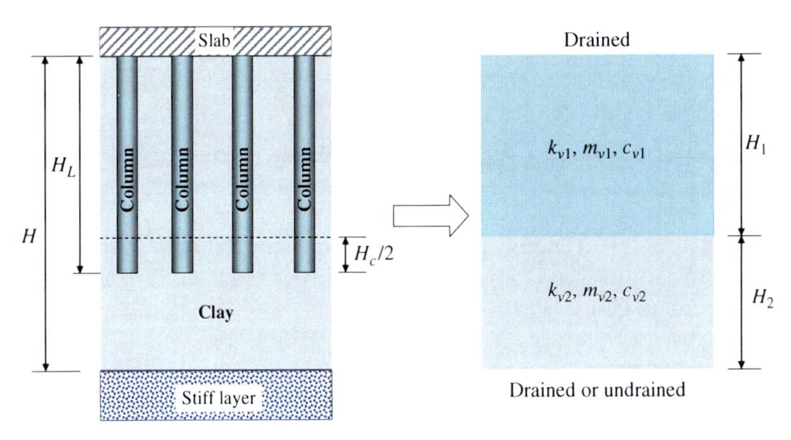

Fig. 5.21 Two-layer model for calculating the degree of consolidation (after Chai and Pongsivasathit 2010)

can be considered as a two-layer system, as shown in Fig. 5.21, and the degree of consolidation can be evaluated by the corresponding theoretical solution for a two-layer system, such as that proposed by Zhu and Yin (1999). In the figure, m_{v1} and m_{v2} are the coefficients of volumetric compressibility of layer-1 and layer-2, k_{v1} and k_{v2} are the vertical hydraulic conductivities of layer-1 and layer-2, c_{v1} and c_{v2} are the vertical coefficient of consolidation of layer-1 and layer-2, respectively. There are two issues which need to be resolved before applying these solutions. One is the representative values of k_{v1} and m_{v1} to be adopted for layer-1 (Fig. 5.21), and another is the thickness of each layer (H_1 and H_2). For a two-layer system, the degree of consolidation is not only influenced by the coefficient of consolidation, but by both values of the hydraulic conductivity (k) and the coefficient of volumetric compressibility (m_v) independently.

Chai and Pongsivasathit (2010) proposed that the value of m_{v1} can be evaluated by using the area weighted average value of the constrained moduli of the column (D_c) and the surrounding soil (D_s) as follows:

$$m_{v1} = \frac{1}{\alpha D_c + (1 - \alpha) D_s} \tag{5.21}$$

Although the hydraulic conductivity of the column can be smaller or larger than the corresponding value of the surrounding soft soil, for most cases there may not be much difference between the two values. However, due to the higher stiffness of the column, the value of the coefficient of consolidation of the column is normally several orders of magnitude higher than that of the soft soil. As a result, there will be radial flow toward the column (Chai et al. 2006), and the column partially functions as a drain. Chai et al. (2001) introduced the concept of the equivalent vertical hydraulic conductivity for a subsoil improved by the installation of prefabricated vertical drains (PVDs), and it is considered that the same concept can also be applied to column-improved subsoil. In such cases the value of k_{v1} can be calculated from the following equation (Chai et al. 2001):

$$k_{v1} = \left(1 + \frac{2.5H_1^2}{\mu d_e^2}\frac{k_h}{k_v}\right)k_v \tag{5.22}$$

where k_v and k_h = the hydraulic conductivity of the soft soil in the vertical and the horizontal directions, respectively, H_1 is the thickness of the layer-1, and μ can be expressed as follows (Hansbo 1981):

$$\mu = \ln\left(\frac{n}{s}\right) + \frac{k_h}{k_s}\ln(s) - \frac{3}{4} + \frac{8H_1^2 k_h}{3d_c^2 k_c} \tag{5.23}$$

where k_c is the hydraulic conductivity of the column material. Note that even for the case where $k_c = k_v$, k_{v1} is larger than k_v due to the effect of assumed radial flow.

When calculating the final consolidation settlement of the soft subsoil improved by the installation of floating columns a portion of the column-improved layer with a thickness H_c (Fig. 5.1) is regarded as an unimproved layer, due to the relative penetration of the columns into the underlying soft soil (Chai et al. 2010). By comparing predictions of the consolidation behaviour with the results of independent finite element analysis (FEA) using a unit cell model, and by trial and error, Chai and Pongsivasathit (2010) proposed that H_1 can be evaluated as:

$$H_1 = H_L - \frac{H_c}{2} \tag{5.24}$$

Another point worth noting is that due to large consolidation strains occurring within the unimproved layer-2, the thickness of layer-2 may be reduced considerably after consolidation, as illustrated schematically in Fig. 5.22. To consider this kind of large deformation phenomenon, at least in an approximate way, it is proposed to use the average thickness before and after the consolidation of layer-2 when calculating the degree of consolidation. For simplicity the total settlement of the system is used as the compression of layer-2, and H_2 is calculated as follows:

$$H_2 = H_u + \frac{H_c}{2} - \frac{s_f}{2} \tag{5.25}$$

where s_f = the final consolidation settlement (compression) of the system and H_u = the original thickness of the unimproved sub-layer.

It is recommended to apply to this problem the double layer consolidation solution published by Zhu and Yin (1999), which can consider linear variations of the total vertical stress increment with depth and time. A special case of Zhu and Yin's solution for cases where the total vertical stress does not change with time has been presented in Eqs. (4.32)–(4.43) for the special case of consolidation under vacuum

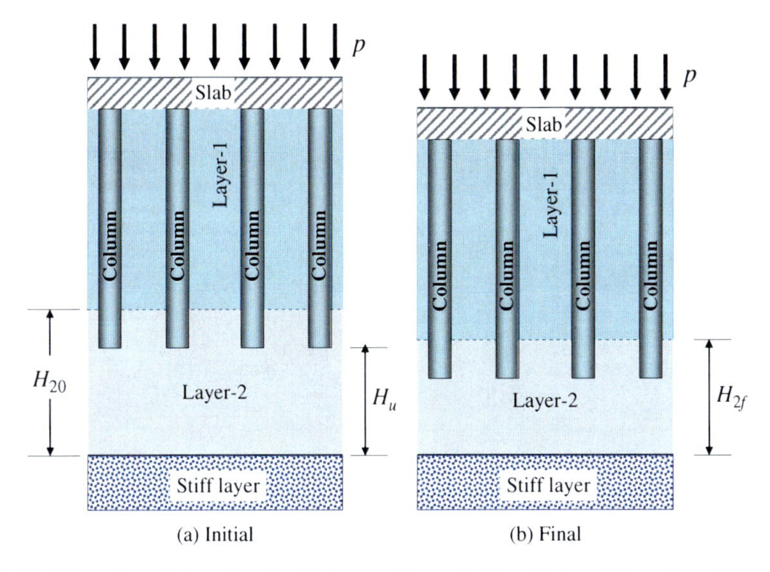

Fig. 5.22 Illustration of changing the thickness of layer-2 (after Chai and Pongsivasathit 2010)

pressure. Here the equations for calculating the degree of consolidation for the general case of a two layer system are given. The variables p, q, α, β, and T_v used in the solution have been defined in Chapter 4 and will not be repeated here. A new parameter, T_c, is defined as follows:

$$T_c = \frac{c_{v1}\, c_{v2}\, t_c}{\left(H_1\sqrt{c_{v2}} + H_2\sqrt{c_{v1}}\right)^2} \tag{5.26}$$

where t_c = the time interval over which the applied load reaches its ultimate magnitude, assuming a linear increase over this period. For two-way drainage boundary conditions, the average degree of consolidation, $U(T_v)$, can be expressed as follows:

$$U(T_v) = \min\left(1, \frac{T_v}{T_c}\right) - \sum_{n=1}^{\infty} \frac{2\left[m_{v1}H_1/\{\alpha \sin(\lambda_n\alpha)\} + m_{v2}H_2/\{\beta \sin(\lambda_n\beta)\}\right] T_n(T_v)}{\lambda_n\left[m_{v1}H_1(\sigma_0+\sigma_1) + m_{v2}H_2(\sigma_1+\sigma_2)\right]} \tag{5.27}$$

$$T_n(T_v) = \begin{cases} \dfrac{b_n}{\lambda_n^3 T_c}\left[1-\exp\left(-\lambda_n^2 T_v\right)\right], & T_v \le T_c \\ \dfrac{b_n}{\lambda_n^3 T_c}\left[1-\exp\left(-\lambda_n^2 T_c\right)\right]\left[\exp\left(-\lambda_n^2 (T_v - T_c)\right)\right], & T_v > T_c \end{cases} \tag{5.28}$$

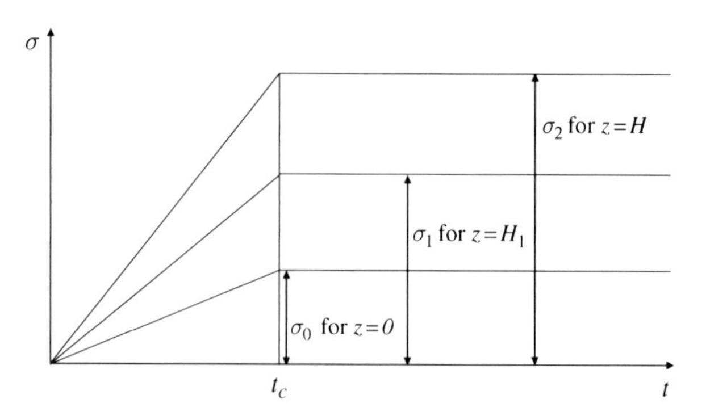

Fig. 5.23 Variation of vertical total stress (after Zhu and Yin 1999)

where σ_0, σ_1, σ_2 = total stresses at the top of layer-1, at the interface between layer-1 and layer-2 and at the bottom of layer-2, respectively. It is also assumed that the variation of total stress within each layer is linear with depth. The variation of total stress with time at the three locations of interest is illustrated in Fig. 5.23 (after Zhu and Yin 1999). In these equations the constant b_n is defined as:

$$b_n = \frac{\left[\frac{m_{v1}H_1}{\alpha \sin(\lambda_n\alpha)}\right]\sigma_0 + \left[\frac{m_{v2}H_2}{\beta \sin(\lambda_n\beta)}\right]\sigma_2 + \left[\frac{m_{v1}H_1}{\alpha^2\lambda_n}\right](\sigma_1 - \sigma_0) + \left[\frac{m_{v2}H_2}{\beta^2\lambda_n}\right](\sigma_1 - \sigma_2)}{m_{v1}H_1/\left[2\sin^2(\lambda_n\alpha)\right] + m_{v2}H_2/\left[2\sin^2(\lambda_n\beta)\right]} \quad (5.29)$$

where λ_n = the nth positive root of the following equation for variable θ:

$$\sin\theta + p\sin(q\theta) = 0 \quad (5.30)$$

For one-way drainage boundary conditions:

$$U(T_v) = \min\left(1, \frac{T_v}{T_c}\right) - \sum_{n=1}^{\infty} \frac{2m_{v1}H_1\, T_n(T_v)}{\lambda_n\alpha\sin(\lambda_n\alpha)\left[m_{v1}H_1(\sigma_0 + \sigma_1) + m_{v2}H_2(\sigma_1 + \sigma_2)\right]} \quad (5.31)$$

The expression for $T_n(T_v)$ is the same as in Eq. (5.28) and the constant b_n is as follows:

$$b_n = \frac{\left[\frac{m_{v1}H_1}{\alpha \sin(\lambda_n\alpha)}\right]\sigma_0 + \left[\frac{m_{v1}H_1}{\alpha^2\lambda_n}\right](\sigma_1 - \sigma_0) + \left[\frac{m_{v2}H_2[\cos(\lambda_n\beta) - 1]}{\beta^2\lambda_n\cos(\lambda_n\beta)}\right](\sigma_1 - \sigma_2)}{m_{v1}H_1/\left[2\sin^2(\lambda_n\alpha)\right] + m_{v2}H_2/\left[2\cos^2(\lambda_n\beta)\right]} \quad (5.32)$$

where λ_n = the nth positive root of the following equation for variable θ:

$$\cos\theta - p\cos(q\theta) = 0 \quad (5.33)$$

5.3.3 Summary

For soft clayey deposits improved by soil-cement columns and under the influence of a surcharge load, the effect of the columns will be not only to reduce the consolidation settlement, but also to increase the degree of consolidation. Although there are several methods available for calculating the degree of consolidation in cases where the columns fully penetrate the soft layer, it is suggested that the solution proposed by Han and Ye (2002) for stone column improvement may be most suitable to evaluate the effects of the columns on the degree of consolidation, because it can consider both the effects of radial drainage and the higher stiffness of the columns on the consolidation response.

In the case of floating columns installed in the soft layer, the interaction behaviour between the columns and the soft layer is more complicated and there is no rigorous solution for calculating the degree of consolidation. A semi-empirical method has been described in this section. It is considered that at the macro level, the subsoil improved by floating soil-cement columns can be regarded as a two-layer system and the degree of consolidation can be calculated by the double soil-layer consolidation theory. Methods for determining the thickness of each layer, the equivalent vertical hydraulic conductivity (k_{v1}) and the coefficient of volumetric compressibility (m_{v1}) of a part of the column improved layer (layer-1) have been proposed.

5.4 Settlement–Time Curve

5.4.1 Method of Calculation

For normally consolidated ground, once the final settlement and the degree of consolidation at any time (t) are known, the settlement–time curve can be easily obtained. However, even soft clayey deposits usually possess a crust layer at the ground surface which is usually over-consolidated or apparently over-consolidated. The presence of this crust should normally be considered when calculating the settlement–time curve. Chai and Pongsivasathit (2010) proposed the following method to deal with this possibility.

The settlement consists of two parts: the compression (s_1) of that part of the column-improved layer with a thickness of $H_{1s}(H_L - H_c)$ and the compression (s_2) of the unimproved layer supplemented by a portion of the improved sub layer of thickness H_c, i.e., the sub layer with a total thickness of H_{2s}, as shown in Fig. 5.24. It is assumed that an average degree of consolidation, $U(t)$, can be used for both layer-1 and layer-2. The equations for calculating the consolidation settlement, $s(t)$, are as follows.

For the layer of thickness H_{1s}:

$$s_1(t) = \sum_{i=1}^{n} \frac{\Delta p_{1i} H_{1i} U(t)}{D_{ci}\alpha + (1-\alpha) D_{si}} \tag{5.34}$$

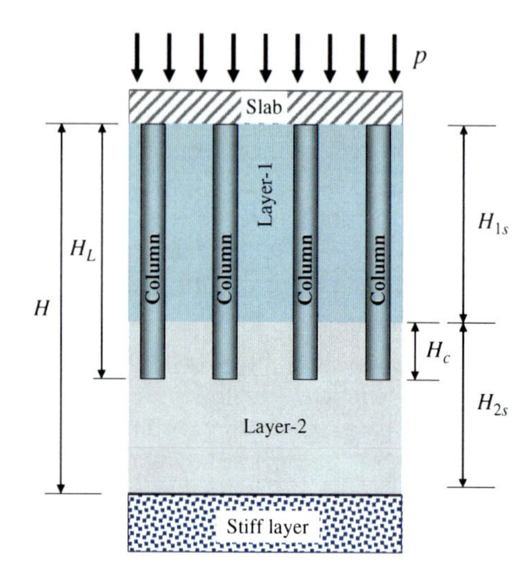

Fig. 5.24 Layer thicknesses for settlement calculation

For the layer of thickness H_{2s}:

$$s_2(t) = \sum_{i=1}^{n} H_{2i} \frac{\lambda_i}{1 + e_{0i}} \ln\left(1 + \frac{\Delta p_{2i}}{\sigma'_{vi}} U(t)\right) \tag{5.35}$$

where H_{1i} and H_{2i} = the thickness of all sub-soil layers contained within the layers with overall thickness of H_{1s} and H_{2s}, respectively, σ'_{vi} = the initial vertical effective stress in the sub-layer H_{2i}, e_{0i} = the initial void ratio, λ_i = the slope of virgin compression line in an $e - \ln p'$ plot (p' is the mean effective consolidation stress), Δp_{1i} and Δp_{2i} = the total vertical stress increments in sub-layers of thickness H_{1i} and H_{2i}, respectively. D_{ci} and D_{si} = the constrained moduli of the column and the surrounding soft soil of the sub-layer of thickness H_{1i} which can be calculated as follows:

$$D_{ci} = \frac{E_i(1 - \nu_i)}{(1 + \nu_i)(1 - 2\nu_i)} \tag{5.36}$$

$$D_{si} = \frac{(1 + e_i)\sigma'_{avi}}{\lambda_i} \tag{5.37}$$

where for each sub-soil layer i, E_i = Young's modulus, ν_i = Poisson's ratio, e_i = void ratio, and σ'_{avi} = the average effective vertical stress in the corresponding sub-soil layer including the incremental stress induced by the embankment loading. In Eqs. (5.35) and (5.37), the slope of the unloading-reloading line in the $e - \ln p'$ plot, i.e., κ_i, should be substituted instead of λ_i in cases where the sub-soil layer is in an over-consolidated state, i.e., where $\left[\sigma'_{vi} + \Delta p_{1i} U(t)\right]$ or $\left[\sigma'_{vi} + \Delta p_{2i} U(t)\right]$ is less than p_{ci}, the consolidation yield stress of the corresponding sub-layer. Finally, the total compression (settlement) at time t, $s(t)$, can be expressed as follows:

$$s(t) = s_1(t) + s_2(t) \tag{5.38}$$

5.4.2 Application to Laboratory Model Tests

In order to demonstrate the validity of the analytical methods presented above for predicting the behaviour of ground improved by the installation of soil-cement columns, a series of laboratory tests will be considered in this section. Further validation of these analytical methods will also be presented in the subsequent section where field case studies are presented. But first attention is confined to laboratory scale testing.

The cylindrical soil model adopted for the laboratory tests described in this section has a diameter of 0.45 m and height of 0.9 m. The test set-up is shown in Fig. 5.25. The soil used was reconstituted Ariake clay with a liquid limit of about 108% and plastic limit of about 59%. The particular cement used in the model tests was US10, which is a typical cement used for the improvement of soft clayey ground in Japan. Non-woven geotextile with a thickness of about 3 mm (under zero confining pressure) was used as drainage medium at the bottom and the top of the soil sample. The geotextile was made of polypropylene with a density of about 131 g/m^2.

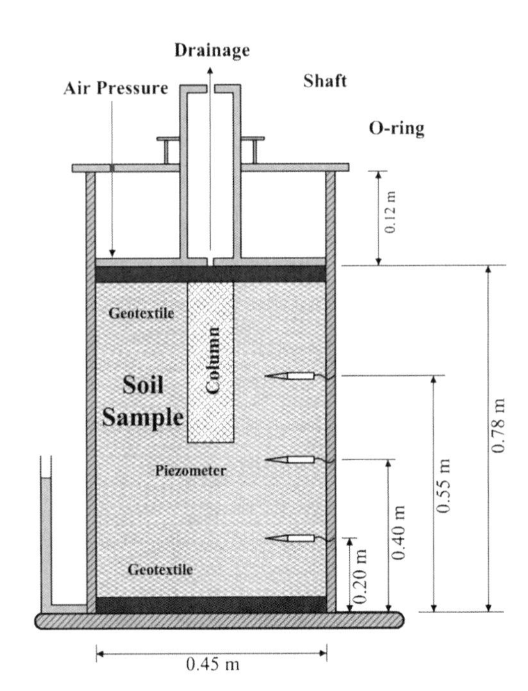

Fig. 5.25 Set-up of the laboratory model test (after Chai and Pongsivasathit 2010)

To accelerate the process of consolidation during preparation of the soil sample, a mini-prefabricated vertical drain (mini-PVD) was installed at the middle of the cylindrical specimen. The mini-PVD was made by folding the geotextile in 3 layers with a cross-section of 30 mm by 9 mm.

The procedures adopted for each test are described in the following sub-sections.

5.4.2.1 Pre-consolidation

Firstly, 3 layers of the geotextile were placed at the bottom of the model to act as a drainage layer. A thin layer of grease was smeared on the inside wall of the cylindrical apparatus to reduce friction. The remoulded Ariake clay with a water content of about 120% was then placed in the model apparatus layer by layer. Piezometers were installed at vertical locations defined by 0.2 m, 0.4 m and 0.55 m height from the bottom of the cylindrical sample. When the thickness of the clay reached about 0.78 m, the mini-PVD was installed by pushing it into the centre of the soil sample with a stainless steel rod. Three layers of the geotextile were then placed on top of the sample. The loading system shown in Fig. 5.25 was then set-up, and air-pressure of 40 kPa was applied to the annular top of the piston resting on the soil sample. Taking into account the plan area of the shaft fixed to the piston, the consolidation pressure experienced by the soil sample was estimated to be about 38 kPa. Consolidation of the soil sample occurred under two-way drainage. During the consolidation process, the settlement at the top of the sample and the excess pore water pressures were monitored. After the degree of primary consolidation reached about 90%, the test was stopped, the loading system was dismantled and the mini-PVD was withdrawn. This means that just before unloading, the average effective vertical stress in the model ground was about 34 kPa if the effect of friction between the model wall and the soil is ignored. Although a thin layer of grease was smeared on the inside wall of the containing cylinder, there may still have been some friction at this boundary. Considering an interface friction angle of 2°, and horizontal earth pressure coefficient (K_0) of 0.5, it is estimated that the friction would reduce the average vertical consolidation stress by about 2 kPa. Therefore, an average vertical effective stress of about 32 kPa can be estimated and this value of initial vertical effective stress was assumed in the subsequent settlement calculations. The thickness of the resulting sample of model ground was about 0.65–0.67 m.

5.4.2.2 Soil-Cement Column Installation

The small hole left after withdrawing the mini-PVD was filled carefully with the same type of clay. At the middle of the model ground specimen, a hole with a diameter of 150–246 mm was made by an auger to a pre-designed depth. The excavated soil was mixed with cement (16.6% by dry weight) and put back into the hole to form a soil-cement column. For ease of mixture, the water content of the excavated soil was adjusted to about 120%. To avoid air-bubbles getting trapped in the column, the soil was carefully compacted layer by layer by a steel-rod. The specially prepared model ground sample containing the soil-cement column was then left for 2 weeks to allow the column to cure before the consolidation test commenced.

5.4.2.3 Consolidation Test

Prior to the consolidation phase of the testing three layers of geotextile were installed on top of the model ground sample and the loading system was set-up again. Air-pressure of 40 kPa (equivalent to the pre-consolidation air-pressure) was applied first and maintained for 2 hours in order to make firm contact between the piston and the model ground and also to reload the model ground to the state it was in at the end of the pre-consolidation phase. The air pressure was then increased to 100 kPa (providing a total consolidation pressure for the soil sample of about 95 kPa) and measurements were made over time of the settlement at the top of the sample and the excess pore water pressures within the soil sample.

5.4.2.4 Test Results and Model Predictions

The cases tested in the laboratory study are listed in Table 5.3. Cases-L1 to L3 were conducted to investigate the effect of the depth improvement ratio β, Case-L4 investigated the effect of the area improvement ratio α and Case-L5 involved a sample without any soil-cement columns.

As explained previously, at the time the pre-consolidation phase was terminated for each test specimen, the vertical effective stress in the model ground can be estimated as approximately 32 kPa. This means that subsequently, after applying an overall surcharge pressure to the soil of 95 kPa, the increment in vertical stress was approximately 63 kPa. If the additional friction effect is then considered, an incremental consolidation pressure of 61 kPa was estimated for calculating the final consolidation settlement of each soil sample.

Based on the results of unconfined compression tests on specimens made using the same soil, cement and preparation method as used for making the soil-cement columns, Young's modulus for the column material was measured as approximately 6×10^4 kPa. Poisson's ratio of the column material was assumed to be 0.2. Taking into account the procedure used to install the columns, it was assumed that there was no smear zone surrounding the column. Using the results of oedometer tests

Table 5.3 Cases studied in laboratory testing

Case	α (%)	β (%)	Model thickness H (m)	Length of column H_L(m)
L1	11	50	0.67	0.34
L2	11	70	0.65	0.48
L3	11	80	0.67	0.54
L4	30	50	0.67	0.34
L5	–	–	0.67	–

for the void ratio, e_0, and the corresponding hydraulic conductivity, k_0, for the the columns composed of Ariake clay and cement, a value of k_c of 3.65×10^{-4} m/day was calculated using Taylor's (1948) equation corresponding to a voids ratio of 3.2 (water content of 120%), with the constant in Taylor's equation $C_k = 0.4e_0$. The values of all other parameters used in the consolidation calculations are listed in Table 5.4. The degree of consolidation was calculated by the double layer theory and the corresponding final settlement by Eqs. (5.34), (5.35), (5.36), and (5.37). The calculated values are compared with the measurements in Figs. 5.26, 5.27, 5.28, and 5.29, which indicate that the proposed method yielded good predictions of the test data. The calculated final consolidation settlement for Case-L2 is less than the measured settlement due to the poor quality of the column material at its tip, which was verified by the post-test investigation. In Fig. 5.28 the case without the column is included to demonstrate that the presence of the columns significantly reduced the final settlement and accelerated the rate of consolidation.

Table 5.4 Parameter values adopted for calculating the degree of consolidation

			m_{v1}	m_{v2}	k_{v1}	k_{v2}		
Case	λ	e_0	($\times 10^{-4}$ m^2/Kn)		($\times 10^{-4}$ m/day)		H_c (m)	s_f (mm)
L1	0.33	2.38	1.31	31.1	1.36	1.02	0.106	42
L2			1.31	31.1	1.38	1.02	0.088	31
L3			1.31	31.1	1.39	1.02	0.104	25
L4			0.49	31.1	1.37	1.02	0.037	37
L5			31.1	31.1	1.02	1.02	–	53

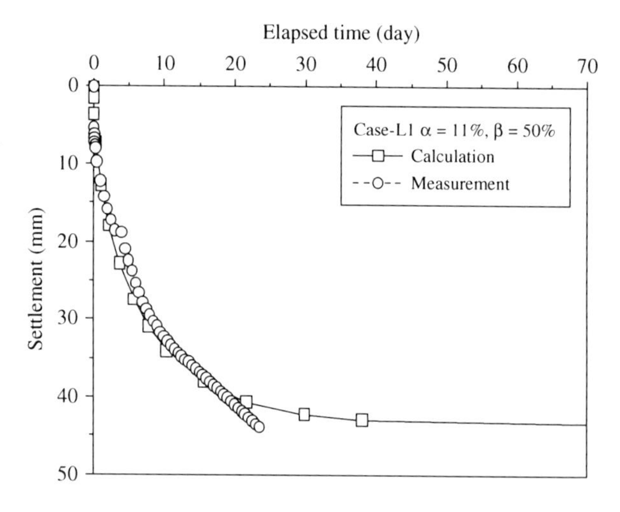

Fig. 5.26 Comparison of the settlement curves for Case-L1 (modified from Chai and Pongsivasathit 2010)

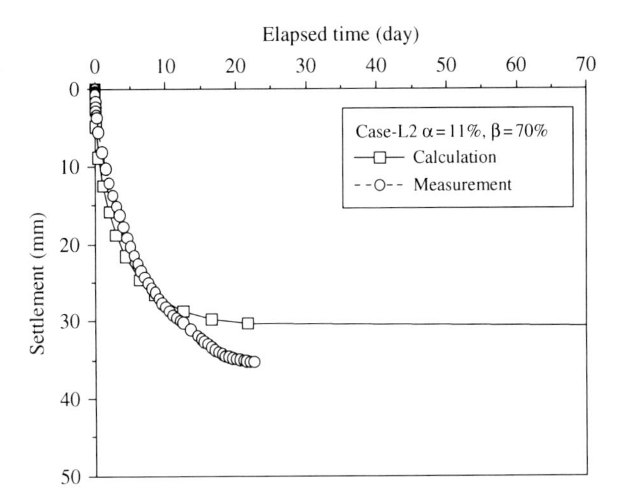

Fig. 5.27 Comparison of the settlement curves for Case-L2 (modified from Chai and Pongsivasathit 2010)

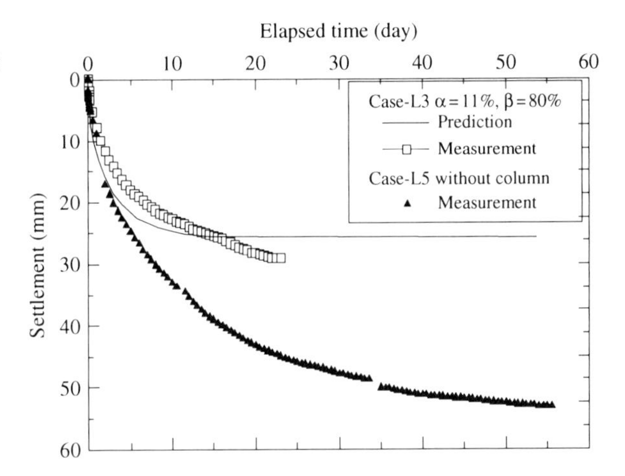

Fig. 5.28 Comparison of the settlement curves for Case-L3 and Case-L5

5.4.3 Application to Case Histories

5.4.3.1 Test Embankments in Fukuoka, Japan

Four test embankments in Fukuoka, Japan (Cases-1, 2 3 and 4) have already been described in detail in Section 5.2.3. The additional parameters required for calculating the degree of consolidation are given in Table 5.5. To demonstrate the effect of the columns, the calculated values of $U(t)$ for Case-1 with and without the columns are compared in Fig. 5.30. This figure clearly shows that the columns should have accelerated the consolidation process. Comparisons of the predicted and measured settlement curves are given in Figs. 5.31, 5.32, and 5.33 for Cases-1 to 3.

Fig. 5.29 Comparison of the settlement curves for Case-L4 (modified from Chai and Pongsivasathit 2010)

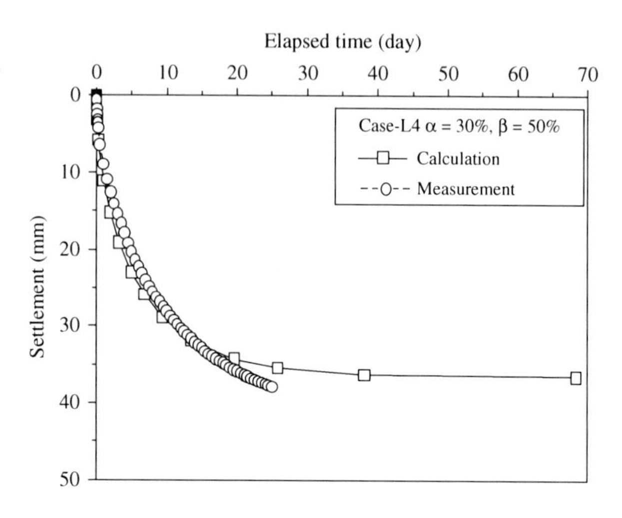

Table 5.5 Parameters for calculating the degree of consolidation

Quantity		Case 1	Case 2	Case 3
H (m)		8.5	6.5	10.6
H_1 (m)		6.10	4.93	4.71
$H_2 = H_{20} - s_f/2$ (m)		2.25	1.43	5.60
σ_0 (kN/m^2)		152.0	157.7	190,0
σ_1 (kN/m^2)		147.0	153.2	187.0
σ_2 (kN/m^2)		140.2	151.0	168.9
α (%)		21.7	9	30
β (%)		76.0	85.0	47.0
E_0		1.53	1.95	1.62
λ		0.20	0.20	0.21
Initial average stress at the middle of upper layer, σ'_{v0}(kN/m^2)		39.02	41.37	41.453
Calculated final settlement s_f (m)		0.306	0.265	0.565
Layer H_1	c_{v1}(m^2/day)	0.596 (0.046)[a]	0.181 (0.036)	1.258 (0.034)
	k_{v1}($\times 10^{-4}$ m/day)	2.81	2.17	4.06
Layer H_2	c_{v2}(m^2/day)	0.035	0.031	0.061
	k_{v2}($\times 10^{-4}$ m/day)	2.67	2.02	1.84

[a]Number in parentheses is for soft soil in layer of thickness H_1 without the columns

Generally, these calculations resulted in slower predicted settlement rates at the early stages compared with the measured rates. A possible reason is that the values of c_v used in the calculations correspond to normally consolidated soil states, but for these 3 cases initially the sub-soil was in an over-consolidated state. Nevertheless,

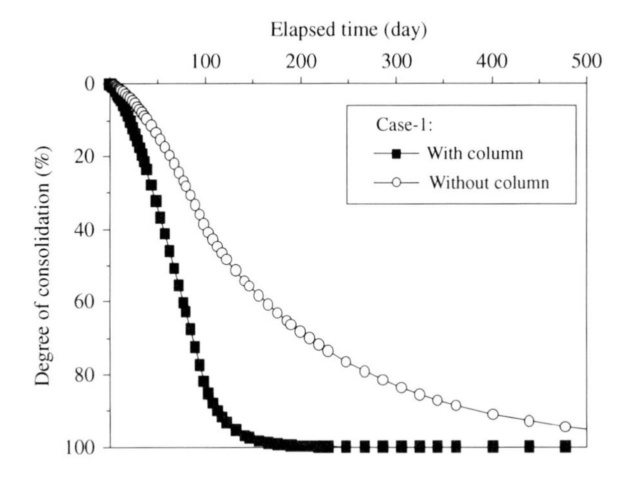

Fig. 5.30 Effect of the columns on the degree of consolidation

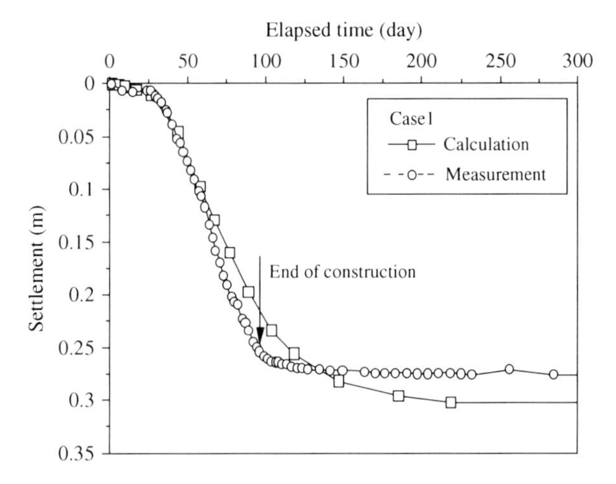

Fig. 5.31 Comparison of settlement curves for Case-1

the proposed method has resulted in a fair prediction of the time-dependent settlements and the proposed calculation method is therefore considered useful as a design tool.

5.4.3.2 Test Embankment in Saga, Japan

(a) Brief Description of the Project

As shown in Fig. 5.34 (after Igaya et al. 2010), a highway around a section of the Ariake Sea, Japan, was planned and construction commenced in 2010 for the section in Saga Prefecture. This road was designed as an access controlled freeway with an embankment height of 5.0–8.0 m for most sections of the road. A deposit of soft Ariake clay exists along the route of the highway, virtually from the ground

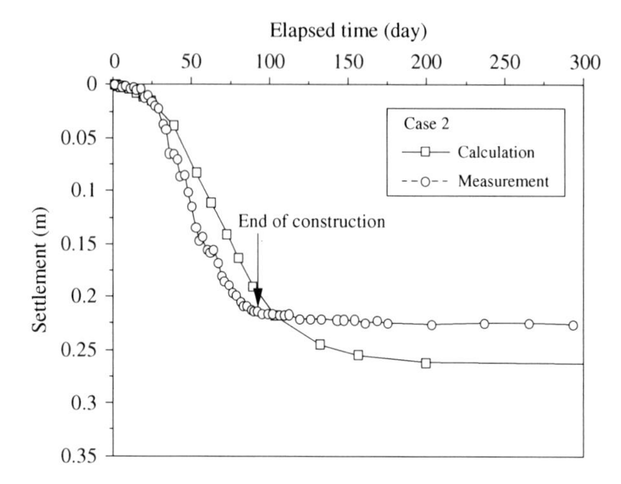

Fig. 5.32 Comparison of settlement curves for Case-2

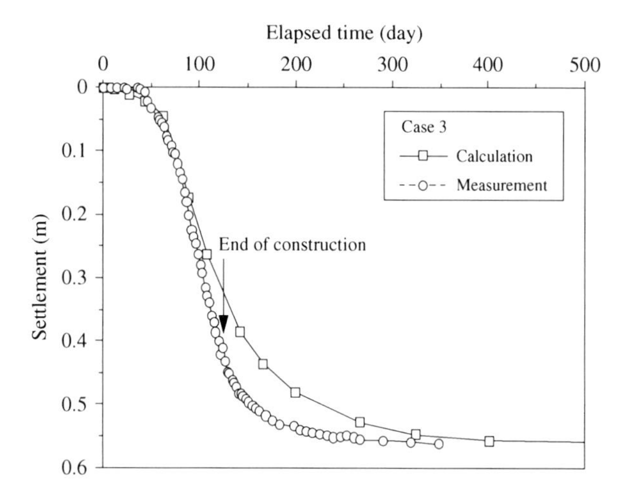

Fig. 5.33 Comparison of settlement curves for Case-3

surface to depths between 10 and 30 m. This deposit is characterized by high compressibility and low strength (Miura et al. 1998a). Below this soft clay layer there is an aquifer (a sand layer). The design for this project required that: (1) the residual settlement of the embankment and road surface should be less than or equal to 0.3m after the completion of construction; (2) the vertical and horizontal displacements along the property boundary of the road should be less than ±50 mm; and (3) the road construction should have no effect on the groundwater quality. In order to satisfy these requirements, one of the construction methods selected involved ground improvement resulting from the installation of floating soil-cement columns. In this case it was anticipated that the mechanical properties of most of the soft clay layer would be improved by the presence of these columns, while leaving an intact clay

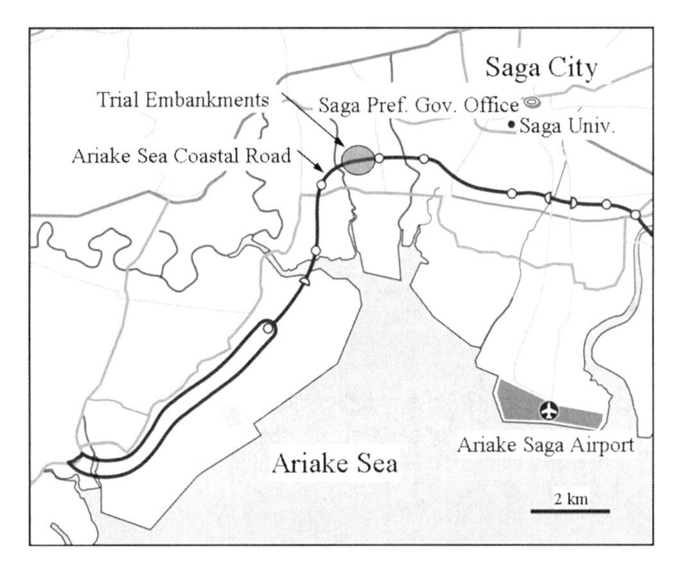

Fig. 5.34 Site of trial embankments (after Igaya et al. 2010)

sub-layer (without improvement) immediately above the aquifer. It was expected that this clay layer would serve as a barrier to prevent hazardous chemicals from the cement treatment entering and possibly contaminating the groundwater. To verify the performance of the proposed method three trial embankments were constructed and the case described here involves just one of them. The location of the trial embankments is indicated in Fig. 5.34.

The soil profile and some engineering properties of the various materials at the test site are summarized in Fig. 5.35 (data from ASCRDO 2009). At this site there exists a surface crust about 1.5 m thick underlain by a thick soft silty clay and/or sandy clay layer (the Ariake clay layer) which is about 8.0 m thick. Below it is a relatively stiff silty clay layer about 1.7 m thick underlain by alternating gravelly sand and silty to sandy clay layers. The groundwater level was about 0.5 m below the ground surface. The soft Ariake clay was initially in a lightly over-consolidated state and its natural water content was generally more than 100% and larger than the corresponding liquid limits. The compression index (C_c) of the layer was about 2.0. The ranges indicated for the coefficients of consolidation (c_v) and hydraulic conductivity (k) correspond to the stress range from the value of the initial vertical effective stress (σ'_{v0}) to σ'_{v0} plus the vertical stress increment due to the weight of the constructed embankment.

Before embankment construction, the ground was improved by the installation of floating soil-cement columns. The columns had a diameter of about 1.2 m and a length of 8.5 m. They were arranged in a square pattern with a spacing of 1.9 m, which resulted in an area improvement ratio (α) of approximately 30%. The columns were constructed by the slurry mixing method, and the amount of cement used was 150 kg/m^3 with a water/cement ratio of 150%. The design unconfined

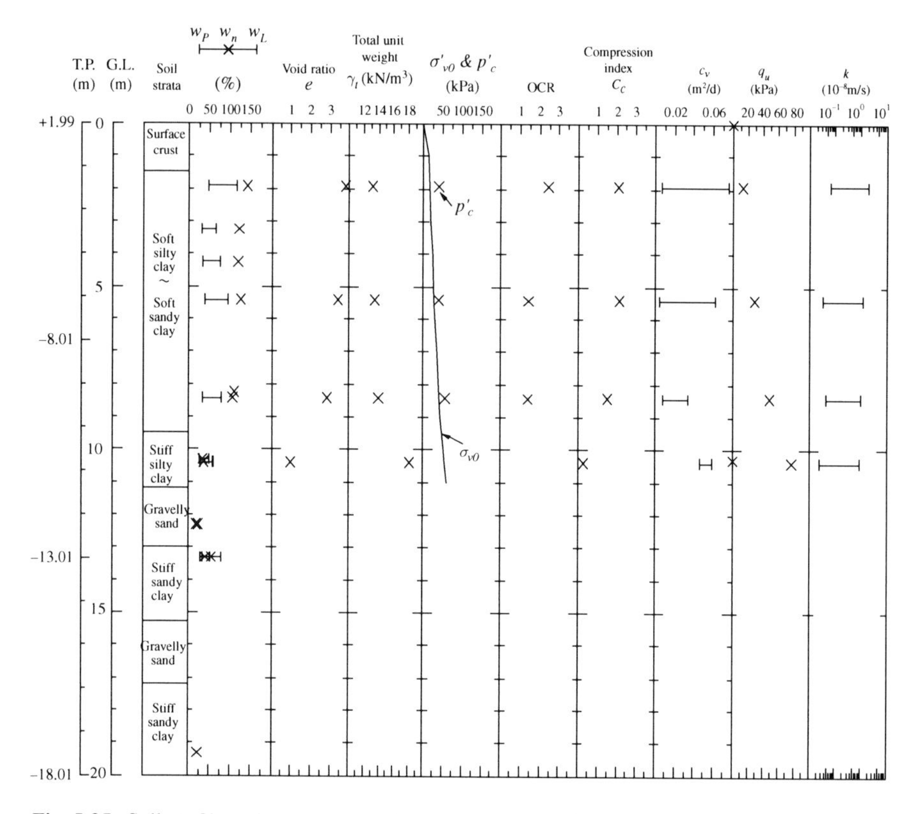

Fig. 5.35 Soil profile and some engineering properties of soils at the test site

compression strength (q_u) of the columns was 600 kPa. However, using core samples retrieved from the columns about 1 month after the field treatment, an average q_u value of more than 1,000 kPa was measured in laboratory tests. Considering the main compression layer extends from the ground surface to 11.2 m depth, the depth improvement ratio (β) was approximately 76%.

The trial embankment considered here had a fill thickness of 6.5 m. The base dimensions (length × width) of the embankment were 56.8 m × 35.4 m and the side slope of the embankment was 1:1.8 (V:H), which resulted in plan dimensions of the top of the embankment of 33.4 m × 12.0 m. Decomposed granite was used as fill material, and the average filling rate was about 0.06 m/day. The average total unit weight of the embankment fill as placed was about 18.2 kN/m³. A cross-section of the embankment and some of the key instrumentation points for measuring settlement, lateral displacement and pore water pressure are shown in Fig. 5.36. In an attempt to reduce the horizontal spreading force applied to the ground surface by the constructed embankment, two layers of diamond-shaped steel wire mesh were placed at the base of the embankment with a vertical spacing of about 0.5 m. The mesh had a grid spacing of 56 mm and the diameter of the wire was 2 mm. All the detailed information presented above is from Igaya et al. (2010).

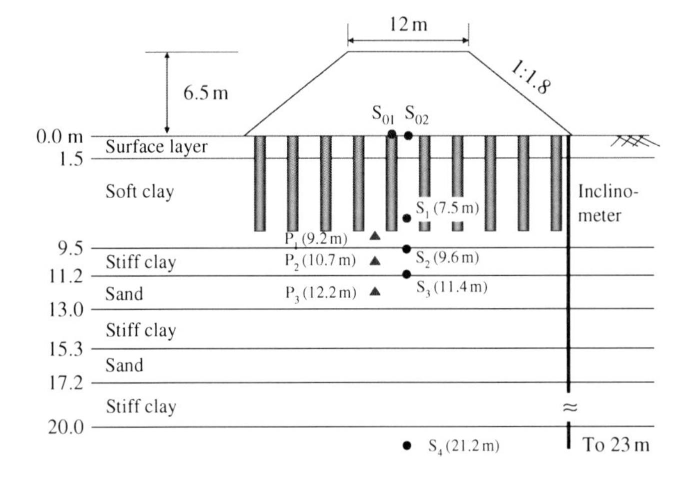

Fig. 5.36 Cross-section of the trial embankment and key instrumentation points

(b) Field-Measured Data

The measured settlement curves are shown in Fig. 5.37. For this case, there was no soil-cement slab constructed on the ground surface prior to construction of the embankment, so that there was significant differential settlement between the columns and the soil around them. Near the embankment centreline, the measured settlement of the original ground surface at the top of the column (S_{01}) was about 100 mm less than that measured on the original soil surface between the columns (S_{02}). The settlement gauge S_1 (as distinct from S_{01}) was located about 7.5 m below the original ground surface and between the columns. Assuming for the moment that the compression of the column is small enough to be ignored, then at 7.5 m

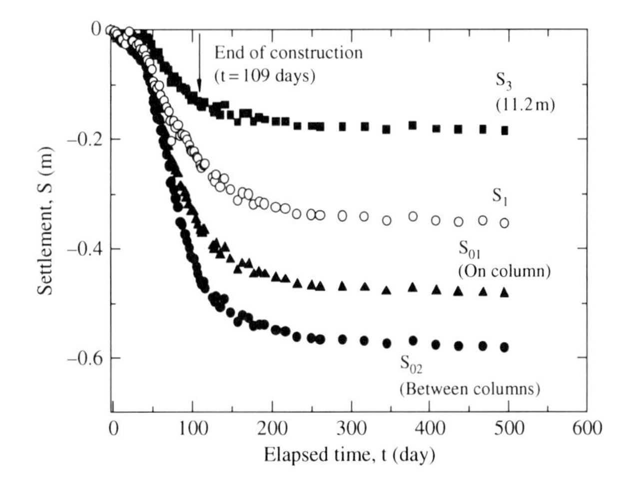

Fig. 5.37 Measured settlement curves

depth the differential settlement between the column and the surrounding soil will be about 131 mm (i.e., the difference between measurements at S_{01} and S_1). On the other hand, if it is assumed that the embankment load is carried entirely by the columns, then if the confining effect of the surrounding soil is ignored a compression of the columns of about 34 mm can be calculated (assuming a Young's modulus of 100,000 kPa for the soil-cement columns). In this case the differential settlement is estimated to be about 97 mm. Therefore, the relative penetration of the columns at 7.5 m depth is expected to be within the range from 97 to 131 mm. It is worth noting that this is one of the rare cases where field measurements have clearly demonstrated that the columns have penetrated into the surrounding soft soil. It is also notable that at this site there was considerable settlement of the stiff soil layers below 11.2 m depth (e.g., see the data for S_3). The differences between S_{01} and S_3 and between S_{02} and S_3 represent the compression of the soil layers from the ground surface to 11.2 m depth. In the next section these measured values of soil compression will be compared with values predicted using the theory presented previously in this chapter.

Although the lateral displacement and the excess pore water pressure are not directly related to the main topic of this chapter, in order to provide complete information on this case study the measured lateral displacement profiles and the measured variations of excess pore water pressure are provided in Figs. 5.38 and 5.39, respectively. It can be seen from Fig. 5.38 that the lateral displacements were generally relatively small, indicating that the method of ground improvement was quite successful in limiting this type of displacement. Clearly these measurements were smaller than the design limit of 50 mm permitted for the lateral displacements at the property boundary. All the data in Figs. 5.37, 5.38, and 5.39 are from Igaya et al. (2010), except the settlement data S_1, which are from ASCRDO (2009).

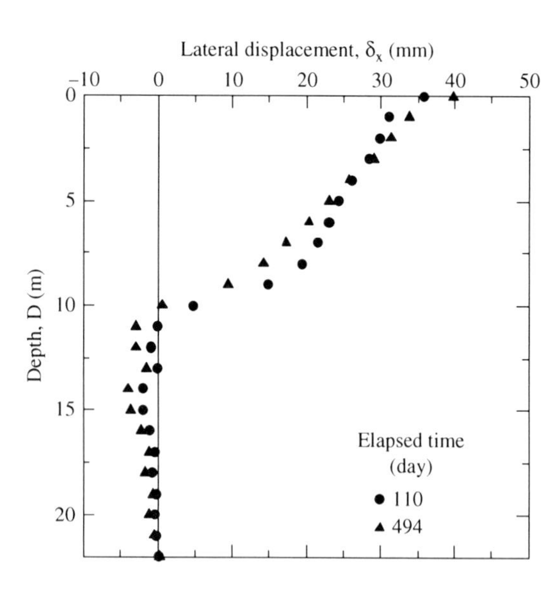

Fig. 5.38 Lateral displacements under the toe of the embankment

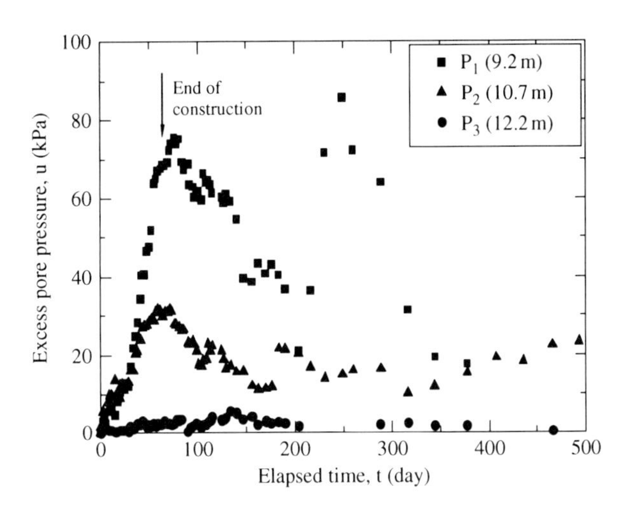

Fig. 5.39 Variations of excess pore water pressures

(c) Comparison of Predicted and Measured Settlement–Time Curves

The parameters adopted for calculating the predictions of final settlement are listed in Table 5.6. They are based on the results of the site investigation, as presented in Fig. 5.35. The values of κ were simply assumed as $\lambda/10$. For the columns, a Young's modulus, $E = 100,000$ kPa (i.e., approximately 100 times the unconfined compression strength, q_u, of about 1,000 kPa) and Poisson's ratio, $\nu = 0.2$ were assumed. Using the method presented in section 5.2.3 of this chapter, the compression of the layers from the ground surface to 11.2 m depth can be estimated as approximately 0.337 m. The measured compression was 0.298 m (i.e., the difference between S_{01} and S_3) and 0.396 m (the difference between S_{02} and S_3) from the top of the column and from the original surface between the columns, respectively. Without a slab on the ground surface, the settlement of the column should normally be less than that for the case with a slab, while the settlement of the ground surface between the columns may be larger than that for the case with a slab. The method used for predicting these settlements was developed assuming the presence of a surface slab spanning the columns, and therefore the outcome described here,

Table 5.6 Parameters adopted for settlement predictions

Depth(m)	Soil strata	λ	e_0	γ_t (kN/m^3)	OCR
0.0–1.5	Surface soil	0.25	1.50	16.0	5
1.5–4.0	Soft clay	0.87	3.10	13.4	2.4
4.0–6.0		0.87	2.81	14.0	1.7
6.0–8.0		0.58	2.58	14.1	1.4
8.0–9.5		0.43	2.49	14.3	1.2
9.5–11.2	Stiff clay	0.15	1.10	18.0	1.2

Table 5.7 Parameters adopted for predicting the rate of consolidation

Parameter			Value
H_1 (m)			8.22
$H_2 = H_{20} - s_f/2$(m)[a]			2.82
σ_0 (kPa)			118.3
σ_1 (kPa)			105.1
σ_2 (kPa)			98.4
Calculated final settlement s_f (m)			0.32
Layer H_1	$c_{v1}(\text{m}^2/\text{day})$	NC	0.819
		OC	5.842
	$k_{v1}(\times 10^{-4}$ m/day)	NC	2.37
		OC	15.94
Layer H_2	$c_{v2}(\text{m}^2/\text{day})$	NC	0.028
		OC	0.052
	$k_{v2}(\times 10^{-4}$ m/day)	NC	0.97
		OC	3.67

[a]Note: $H_1 + H_2 = H - s_f/2$

where the predicted value lies between the two measured values, is considered to be reasonable.

The parameter values adopted for calculating the degree of consolidation using the double layer consolidation theory are listed in Table 5.7. These values were determined based on the laboratory consolidation test results on undisturbed soil samples. In calculating the values of k_{v1} and c_{v1}, the ratio of the hydraulic conductivities in the horizontal and vertical directions (k_h/k_v) of 2 was assumed (Chai and Miura 1999), no smear effect was considered and the hydraulic conductivity of the columns was assumed to be the same as the surrounding soil. Within the range of the vertical effective stress considered, two sets of the values of c_{v1}, c_{v2}, k_{v1} and k_{v2}, were evaluated corresponding to a normally consolidated state (NC), and an over-consolidated state (OC) in the Ariake clay. Accordingly, two calculations were carried out using these two sets of parameters. A comparison of the measured settlement curves with those calculated using the theory presented in sections 5.3.2 and 5.4.1 is shown in Fig. 5.40. It can be seen that for much of the elapsed time the two calculated curves lie between the two measured curves indicating that both predictions are probably acceptable. However, the predicted final settlement is closer to the measured value of (S_{01} – S_3), and from this point of view it could be argued that adoption of the NC parameters has resulted in a predicted settlement performance that is closer to the measured settlement rate. This outcome is considered to be quite reasonable because the ground was initially in a lightly over-consolidated state but for most of the loading increment arising from embankment construction it should have been in the normally consolidated range of behaviour.

Again, the results of this case history have indicated that the proposed method for predicting the behaviour of soft clay soils reinforced by soil-cement columns can be adopted in practice as a useful design tool.

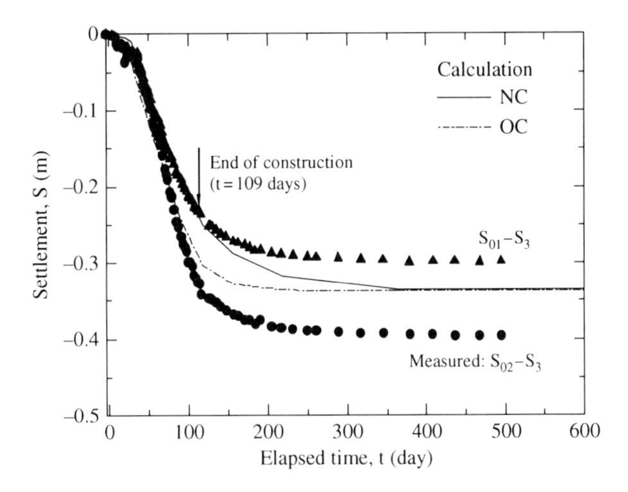

Fig. 5.40 Comparison of measured and predicted settlement-time curves

5.5 Deformations Induced by Column Installation

5.5.1 Introduction

During the installation of soil-cement columns either cement slurry (wet mixing) or cement powder (dry mixing) is injected into the ground under pressure and the application of this pressure may cause lateral deformation of the surrounding subsoil. In urban environments this deformation may be very significant because it can affect nearby structures. Therefore, predicting and controlling the lateral displacements induced by soil-cement column installation are important design considerations.

Installing soil-cement columns is a complicated process, and even with numerical analysis there are significant difficulties in simulating the entire process (Shen et al. 1999). The main mechanisms causing lateral displacement during column installation are expansion and possible hydraulic fracturing in the ground due to the injection pressure and the addition of extra material into the ground (Miura et al. 1998b). It may be assumed that the expansion caused by soil-cement column installation is analogous to a cavity expansion. Indeed, it would appear that the only method of analysis available to date for this problem is the one proposed by Chai et al. (2005, 2007, 2009) which is based on the theory of cavity expansion. This method will be explained in the following sections.

5.5.2 Lateral Deformation

5.5.2.1 Cavity Expansion Solutions

The theory of cavity expansion in an infinite medium proposed by Vesic (1972), has been used to determine the equations for calculating the lateral displacement induced by a cavity expansion, as summarized in Table 5.8. Equations have been

Table 5.8 Equations for calculating lateral displacements caused by cylindrical and spherical cavity expansion

Eq. No	Item	Cylindrical (a)	Spherical (b)	Reference
Eq. (5.39)	Radius of plastic zone, R_p	$\frac{R_p}{R_u} = \sqrt{I_r}$ $I_r = \frac{E}{2(1+\nu)S_u} = \frac{G}{S_u}$	$\frac{R_p}{R_u} = \sqrt[3]{I_r}$ $I_r = \frac{E}{2(1+\nu)S_u} = \frac{G}{S_u}$	Vesic (1972)
Eq. (5.40)	Limiting cavity pressure, p_u	$p_u = S_u \cdot F_c + p_0 \cdot F_q$ $F_q = 1.0$ $F_c = \ln(I_r) + 1$	$p_u = S_u \cdot F_c + p_0 \cdot F_q$ $F_q = 1.0$ $F_c = \frac{4}{3}\left[\ln(I_r) + 1\right]$	
Eq. (5.41)	Pressure at $r = R_p$, σ_p	$\sigma_p = p_u - 2S_u \ln\left(\frac{R_p}{R_u}\right)$	$\sigma_p = p_u - 4S_u \ln\left(\frac{R_p}{R_u}\right)$	
Eq. (5.42)	Displacement at $r = R_p$, δ_p	$\delta_p = \frac{1+\nu}{E} R_p(\sigma_p - p_0)$ $= \frac{1+\nu}{E} R_p S_u$	$\delta_p = \frac{1+\nu}{2E} R_p(\sigma_p - p_0)$ $= \frac{2(1+\nu)}{3E} R_p S_u$	Chai et al. (2005, 2007) for cylindrical case. Equations for spherical case have been derived.
Eq. (5.43)	Displacement in plastic zone, δ $(R_u \le r \le R_p)$	$\delta \approx \frac{2R_p + \delta_p}{2r + \delta_p R_p / r}\delta_p$	$\delta = \frac{1}{R_p^2/R_u^2 - 1}\left[\frac{R_p^2}{R_u^2}\delta_p - \delta_u + (\delta_u - \delta_p)\frac{R_p^2}{r^2}\right]$	
Eq. (5.44)	Displacement in elastic zone, $\delta(r \ge R_p)$	$\delta = \frac{R_p}{r}\delta_p = \frac{R_u^2}{2r}$	$\delta = \frac{R_p^2}{r^2}\delta_p = \frac{R_u^3}{3r^2}$	

Note: R_u = the radius of a cavity; δ_u = displacement at R_u; r = radius from the centre of the cavity; E = Young's modulus of the soil; G = shear modulus of the soil; S_u = undrained shear strength of the soil; ν = Poisson's ratio of the soil; p_0 = initial mean stress in the ground; F_c, F_q = cavity expansion factors; and I_r = rigidity index of the soil

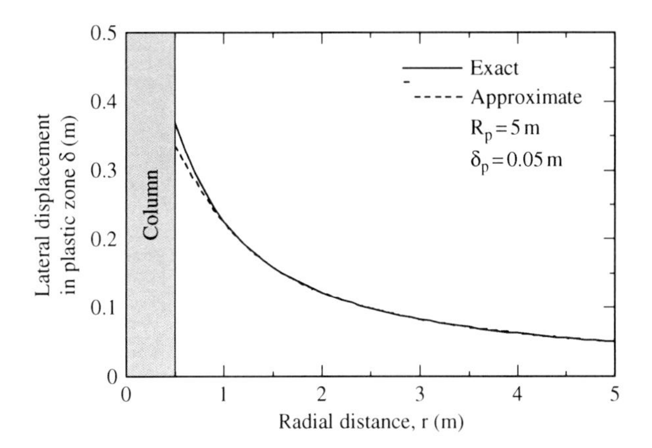

Fig. 5.41 Comparison of lateral displacements occurring in the plastic zone of soil surrounding an expanding cylindrical cavity (after Chai et al. 2009)

presented for both cases of cylindrical and spherical cavities and these are numbered as Eqs. (5.39a, b) to (5.44a, b). They correspond to the specific case of an undrained soil for which it is assumed the friction angle, $\phi = 0$, and for which the volumetric strain in the plastic zone, $\Delta = 0$. It should be noted that Eq. (5.43a) is not a rigorous expression. A precise expression can be obtained by solving a quadratic equation, but in order to obtain an explicit equation for the lateral displacement caused by installing a row of columns, an approximation form (Eq. (5.43a)) has been adopted. The magnitude of the errors introduced by making this approximation depends on the specific values of the parameters r, R_p, and δ_p. The smaller the value of the radial position r and the larger the values of the plastic radius R_p and the displacement at this radius δ_p, the larger the error. As an example, for the field cases reported by Chai et al. (2005), R_p was in a range from about 1.5 m to 5 m, and δ_p from about 0.015 m to 0.05 m (the larger values are for the wet jet mixing method - WJM). Using $R_p = 5.0$ m, $\delta_p = 0.05$ m and $r \geq 0.5$ m, Fig. 5.41 shows a comparison of the calculated lateral displacements in the plastic zone. It can be seen that the maximum error is about 9%. Further, both Eqs. (5.43a, b) have been derived using initial values of the parameters R_u, R_p and r, so that for cases involving large displacements, there will be some additional error involved. It is also noted that Eq. (5.43b), which describes the solution for the spherical case, is in fact a rigorous solution.

5.5.2.2 Infinite Columns

The additional Eqs. (5.45), (5.46), and (5.47), which are based on application of the theory of cylindrical cavity expansion (Chai et al. 2007), have been obtained to calculate the lateral displacement at point A in the x-direction caused by installing a single row of infinitely long soil-cement columns in clayey soil (Fig. 5.42). These equations are listed in Table 5.9. In deriving these equations the method of superposition has been adopted. Strictly speaking, superposition is applicable for linear materials only. However, the soil model assumed in the analysis is non-linear, i.e.,

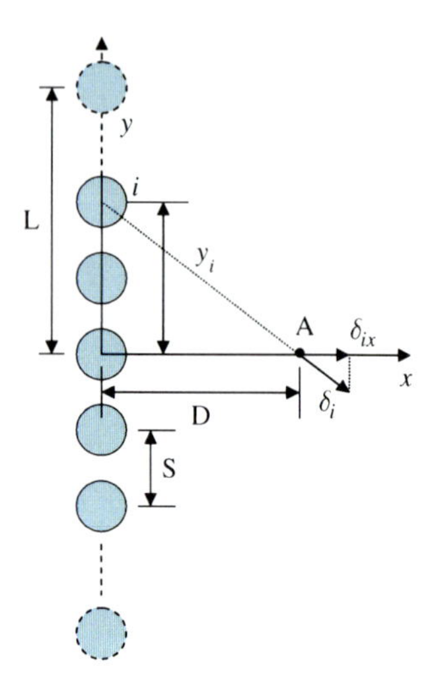

Fig. 5.42 Plan section illustration of the lateral displacement in the x direction caused by installation a single row of soil-cement columns (after Chai et al. 2009)

Table 5.9 Equations for calculating lateral displacement caused by installation of a single row of columns of infinite length

Eq. No.	Condition	Lateral displacement
Eq. (5.45)	For $D < R_p$ and $D^2 + L^2 > R_p^2$	$\delta_{xA} = \frac{2D(2R_p+\delta_p)\delta_p}{S\sqrt{4D^2+2R_p\delta_p}} \cdot \tan^{-1}\sqrt{\frac{2\left(R_p^2-D^2\right)}{2D^2+R_p\delta_p}} + \frac{2R_p}{S}\delta_p\left(\tan^{-1}\frac{L}{D} - \tan^{-1}\sqrt{\frac{R_p^2}{D^2} - 1}\right)$
Eq. (5.46)	$D < R_p$ and $D^2 + L^2 \leq R_p^2$	$\delta_{xA} = \frac{2D(2R_p+\delta_p)\delta_p}{S\sqrt{4D^2+2R_p\delta_p}}\tan^{-1}\sqrt{\frac{2L^2}{2D^2+R_p\delta_p}}$
Eq. (5.47)	For $D \geq R_p$	$\delta_{xA} = \frac{2R_p}{S}\delta_p\left(\tan^{-1}\frac{L}{D}\right)$

Note: D = offset distance from the centre of a row of columns to the point of interest; S = spacing between two adjacent columns in a row; L = half length of a row. Equations in this table assume that the point of interest is on the perpendicular bisector of a row of columns. If the point of interest is not on the bisector of the row, two calculations are needed, i.e., using two different values of L and then averaging the results

linear elastic-perfectly plastic, and so using superposition in the plastic zone will also induce some error, possibly a slight over-prediction of the lateral displacements. It is also noted that the method of superposition has been used to consider the effect of an installation composed of multiple rows of soil-cement columns.

It can be seen from the governing equations that in addition to the geometric conditions, the parameters controlling the lateral displacement are R_p and δ_p. Both R_p and δ_p are functions of I_r and R_u (Eqs. (5.39) and (5.42) in Table 5.8). The rigidity index, I_r, is a basic soil parameter, being a measure of the soil's stiffness to strength ratio. R_u, the radius of a cavity, is also a crucial parameter controlling the amount of lateral displacement that will occur. The installation of a soil-cement column is neither a displacement controlled nor a pressure controlled process. It is a process of injecting a pre-determined amount of cement slurry/powder into the ground by pre-determined pressures (which may not be precisely controlled). For example, injecting the same amount of admixture into the ground by the cement deep mixing (CDM) method and by the wet jet mixing (WJM) method, which normally uses a higher injection pressure, will in general cause different lateral displacements. Therefore, several factors will affect the value of R_u. The first of these is the injection pressure (p) of the admixture and the second is the effective volume of admixture injected into the subsoil per unit length of a column (Δvol), which in some cases may include the amount of water injected into the ground prior to injecting the admixture. The effective volume means the total volume injected into the ground, which is obtained after subtracting the volume of returned spoil (slurry and clay mixture). The third and fourth factors affecting the value of R_u are the undrained shear strength (S_u) and the modulus (E) of the subsoil. At present, there are insufficient data available to derive a meaningful theoretical expression for R_u in terms of all these factors. Instead, empirical equations have been proposed to evaluate R_u (Chai et al. 2005).

The effect of the injection pressure p is incorporated indirectly in the back-fitted radius of the cavity, R_{u0}, which also depends on the amount of injected admixture and the soil conditions (specifically the modulus E_0). An empirical equation has been proposed to consider the effects of S_u and E on R_u. Since for most soft clayey deposits, E can be estimated as a linear function of S_u, only E is required as a variable in this empirical expression, and consequently the following simple power function has been selected:

$$R_u = R_{u0}\left(\frac{E_0}{E}\right)^{1/3} \tag{5.48}$$

where R_{u0} = the radius of the cavity corresponding to a modulus of E_0. There are three main methods used for soil-cement column installation, namely, the cement deep mixing (CDM) method, the dry jet mixing (DJM) method, and the wet jet mixing (WJM) method. The injection pressures used in these methods are usually different. By back-fitting the field measured data from three sites in Saga, Japan, Chai et al. (2005) proposed values of R_{u0} for these three methods, as listed in Table 5.10, together with the corresponding values of p, Δvol and E_0.

For a given installation method, the effect of varying Δvol on R_{u0} can be estimated under equivalent volume conditions. Suppose the value of R_{u0} corresponding to Δvol_1 is R_{u01}, then for Δvol_2, the corresponding value R_{u02} can be approximated as:

$$R_{u02} = R_{u01} - \frac{\Delta vol_1 - \Delta vol_2}{2\pi R_{u01}} \tag{5.49}$$

Table 5.10 Back estimated empirical parameters

Mixing method	Injection pressure p (MPa)	Injected volume Δvol (m^3/m)	Radius R_{u0} (m) ($E_0 = 2250$ kPa)
CDM	0.1–0.2	0.146	0.21
DJM	0.5–0.7	0.036	0.46
WJM	~20	0.146	0.58

5.5.2.3 Lateral Deformation Caused by Columns of Limited Length

The equations in Table 5.9 are for calculating the lateral displacement caused by installing an infinitely long column in an infinite soil medium. However, actual soil-cement columns have limited length, and the equations in Table 5.9 generally over-predict the lateral displacement near the ends of the columns. A correction function is proposed to take into account the effect of limited length of the columns (Chai et al. 2009). This correction function has been developed by comparing the solution based on the theory of a spherical cavity expansion with that of a cylindrical cavity expansion for a single column installation. The cavity expansion theory used to derive the equations in Table 5.9 is not a half space solution, which implies that the method does not satisfy the stress conditions at the free surface and it ignores the vertical displacement of the free surface during soil-cement column installation. Considering the simplicity of the resulting equations (Eqs. (5.45) to (5.47) in Table 5.9) and the fact that one of the key parameters, R_u, is determined empirically, this violation of the free surface boundary condition is usually accepted as a reasonable approximation.

However, when simulating the installation of a column of limited length as a series of spherical cavity expansions, the effect of the free surface has to be considered. The theoretical solution for the expansion of a single spherical cavity in a half-space has been presented by Keer et al. (1998). For the case of a series of spherical cavity expansions an approximate solution can be obtained approximately by adopting the method of superposition. In practice it is difficult to obtain an explicit closed form solution for this case because of difficulties with the required integration. For simplicity, it is proposed to take into account the effect of the free surface in an approximate manner by using a virtual image approach, as shown in Fig. 5.43. Using the theory of spherical cavity expansion in an infinite space implies that there is an infinitely thick soil layer existing above the ground surface, which will restrain the lateral displacement induced by the spherical cavity expansion. In reality, no such restraint exists. Therefore, the main propose of using the virtual imagine approach is to eliminate this 'virtual restriction' on lateral displacement. The lateral displacements caused by the expansion of both the real and the imaginary spherical cavities in an infinite space are superposed. In this case, zero vertical displacement at the free surface is imposed.

As shown in Fig. 5.43, it is assumed that a soil-cement column is formed by a series of spherical cavity expansions with a spacing of S_s along a straight line. The lateral displacement at point A in the x direction (δ_{ix}) caused by the ith and the imth

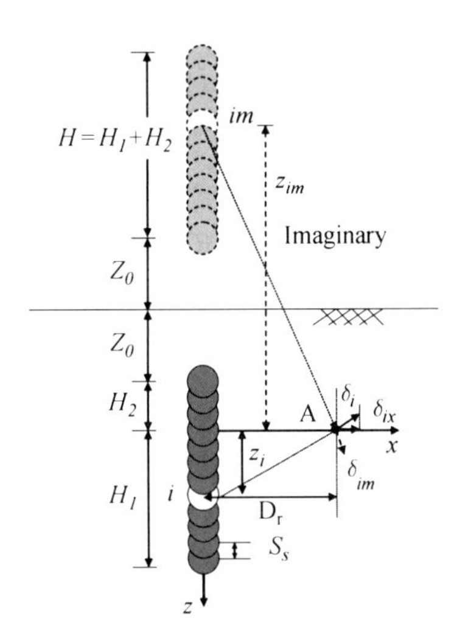

Fig. 5.43 Vertical section illustration of the lateral displacement in the x direction caused by installation of a single soil-cement column (after Chai et al. 2009)

spherical cavity expansions can be calculated as follows:

$$\delta_{ix} = \delta_i \frac{D_r}{\sqrt{D_r^2 + z_i{}^2}} + \delta_{im} \frac{D_r}{\sqrt{D_r^2 + z_{im}^2}} \tag{5.50}$$

where D_r = the horizontal offset (radial) distance from the centre of the column to point A, z_i and z_{im} = respectively the vertical distances between point A and the ith spherical cavity and the corresponding imth imaginary cavity, δ_i and δ_{im} = the displacement components at point A in the corresponding radial directions. The overall lateral displacement at point A in the x direction (δ_{xA}) caused by installation of a single column can then be calculated from the appropriate integration, as follows:

$$\delta_{xA} = \frac{D_r}{S_s} \int_{-H_2}^{H_1} \frac{\delta}{\sqrt{D_r^2 + z^2}} dz + \frac{D_r}{S_s} \int_{2Z_0+H_2}^{2Z_0+H_2+H} \frac{\delta}{\sqrt{D_r^2 + z^2}} dz \tag{5.51}$$

where S_s = the centre-to-centre spacing of two adjacent spherical cavities in a column, H = the length of the column, H_1 and H_2 = the vertical distances from point A to the bottom and the top of the column, and Z_0 = the vertical distance from the ground surface to the top of the column. When the radial distance $R = \sqrt{(D_r{}^2 + z^2)} < R_p$, the parameter δ in Eq. (5.51) can be calculated from the elastic solution (Eq. (5.44b) in Table 5.8), and the result of the integration is as follows:

$$\delta_{xA} = \frac{R_p^2\delta_p}{S_s \cdot D_r}\left(\frac{H_1}{\sqrt{D_r^2+H_1^2}} + \frac{H_2}{\sqrt{D_r^2+H_2^2}} + \frac{2Z_0+H_2+H}{\sqrt{D_r^2+(2Z_0+H_2+H)^2}} - \frac{2Z_0+H_2}{\sqrt{D_r^2+(2Z_0+H_2)^2}}\right) \quad (5.52)$$

If the length of a column (H) approaches ∞, Eq. (5.52) reduces to:

$$\delta_{xA} = \frac{2R_p^2\delta_p}{S_s \cdot D_r} \quad (5.53)$$

If it is assumed that the radii of the plastic zones (R_p) created around a cylindrical and a spherical cavity must be the same, and if the lateral displacement expressed in Eq. (5.53) is equated to the displacement expressed by Eq. 5.44a in Table 5.8, then an expression for the equivalent spacing, S_s, is derived as follows:

$$S_s = \frac{4}{3}R_p \quad (5.54)$$

Note that in this derivation, $D_r = r$, and $(\delta_p)_{\text{spherical}} = 2(\delta_p)_{\text{cylindrical}}/3$ (refer to Eqs. (5.42a and b) in Table 5.8) have been adopted, where the subscripts '*spherical*' and '*cylindrical*' refer to the values for a spherical and a cylindrical cavity expansion, respectively. The process for obtaining S_s is illustrated in Fig. 5.44.

As indicated by Eq. (5.39) in Table 5.8, the ratio of the radius of the plastic zone and the cavity radius is different for cylindrical and spherical cavities. Assuming the

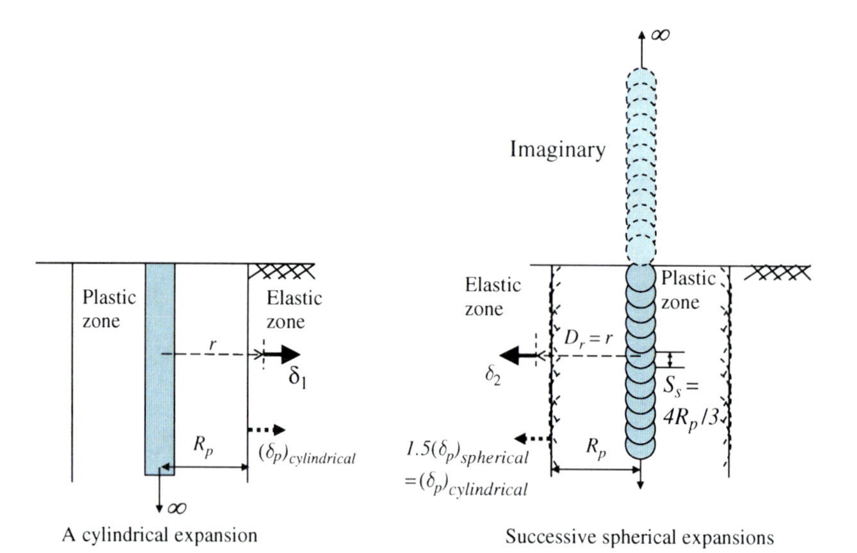

Fig. 5.44 Illustration of the conditions for obtaining the expression for S_s (after Chai et al. 2009)

same radius of the plastic zone for each case implies that the radius of the spherical cavity must be larger than the radius of the cavity in the cylindrical case. In this case the ratio of the radii of the spherical and cylindrical cavities is designated as R_{sc}, which can be expressed as a function of the rigidity index (I_r) of soil:

$$R_{sc} = I_r^{1/6} \tag{5.55}$$

As shown in Fig. 5.45, R_{sc} increases with I_r, and for I_r in the range from 10 to 200, the value of R_{sc} is about 1.5–2.4.

If the expression for S_s given in Eq. (5.54) is substituted into Eq. (5.52) and the resulting expression for the lateral displacement is compared with Eq. (5.44a) in Table 5.8, the ratio of lateral displacement (*RLD*) caused by a column with limited length to that predicted by an infinitely long cylindrical cavity expansion can be evaluated as follows:

$$RLD = \frac{1}{2}\left(\begin{array}{l} \frac{H_1}{\sqrt{D_r^2+H_1^2}} + \frac{H_2}{\sqrt{D_r^2+H_2^2}} + \frac{2Z_0+H_2+H}{\sqrt{D_r^2+(2Z_0+H_2+H)^2}} \\ -\frac{2Z_0+H_2}{\sqrt{D_r^2+(2Z_0+H_2)^2}} \end{array}\right) \tag{5.56}$$

For the case of a single column, Eq. (5.56) indicates the effects of the horizontal offset radial distance (D_r), the length of the column (H) and the location of the point interested (Z_0, H_1, H_2) on *RLD*. Figure 5.46 shows the variation of *RLD* with depth for two assumed cases at a location 5 m horizontally from the centre of the column. Case-1 is a situation where the column is installed at depths between 5 and 15 m, and Case-2 is a column installed to 15 m depth from the ground surface. For Case-1, *RLD* has its largest value around the middle of the column and reduces towards the two ends, and for Case-2, *RLD* reduces with depth. Figure 5.47 shows the variation of *RLD* with distance from the column for a point at the middle of the 15 m length of the column assumed in Case-2. From Fig. 5.47, it can be seen that *RLD* reduces significantly with the offset (radial) distance from the column. It is noted that Eq. (5.56) was obtained assuming that the point of interest lies within

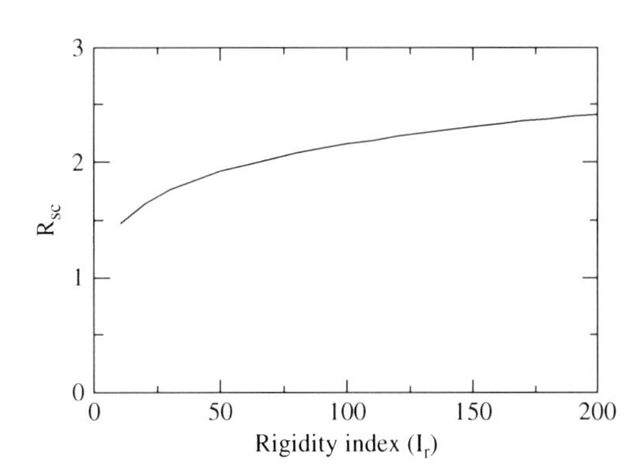

Fig. 5.45 Variation of R_{sc} with I_r (after Chai et al. 2009)

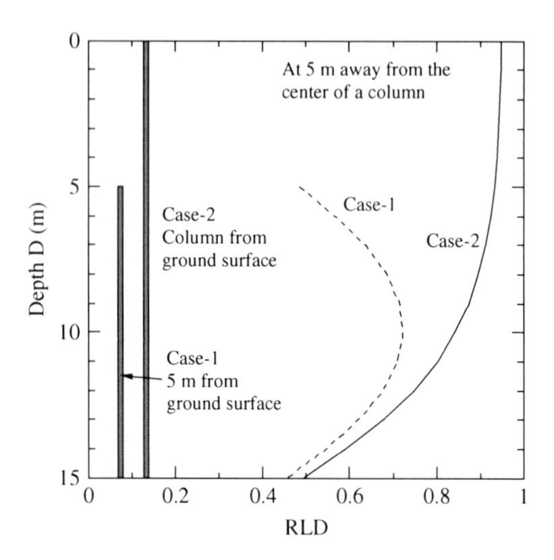

Fig. 5.46 Example variation of RLD with depth for a single column case (after Chai et al. 2009)

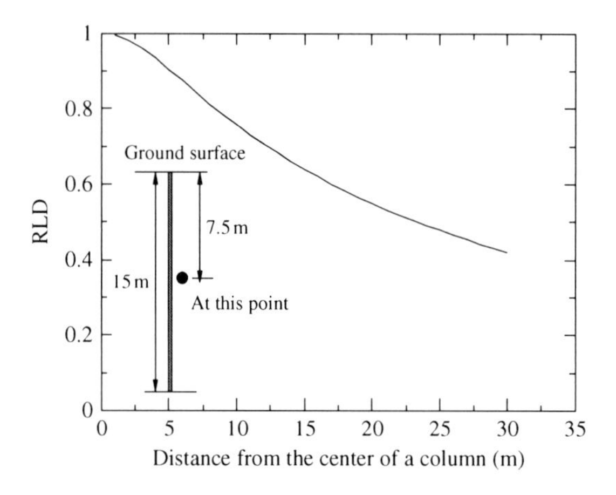

Fig. 5.47 Example variation of RLD with radial distance for a single column case (after Chai et al. 2009)

the elastic zone of the soil around the column. Considering that Eq. (5.56) simply provides a geometrical correction factor, for analytical convenience it is assumed that this equation can also be used to sufficient accuracy for points located within the plastic zone.

In engineering applications, soil-cement columns are normally installed in a single row or in multiple rows. In the case of a row of columns, the offset (radial) distance from each column to the point of interest is different and Eq. (5.56) cannot be directly applied to the equations shown in Table 5.9. Using the principle of superposition, the lateral displacement induced by installing a row of columns with limited length can be expressed as follows:

$$\delta_{xA} = \frac{2D}{S} \int_0^L \frac{\delta}{\sqrt{D^2 + y^2}} RLD(\sqrt{D^2 + y^2}) dy \qquad (5.57)$$

where $RLD\sqrt{(D^2+y^2)}$ represents the function defined in Eq. (5.56) with the variable of $D_r = \sqrt{(D^2+y^2)}$. The meanings of the other parameters are indicated in Fig. 5.42. Integration of Eq. (5.57) becomes complicated and explicit forms have not been obtained. However, the integration of Eq. (5.57) can be easily done numerically. The values of *RLD* (varying with depth) for a single row of columns can be evaluated by using the numerical result of Eq. (5.57) and the lateral displacement calculated by the equations in Table 5.9 for a single row of infinite cylindrical cavity expansions. L is defined as the length from the point of interest to one end of a row. For cases where the point of interest is located on the perpendicular bisector of a row, L is half the total length of the row. For cases where the point of interest does not lie on the bisector of the row, two separate calculations with different values of L are needed, and then the results must be averaged. Numerical results indicate that the representative value of D_r in the elastic zone from the integration of Eq. (5.57) depends mainly on the values of D and L, but also on the length of the columns (H), and the location of the target point (H_1, H_2, and Z_0). The results given in Fig. 5.48 were calculated assuming $H = 10$ m, $H_1 = H_2 = 5$ m and $Z_0 = 0$. Assuming H varies from 5 to 30 m, and Z_0 from 0 to 10 m, and with the D_r values given in Fig. 5.48, limited numerical investigation indicates that the maximum error in estimating D_r is about 15% for $D = 1.0$ m, and the corresponding error in *RLD* is about 3%. This error reduces rapidly with increasing values of D.

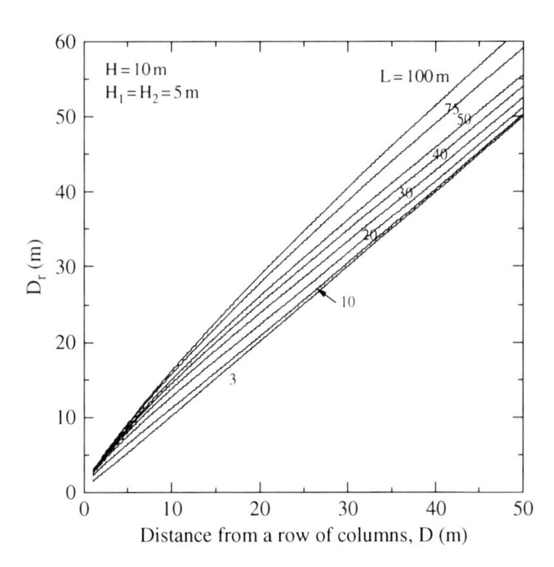

Fig. 5.48 Radial distance D_r as a function of D and L (after Chai et al. 2009)

Taken together, Eq. (5.57) and the representative offset (radial) distance D_r obtained from Fig. 5.48 form the correction function. Multiplying the correction function by the lateral displacement values obtained from Eqs. (5.45), (5.46), and (5.47) (Table 5.9) forms the method for predicting the lateral displacement caused by installation of soil-cement columns with limited length in soft deposits.

5.5.3 Case History in Clayey Ground

Shen et al. (1999) reported a case where three field tests involving installation of soil-cement columns using the CDM, DJM and WJM methods were conducted along the banks of the Rokkaku River in Saga, Japan. All these cases have been analyzed previously by Chai et al. (2005). At this site, soft Ariake clay is deposited with a thickness of about 15 m. The natural water contents of the soft Ariake clay are 80–120% and these values are generally slightly higher than the corresponding liquid limits. The undrained shear strength (S_u), derived from unconfined compression test results, is within a range of 15–30 kPa. For each installation method, three rows of columns with 5 columns per row were constructed under the toe of an embankment. At the site where the DJM method was used the columns were installed from the ground surface to a depth of 13 m with a diameter of about 1.0 m. The layout of the columns and the location of inclinometer casing are illustrated in Fig. 5.49. For this case the lateral displacements caused by installation of each row of columns were measured, and these are compared here to the values predicted by the method suggested above. For this site, the ratio E/S_u of about 150 is adopted, and the back fitted value of R_{u0} is 0.46 m, corresponding to a value of E_0 of 2250 kPa and (dry) mass injected Δvol of 0.036 m^3/m (Table 5.10). Other test details can be found in Chai et al. (2005). $E/S_u = 150$ was selected based on the reported values of the ratio E_{50}/S_u (E_{50} is the secant modulus at the stress level equivalent to 50% of peak strength) of 100–200 for soft Ariake clay (Fujikawa et al. 1996).

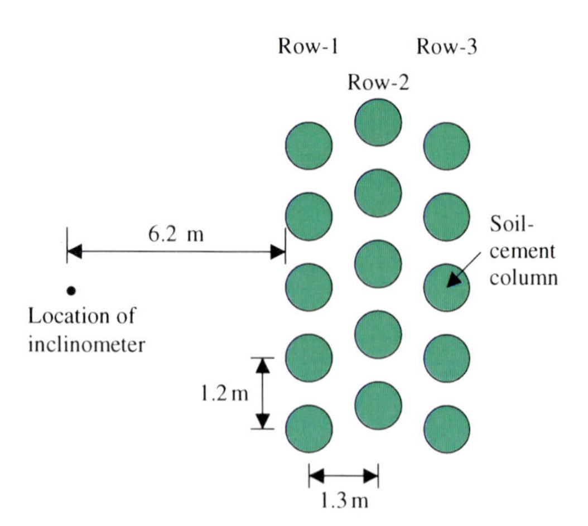

Fig. 5.49 Plan layout of columns and the location of inclinometer of Saga case (after Chai et al. 2009)

The predicted and the measured lateral displacements at the ground surface are compared in Fig. 5.50. It can be seen that the calculated values almost match the measured data. Comparison of the variation of the lateral displacement with depth at 6.2 m from the edge of the columns is given in Fig. 5.51. Although the calculations over-predicted the lateral displacements within the top 5 m, generally the method yields fair predictions. Note that the measured surface lateral displacements given in Figs. 5.50 and 5.51 at a distance of 6.2 m from the edge of the improved zone are slightly different. The data given in Fig. 5.50 were measured from a surface mark and are larger than the corresponding inclinometer readings (Fig. 5.51).

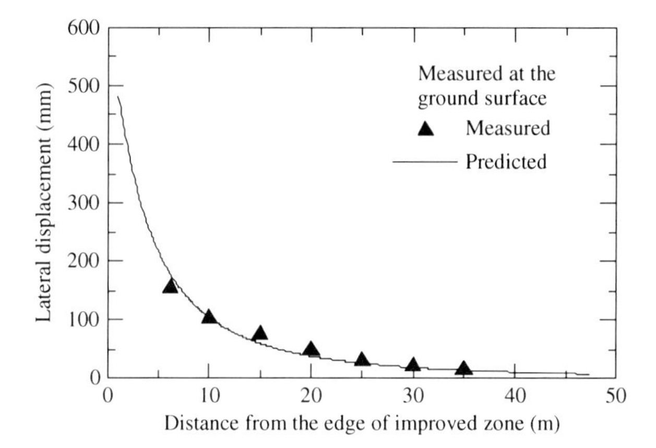

Fig. 5.50 Comparison of lateral displacements at the ground surface of Saga case (after Chai et al. 2009)

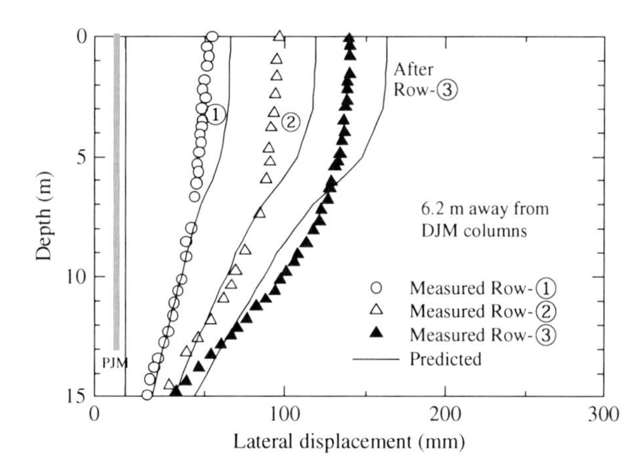

Fig. 5.51 Lateral displacement increments caused by each row of DJM-installed columns of Saga case (after Chai et al. 2009)

5.5.4 Case History in Sandy Ground

5.5.4.1 Modifications

The method described previously for calculating the effects of column installation was specifically for clayey soil deposits for which the assumptions of undrained soil behavior and a friction angle $\phi = 0$ were made. However, the method can be extended to a cohesive-frictional deposit, such as sandy ground. The main modification required relates to the equations used for calculating the lateral displacement (δ_p) at the location of the boundary between the plastic and elastic soil zones. Assuming a Mohr-Coulomb failure criterion and an effective stress approach, it is not difficult to show that the corresponding equations are as follows (re-arranged from Vesic 1972).

For a cylindrical cavity expansion:

$$\delta_p = \frac{1+\nu}{E} R_p \left(c' \cdot \cot\phi' + p_0\right) \sin\phi' \tag{5.58a}$$

and for a spherical cavity:

$$\delta_p = \frac{1+\nu}{2E} R_p \left(c' \cdot \cot\phi' + p_0\right) \frac{4\sin\phi'}{3-\sin\phi'} \tag{5.58b}$$

The effective stress strength parameters of the soil are the cohesion, c', and friction angle, ϕ'. With these modifications, the expression for the spherical cavity spacing, S_s (Eq. 5.54), becomes:

$$S_s = R_p \frac{4\sin\phi'}{\left(3-\sin\phi'\right)\sin\phi'} \tag{5.54a}$$

For values of ϕ' of 1°–45°, Eq. (5.54a) gives $S_s = (1.34–1.74)R_p$. However, the resulting expression for *RLD* (Eq. (5.56)) remains the same.

In the analysis presented previously for clay soils it was assumed that the soil behaved in an undrained manner, so that no volumetric strain would occur in the plastic zone ($\Delta = 0$). However, during the injection of cement slurry into sandy ground, some drainage can occur in the soil resulting in finite volumetric strain in the plastic zone surrounding the column. In order to consider the effect of a finite value of Δ, the reduced rigidity index (I_{rr}) should be used instead of I_r, as recommended by Vesic (1972). The relevant expressions for I_r and I_{rr} are as follows:

$$I_r = \frac{E}{2(1+\nu)\left(c' + p_0 \tan\phi'\right)} \tag{5.59}$$

so that for a cylindrical cavity:

$$I_{rr} = \frac{I_r}{1 + I_r \Delta \sec\phi'} \tag{5.60a}$$

and for a spherical cavity:

$$I_{rr} = \frac{I_r}{1 + I_r \Delta} \tag{5.60b}$$

As indicated in Eqs. (5.60a and b), for a given value of Δ and for cases where $\phi' > 0$, the value of the reduced rigidity index I_{rr} for the cylindrical case will be smaller than for the spherical case. Conversely, a larger value of Δ would be required in the spherical case to result in the same value of reduced rigidity index for both the cylindrical and spherical cases. This latter condition is adopted when applying Eq. (5.56) to sandy soils.

It is also noted that Eq. (5.39) (in Table 5.8) remains the same for the spherical case but has to be changed for the cylindrical case in sandy soils, following the recommendation of Vesic (1972), i.e., in this case:

$$\frac{R_p}{R_u} = \sqrt{I_{rr} \sec \phi'} \tag{5.39a}$$

Equation (5.43) in Table 5.8 has been derived based on the assumption of no volumetric strain in the plastic zone. For cases where $\Delta > 0$, Eq. (5.43) will predict smaller lateral displacement in the plastic zone, especially near the cavity boundary. It is considered that this error has no great practical significance and therefore Eq. (5.43) has not been modified for the case of sandy soils.

5.5.4.2 Case History

Minami et al. (2007) and Sato et al. (2007) reported a case history that involved installation of soil-cement columns in loose sandy ground to form walls arranged in grids in plan-view. The purpose of these walls was to prevent potential liquefaction of the sandy ground contained within them. The profiles of grain size composition, natural water content, voids ratio, total unit weight and standard penetration test (SPT) N-values for this site are given in Fig. 5.52. The grain size distribution curves for the samples from about 12.3 m and 13.3 m depths are given in Fig. 5.53. Except for the SPT N-values (from Sato et al. 2007), all other data plotted in Figs. 5.52 and 5.53 were provided by Dr. E. Sato of the Takenaka Corporation, Japan. It can be seen that D_{50} is about 0.1–0.2 mm and the sand layer is potentially liquefiable. At the site, the groundwater level was about 1.0 m below the ground surface. For the sandy soil layers above 16 m depth, there are no measured data on total unit weight. Using the measured initial water contents, values of total unit weight of 17.5, 17.2 and 18.2 kN/m^3 have been estimated for the soil layers from 0 to 4.5 m, 4.5 to 7.0 m and 7.0 to 16.0 m depth, respectively.

The depth of improved soil was 16.0 m from the ground surface (as indicated in Fig. 5.52). The plan-view of the improved areas and the construction sequence are shown in Fig. 5.54. The machine used for this ground improvement project is capable of constructing 2 soil-cement columns with a diameter of about 1.0 m at the same time, with a centre-to-centre spacing of 0.8 m. The construction speed was approximately 1.0 m/min. in the vertical direction (personal communication with

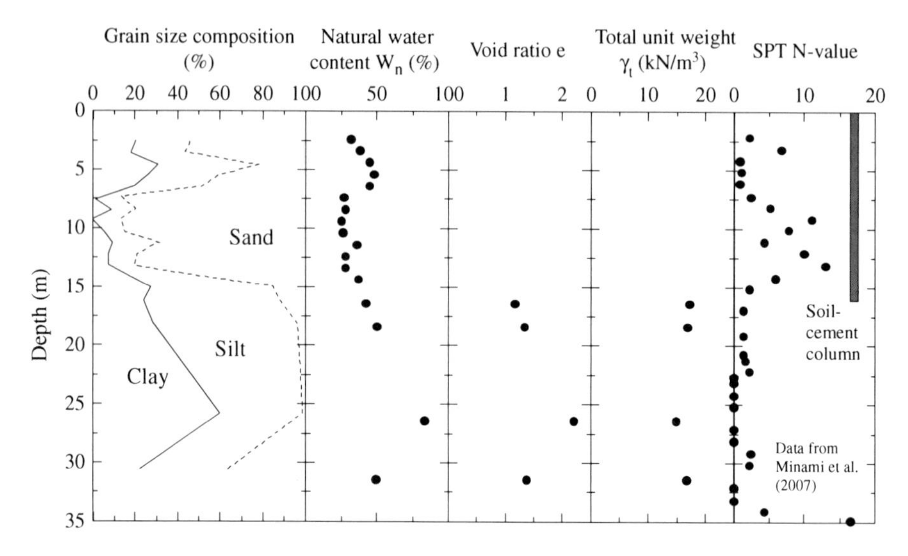

Fig. 5.52 Profiles of grain size composition, natural water content, voids ratio total unit weight, and SPT N-value of the sandy ground (after Chai et al. 2009)

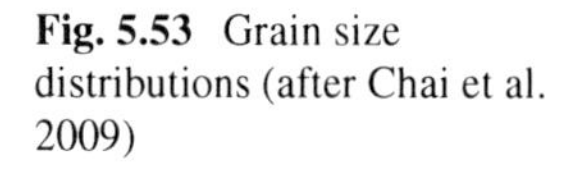

Fig. 5.53 Grain size distributions (after Chai et al. 2009)

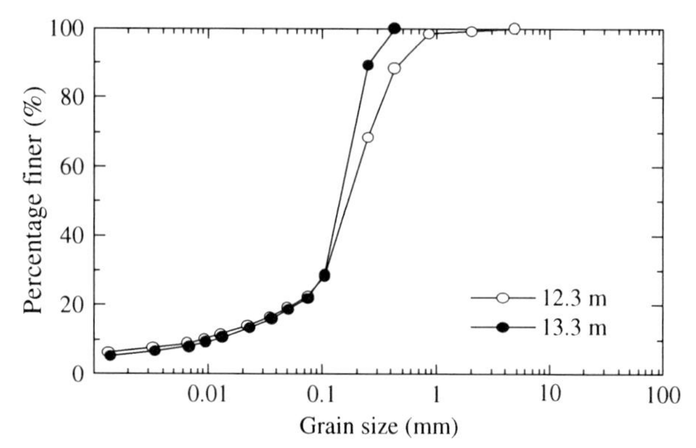

Dr. E. Sato). The adjacent columns were deliberately overlapped to form a continuous wall, and the average centre-to-centre spacing between the adjacent columns was about 0.8 m. With different design requirements for the strengths of the columns on the perimeter and on the inside, the amount of cement injected into each type of column was different. For the pairs of columns designated as Nos 1–17 and No. 31, the amount of cement used was about 210 kg/m^3, and for column Nos 18–30, and Nos 32–44, it was 159 kg/m^3. The water/cement ratio of the slurry by weight was 70% (Minami et al. 2007). Considering the density of cement particles as 3 Mg/m^3, the volume of slurry injected into the ground can be evaluated as 0.216 m^3/m^3 and 0.165 m^3/m^3 for the cement contents of 210 kg/m^3 and 159 kg/m^3, respectively, and

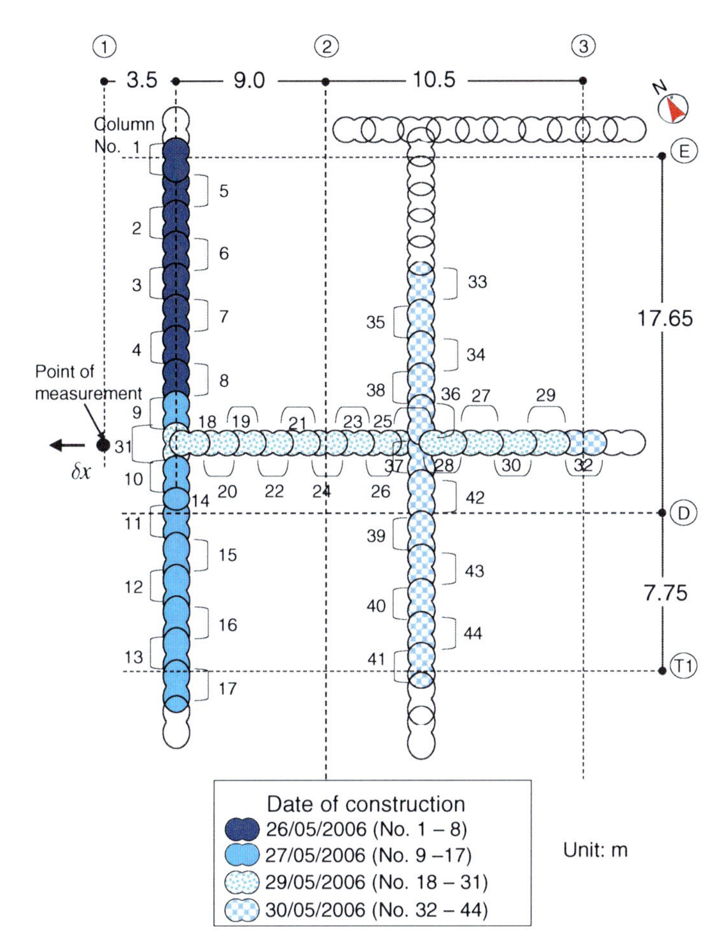

Fig. 5.54 Plan view of the arrangement of the columns and construction schedule of sandy ground case (after Chai et al. 2009; data from Minami et al. 2007)

for a column with a diameter of 1.0 m, the corresponding volume of slurry injected for 1.0 m length will be 0.17 m^3/m and 0.13 m^3/m, respectively.

For the case reported by Minami et al. (2007) and Sato et al. (2007), there are no test data for the values of c' and ϕ' for the soil strata. The ϕ' values listed in Table 5.11 were determined from SPT N-values using the equation proposed by Japanese Railway (JSCE 1986), i.e.,

$$\phi' = 1.85^\circ \cdot \left(\frac{N}{0.7 + {\sigma'}_v / p_a} \right)^{0.6} + 26^\circ \tag{5.61}$$

where ${\sigma'}_v$ = effective vertical stress and p_a = atmospheric pressure. The effective stress c' values shown in Table 5.11 have been determined purely empirically,

Table 5.11 Assumed c' values and estimated ϕ' values from SPT-N values

Soil layer	Depth (m)	c' or S_u (kPa)	ϕ' (°)
Surface crust	0–1	20	30
Surface layer	1–3	10	30
Clayey sand	3–7	10	25
Fine sand	7–15	2	35
Clayey silt	15–22	10	25
Silty clay	22–30	$0.25\sigma'_{v0}$[a]	0

[a]Ladd (1991) for undrained shear strength S_u

based on experience. In selecting these values it was noted that the undrained shear strength of soft clay is typically 10–30 kPa. These parameters were used to calculate the 'undrained' shear strength according to the initial effective stress state. For the 1.0 m thick surface crust, a relatively high effective cohesion was assumed to account for its higher strength (and stiffness). 'Undrained' shear strength (S_u) was calculated as follows for all strata except for the silty clay layer:

$$S_u = c' + p_0 \tan \phi' \tag{5.62}$$

For the silty clay layer S_u was calculated by the equation proposed by Ladd (1991).

Tatsuoka et al. (1986) reported drained strength-deformation characteristics of saturated fine sand under plane strain conditions. For loose fine sand samples with a voids ratio (e) of about 0.8, from the representative values of σ'_1/σ'_3, σ'_3 and ε_a (σ'_1, σ'_3 are major and minor effective mean stresses and ε_a is vertical strain) at stress levels of 50–75% of the corresponding strength under confining stress of 4.9–98.0 kPa, and assuming an effective stress Poisson ratio $\nu' = 0.3$, the ratios of E'/τ_f (E' is effective stress secant modulus and τ_f is drained shear strength) of about 100–350 can be estimated. The corresponding values of I_r are about 38–135. Vesic (1972) suggested that $I_r = 70$–150 for sand from loose to dense states, and 10–300 for clay from very soft to stiff consistency. Referring to these suggestions and considering the fact that the clayey layer at this site is relatively stiff (and strong), the values of Young's modulus (E) were calculated as 200 times the corresponding S_u values for both the sandy and clayey layers. The assumed value of Poisson's ratio was $\nu = 0.45$ (to allow for some volumetric strain in the elastic region of the soil). With all these parameters, a rigidity index (I_r) of about 68 can be calculated. A value of 68 is almost the lower boundary for sand corresponding to a very loose state as suggested by Vesic (1972). For clay, a value of 68 corresponds to a soft state.

Regarding the radius of the cavity (R_u), no suitable data have been proposed for sandy ground. Since the construction method is CDM, the values listed in Table 5.10 for CDM were used directly to investigate their applicability to this case. For the pairs of columns designated as Nos 1–17 and No. 31, a value of R_{u0} of 0.23 m ($E_0 = 2250$ kPa) was evaluated via Eq. (5.49). Similarly for column Nos 18–30, and Nos 32–44 a value of 0.2 m was evaluated.

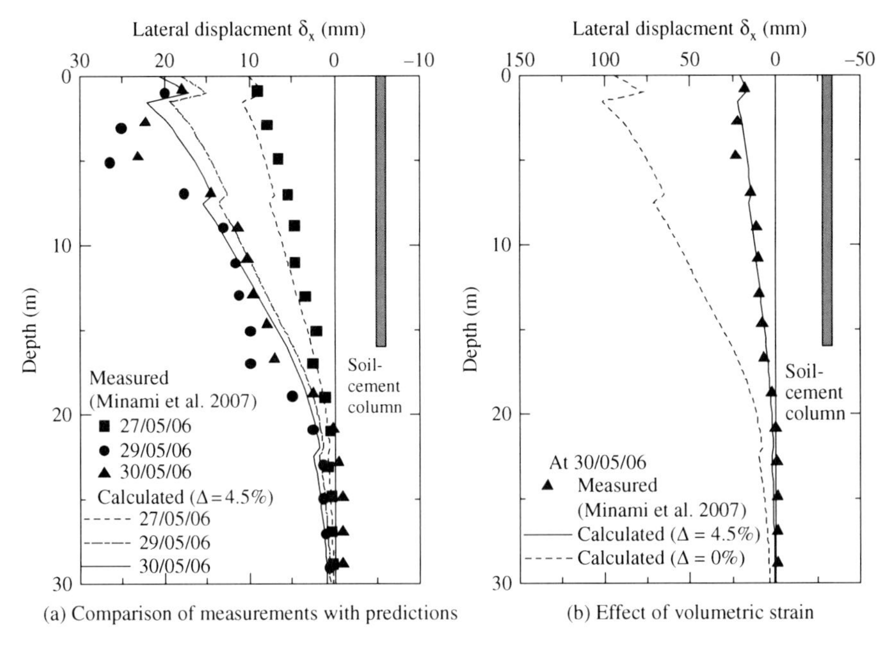

(a) Comparison of measurements with predictions (b) Effect of volumetric strain

Fig. 5.55 Comparison of measured and calculated lateral displacements of sandy ground case (after Chai et al. 2009). (**a**) Comparison of measurements with predictions; (**b**) Effect of volumetric strain

Two calculations of the lateral soil displacements were conducted. One assumed no volumetric strain in the plastic zone ($\Delta = 0$) while the other assumed an average volumetric strain of $\Delta = 4.5\%$ (the back-fitted value) in the plastic zone. The results are compared in Fig. 5.55a, b, from which two observations are made. First, if $\Delta = 0$ is assumed for this sandy ground the calculated displacements are about four times the measured values (Fig. 5.55b), whereas the field data are approximated quite well if $\Delta = 4.5\%$ is assumed. The second observation is that superposition of the lateral displacement is not strictly applicable to cases where subsequent consolidation of the soil causes a considerable reduction in the lateral displacement induced by the cavity expansion. From observations made in the period from May 27 to May 29, 2006, superposition generally appears to work well, although on May 27, 2006, the calculated values are larger than the measured data over the depth range 0–10 m, and on May 29, 2006, the calculated displacements are less than the measured data for most depths. However, the lateral displacements of May 30, 2006, calculated using the superposition method, are generally larger than those of May 29, 2006, while the measured data show the reverse tendency (Fig. 5.55a). Constructing the pair of columns designated as No. 31 caused a large increase in the lateral displacement at the measuring point, and subsequent consolidation (due to dissipation of the excess pore water pressures induced by column installation) appears to have caused a significant reduction in the lateral displacement.

Since the values of c' listed in Table 5.11 were assumed and since a value of $\Delta = 4.5\%$ was back fitted, the calculation procedure described above is not entirely rigorous. However, there are very few, if any, reported cases of predicting lateral displacements induced by soil-cement column installation in sandy ground, and it is believed that at the very least the data presented here may serve as a reference case for future studies of this problem.

5.6 Summary

The theories and methods for calculating consolidation settlements of soft soil deposits improved by the installation of soil-cement columns and the lateral displacements of the ground induced by column installation have been described. The methods have been applied to several case histories reported in the literature. By comparing the predicted values with the field measurements, the usefulness of these methods has been demonstrated.

The theories for predicting consolidation settlements can be divided into two groups, namely, for the case where the columns fully penetrate the soft deposit and for the case where the columns only partially penetrate the deposit (floating columns). For the fully penetrating case, the final consolidation settlement can be calculated by the equilibrium method if focusing on the settlement of the soft soil surrounding the columns, or the composite modulus method if the aim is to predict the average ground settlement for the columns and the surrounding soil. It has been recommended that the degree of consolidation for the fully penetrating case can be evaluated by Han and Ye's (2002) solution for stone-column improvement, which can consider both the higher stiffness and enhanced drainage properties of the column material.

For the floating column case, a method considering the influence of both the improvement area ratio (α) and the depth improvement ratio (β), i.e., the so-called α–β method, has been recommended and described for calculating the final consolidation settlement. At the macro level, a soft soil layer improved by the installation of soil-cement columns forms a double layer system. It has been suggested that the consolidation theory for double layer systems can also be used to calculate the degree of consolidation for the problem of floating columns. Methods for determining the representative consolidation parameters of the column-improved layer have been given. Four case histories in Fukuoka and one from Saga in Japan have been described and analyzed with satisfactory agreement being obtained between the predicted and measured performance in all cases.

Finally, a method for predicting the lateral displacements in the ground caused by the installation of soil-cement columns has been presented. Explicit equations were derived to describe the soil response using the theories of both cylindrical and spherical cavity expansion as well as the principle of superposition. The method was applied successfully to two case histories reported in the literature, one involving clay soils and the other involving sandy soils.

References

Aboshi H, Ichmoto E, Enoki M, Harada K (1979) The composer – A method to improve characteristics of soft clay by inclusion of large diameter sand columns. Proceedings of the international conference on soil reinforcement: reinforced earth and other techniques, vol 1, Paris, pp 211–216

Ariake Sea Coastal Road Development Office (ASCRDO), Saga Prefecture (2009) Research report on the techniques for building high embankment on soft Ariake clay deposit (in Japanese)

Barron RA (1948) Consolidation of fine-grained soils by drain wells. Trans Am Soc Civil Eng 113:718–742

Bergado DT, Chai J-C, Alfaro MC, Balasubramaniam AS (1994) Improvement techniques of soft ground in subsiding and lowland environment. Balkema, Rotterdam

Broms BB, Boman PO (1979) Lime columns – a new foundation method. J Geotech Eng Div ASCE 105(GT4):539–556

Carrillo N (1942) Simple two and three dimensional cases in the theory of consolidation of soils. J Math Phys 21:1–5

Chai J-C, Miura N (1999) Investigation on some factors affecting vertical drain behavior. J Geotech Geoenviron Eng ASCE 125(3):216–226

Chai J-C, Pongsivasathit S (2010) A method for predicting consolidation settlements of floating column improved clayey subsoil. Front Archit Civil Eng China 4(2):241–251. doi: 10.1007/s11709-010-0024-3 (Higher Education Press and Springer, Heidelberg)

Chai J-C, Shen S-L, Miura N, Bergado DT (2001) Simple method of modeling PVD improved subsoil. J Geotech Geoenviron Eng ASCE 127(11):965–972

Chai J-C, Miura N, Koga H (2005) Lateral displacement of ground caused by soil-cement column installation. J Geotech Geoenviron Eng ASCE 131(5):623–632

Chai J-C, Onitsuka K, Hayashi S (2006) Discussion on "Physical modeling of consolidation behavior of a composite foundation consisting of a cement-mixed soil column and untreated soft marine clay" (by Yin JH, Fang Z Géotechnique 56(1):63–68). Géotechnique 56(8):579–82

Chai J-C, Miura N, Koga H (2007) Closure to "Lateral displacement of ground caused by soil-cement column installation." J Geotech Geoenviron Eng ASCE 133(1):124–126

Chai J-C, Carter JP, Miura N, Zhu HH (2009) Improved prediction of lateral deformations due to installation of soil-cement columns. J Geotech Geoenviron Eng ASCE 135(12):1836–1845

Chai J-C, Miura N, Kirekawa T, Hino T (2010) Settlement prediction for soft ground improved by columns. Ground improvement. Proc Inst Civil Eng UK 163(2):109–119

Fujikawa K, Miura N, Beppu I (1996) Field investigation on the settlement of low embankment road due to traffic load and its prediction. Soils Found 36(4):147–153 (in Japanese)

Fukuoka National Highway Office (FNHO) (2003) Report on the field monitoring of test embankments at Shaowa-Biraki, Road No. 208, Fukuoka Part-2, Kyushu Regional Development Bureau, Ministry of Land, Infrastructure, Transport and Tourism (in Japanese)

Gray H (1936) Stress distribution in elastic solids. Proceedings of 1st international conference on soil mechanics and foundation engineering, Harvard, Mass, vol 2, p 157

Han J, Ye S-L (2001) A simplified solution for the consolidation rate of stone column reinforced foundations. J Geotech Geoenviron Eng ASCE 127(7):597–603

Han J, Ye S-L (2002) A theoretical solution for consolidation rates of stone column-reinforced foundations accounting for smear and well resistance effects. Int J Geomech 2(2):135–151

Hansbo S (1981) Consolidation of fine-grained soils by prefabricated drains. Proceedings 10th international conference on soil mechanics and foundation engineering, vol 3, Stockholm, pp 677–682

Igaya Y, Hino T, Chai J-C (2010) Behavior of trial embankments on soft Ariake clay deposits with and without column improvement. Proceedings of international symposium on geotechnical and geosynthetics engineering: challenges and opportunities on climate change, Bangkok, Thailand, pp 354–362

Japan Institute of Construction Engineering (JICE) (1999) Flexible foundation, foundation structure part, design code for flexible box culvert – II (Chapter 4). San-kai-dou Press, Tokyo, pp 233–248 (in Japanese)
Japanese Society of Civil Engineering (JSCE) (1986) Interpretation of the design code for Japanese railway structures, earth pressure retaining structures and foundations, JSCE, Tokyo, Japan
Keer LM, Xu Y, Luk VK (1998) Boundary effects in penetration or perforation. J Appl Mechan ASME 65:489–496
Kitazume M (1996) Deep mixing method. Found Eng Equip 24(7):14–19 (in Japanese)
Ladd CC (1991) Stability evaluation during staged construction. J Geotech Eng Div ASCE 117(4):541–615
Lorenzo GA, Bergado DT (2003) New consolidation equation for soil-cement pile improved ground. Can Geotech J 40:265–275
Miao L, Wang X, Kavazanjian E (2008) Consolidation of a double layered compressible foundation partially penetrated by deep mixed column. J Geotech Geoenviron Eng ASCE 134(8):121–1214
Minami T, Sato E, Namikawa T, Aoki M (2007) Evaluation of ground deformation during construction of deep mixing method – part I: measurement of ground deformation during construction. Proceedings of the 42nd annual meeting of Japanese Geotechnical Society, Tokyo, Japan, pp 905–906 (in Japanese)
Miura N, Chai J-C, Hino T, Shimoyama S (1998a) Depositional environment and geotechnical properties of Ariake clay. Ind Geotech J 28(2):121–146
Miura N, Shen S-L, Koga K, Nakamura R (1998b) Strength change of clay in the vicinity of soil-cement column. J Geotech Eng, JSCE No. 596/III-43:209–221 (in Japanese)
Osterberg JO (1957) Influence values for vertical stresses in semi-infinite mass due to embankment loading. Proceedings of 4th international conference on soil mechanics and foundation engineering, London, UK, vol 1, p 393
Sato E, Minami T, Namikawa T, Aoki M (2007) Evaluation of ground deformation during construction of deep mixing method – part II: effect of deformation control method. Proceedings of the 42nd annual meeting of Japanese Geotechnical Society, pp 907–908 (in Japanese)
Shen S-L, Miura N, Koga H (1999) Lateral displacement of the surrounding ground during installation of DM columns. In: Wang CW, Lee KH, Ang KK (eds) Proceedings of APCOM'99, the 4th Asia-Pacific conference on computational mechanics, vol 2. Elsevier, London, pp 773–778
Shen S-L, Chai J-C, Miura N (2001) Stress distribution in composite ground of column-slab system under road pavement. Proceedings of 1st Asian-Pacific congress on computational mechanics, Elsevier, London, pp 485–490
Tatsuoka F, Sakamoto M, Kawamura T, Fukushima S (1986) Strength and deformation characteristics of sand in plane strain compression at extremely low pressure. Soil Found 26(1):65–84
Taylor DW (1948) Fundamentals of soil mechanics. Wiley, New York, NY
Terzaghi K, Peck RB (1967) Soil mechanics in engineering practice. Wiley, New York, NY
Vesic AS (1972) Expansion of cavities in infinite soil mass. J Soil Mechan Found Eng ASCE 98(SM3):265–290
Zhu G, Yin JH (1999) Consolidation of double soil layers under depth-dependent ramp load. Géotechnique 49(3):415–421

Chapter 6
Concluding Remarks

Abstract Many of the techniques for deformation analysis presented in the preceding chapters involve approximations of reality, sometimes quite crude approximations. These have been clearly and deliberately identified. In many cases it has been demonstrated, by presenting comparisons of predictions with field data, that these approximations provide a reasonable balance between tractability of the problem at hand and the accuracy and reliability of the resulting deformation predictions. However, there is still a need for considerable further research in this area of geotechnology. Much still needs to be done to increase our predictive capabilities and particularly the accuracy and reliability of our predictions of ground deformations associated with common ground improvement techniques. It is anticipated that this need will only continue as further advances are made in the technology of ground improvement. This chapter provides some suggestions for future research directions, which hopefully will address some of the present shortcomings or gaps in our current predictive abilities.

6.1 What Else Needs to Be Done?

The aim of this book has been to present up-to-date methods for the analysis of ground deformations associated with commonly used techniques of soft ground improvement. Particular emphasis has been placed on those methods involving preloading soft ground whose drainage characteristics have been improved by the installation of prefabricated vertical drains (PVDs). The preloading could be applied directly to the ground surface either as a surcharge load (e.g., embankment fill, etc.) or by the imposition of a vacuum pressure. An entire chapter has also been devoted to the topic of improving soft ground by the installation of soil-cement columns.

All of the techniques for deformation analysis presented in the preceding chapters have involved approximations of reality, in some form or other. Sometimes these have been quite crude approximations, but it was always the intention that they should be clearly identified. It is to be hoped, and in many cases it has in fact been demonstrated by comparisons of predictions with field data, that these approximations provide a reasonable balance between tractability of the problem at hand and the accuracy and reliability of the resulting deformation predictions.

J. Chai, J.P. Carter, *Deformation Analysis in Soft Ground Improvement*, Geotechnical, Geological and Earthquake Engineering 18, DOI 10.1007/978-94-007-1721-3_6,

However, there remains a need for considerable further research in this area of geotechnology. Much still needs to be done to increase our predictive capabilities and particularly to improve the accuracy of our predictions of ground deformations associated with common ground improvement techniques. It is anticipated that this need will only continue as further advances are made in the technology of ground improvement. The following are some suggestions for future research directions, which if followed should address some of the present shortcomings and gaps in our current predictive abilities.

6.1.1 Constitutive Models for Soft Clays

Constitutive models are essential in any rational theoretical modelling in geotechnics. How the stress-strain response of soil is represented in these models is usually the key to successful prediction of the behaviour of geotechnical structures. However, the important details of these models, particularly the idealizations that are made are sometimes poorly understood or even ignored, sometimes at significant cost to the unwary analyst. Indeed, the capabilities and the shortcomings of these models, especially the more advanced models, are not always easy to ascertain. In some cases determination of the input parameters is not straightforward. Consequently, in general, it may be difficult to determine which constitutive model to select for a particular analytical or numerical task.

Clearly, this suggest that one task requiring our attention as a geotechnical profession is a continuing education program, with the ability to effectively disseminate the capabilities and especially the limitations of the theoretical models used for estimating the deformations associated with current ground improvement technologies. Such programs should include an emphasis on how the constitutive behaviour of the soil and other relevant materials is represented in these models. It is the humble hope of the authors that this book will be a small step in this direction.

It is also worthwhile addressing the specific question of what are the major limitations of the constitutive models for soft clay soils considered already in Chapter 2 and used in the development of numerical solutions to the various problems of ground improvement. Are there any features of soft soil behaviour missing in current models, which perhaps should be included in order to provide more accurate representation of the ground undergoing improvement?

It is useful to commence this assessment by reviewing the major capabilities of the soil stress-strain models described in Chapter 2. Such an assessment is included in Table 6.1. This table indicates what is possibly an incomplete list of the desired features for stress-strain models of soft soils, and also an indication whether each of the three different models described in Chapter 2 possess the listed features or capabilities.

It is clear from the details presented in Table 6.1 that none of the models considered in Chapter 2 possesses all the 'desirable' features, and so potentially they may not capture all the critical aspects of soft-clay behaviour.

Table 6.1 Features of constitutive models for soft soils

Feature	Modified Cam clay model	Sekiguchi-Ohta model	Generalized two surface 'bubble' model
Non-linear, stress-dependent elasticity	Yes	Yes	Yes
Non-linear, strain-dependent (small strain) elasticity	No	No	Yes
Pre-failure (work hardening/softening) plasticity	Yes	Yes	Yes
Critical state failure criterion in the triaxial effective stress plane	Yes	Yes	Yes
Variable shear strength in the π plane of effective stress space	No	Yes	Yes
Non-associated plastic flow rule	No	No	Yes
Variable plastic dilatancy /contraction including critical state response at large strains	Yes	Yes	Yes
Inherent anisotropy in stiffness	No	Yes	Yes
Inherent anisotropy in strength	No	No	Yes
Developing (stress-dependent) anisotropy in stiffness	No	Yes	Yes
Developing (stress-dependent) anisotropy in strength	No	No	Yes
Inclusion of material creep and relaxation	No	Yes	No

This simple analysis indicates that more model development may be required to capture some of the missing features in the stress-strain models that should be adopted for soft ground when analyzing the implications of the various ground improvement technologies. In particular, it will be important for any future research of this type to identify which of the listed features are critical and which may be relatively unimportant. Presumably the latter may therefore be safely ignored in routine analysis. Answering this particular question is probably the greatest challenge in this area of endeavour.

6.1.2 Electrical and Thermal PVD Improvement

Research into the effects of ground improvement using mechanical means, including the installation of PVDs, has been quite intense and comprehensive over the past 30 years. The basic consolidation theories and numerical modelling methods are now well established.

However, in recent years there have been a number of important technological innovations in the area of pre-fabricated drains. Examples include the use of electrical PVDs (Jones 2002; Chew et al. 2004) and thermal PVDs (Pothiraksanon et al. 2008). The clear aim of these endeavours has been to utilize the effects of both electric fields and temperature to enhance the effectiveness of the PVD as a drainage medium when installed in soft ground. To further develop these technologies and to assist their implementation in engineering practice, advances are now required in our understanding of the underlying physical phenomena, as well as the theories to allow rational predictions to be made of their effects on soft ground behaviour, including the important task of predicting any consequent deformations of the ground. In particular, a consolidation theory that can include the effects of electrical gradients (voltage) and thermal gradients (temperature) is needed. A special requirement is also the associated need for methods for determining soil parameters related to electrically and thermally induced soil consolidation. These have yet to be reliably established.

6.1.3 Soil-Cement Slab and Column Interaction

The method for calculating the consolidation settlement of soft subsoil improved by the installation of floating (i.e., partially penetrating) soil-cement columns, together with an overlying soil-cement slab, has been developed and presented in Chapter 5 of this book. However, implicit in this method is the assumption that the top slab will not fail structurally under the loading imposed on it. In other words, the structural interaction between the slab and the column has not really been investigated to any extent.

Soil-cement slabs are generally relatively strong in compression but much weaker in tension. Indeed, the tensile strength of the soil-cement mix is typically about 1/8 of its compression strength (Takakura et al. 2002). As shown in Fig. 6.1, under a distributed surcharge load the sections of the slab spanning between the columns will tend to settle more than those immediately above the columns, so that tensile stress can develop at locations such as A and B in Fig. 6.1. Hence there is the potential for tensile failure to occur in the slab, especially for cases where the

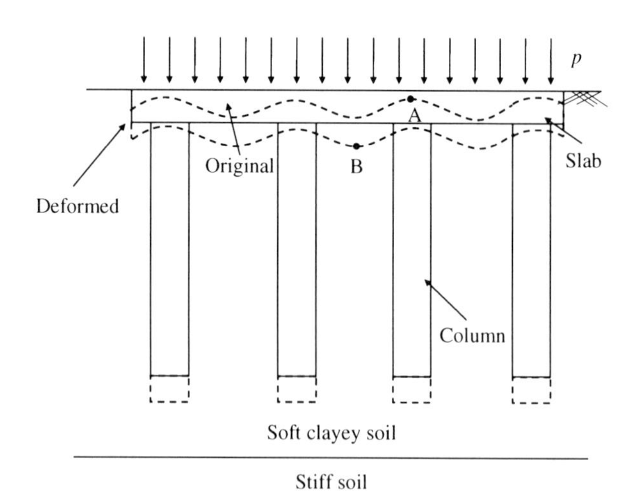

Fig. 6.1 Locations of possible tensile cracks in the soil-cement slab

installed columns provide a low area replacement ratio (α). When a slab fails in tension in this manner it will influence the structural interaction between the column and the surrounding soft soil, and therefore the consolidation settlement of the the overall ground improvement system.

It is suggested that the structural interaction between the slab, the column and the surrounding soft soil needs to be further investigated and a method for designing slab-column soft ground improvement systems taking into account the finite tensile strength of the slab, and possible tensile failure in the slab, needs to be developed.

There have been reported cases where soft soil deposits have been improved by the installation of floating columns alone, without the presence of a slab at the ground surface (e.g., Igaya et al. 2010, as described in Chapter 5). In these cases, there may be some relative settlement between the column and the soft soil near the ground surface and this will influence the consolidation and deformation characteristics of the system. Further study is required to include this possibility in the deformation analysis of soft soils improved by the installation of soil-cement columns without an overlying slab.

6.1.4 Combined Vacuum Pressure and Surcharge Loading

Combining embankment loading with the application of a vacuum pressure to consolidate soil has several advantages, such as increasing the overall consolidation pressure exerted on the soft soil, increasing the factor of safety (FS) of the embankment during construction, as well as minimizing lateral displacement of the ground. This particular technique has already been adopted in several engineering projects (e.g., Bergado et al.1998).

As previously noted, embankment loading can induce shear stress increments in the ground and thus cause immediate outward lateral displacement of the ground. On the other hand vacuum pressure is an isotropic consolidation pressure which will tend to induce inward lateral displacement of the ground during the consolidation process (Chai et al. 2006). Conceptually, it is possible to control or minimize the lateral displacement of the ground by the combination of embankment loading and the application of vacuum pressure (Fig. 6.2). However, there is still no well established method to predict the lateral displacements for this particular combination of loading applied to the ground.

Theoretically, the ground deformation under the combination of embankment load and vacuum pressure can be simulated or predicted by finite element analysis (FEA). However, in engineering practice even predicting the lateral displacement due to embankment loading alone is a challenging task. As a general tendency, when an embankment has a relatively higher FS, finite element analysis will generally over-predict the lateral displacement of the ground (Poulos et al. 1989). Conversely, when the FS is low and approaches unity, FEA will tend to under predict the lateral displacement (Chai et al. 2002). In addition, although nowadays FEA is used

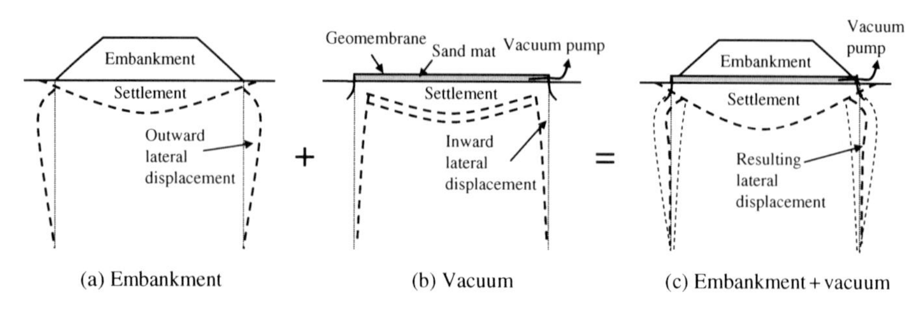

Fig. 6.2 Illustration of the principle of minimizing lateral displacement by the combination of embankment loading with vacuum pressure

increasingly in routine design of geotechnical projects, the results of the numerical analysis still depend on several factors including the analyst's judgment and skill in using a particular computer program. Indeed, there can be no guarantee of a unique result. It is considered desirable that a reliable and easy to use method for engineering design should be developed for this particular problem.

6.2 Hybrid Soft Ground Improvement Techniques

For soft ground improvement to become more economical, effective and efficient, it is almost inevitable that hybrid techniques will need to be developed. If this prophesy is realized, then the corresponding methods of deformation analysis will also need to be established. Combining the installation of soil-cement columns with vacuum consolidation is one of these possible hybrid methods.

For the case of embankment construction on soft clayey soils, improving the soft subsoil by first installing soil-cement columns is an effective technique, but generally it is more expensive than the preloading method. However, preloading is very likely to induce greater settlement and possibly greater lateral displacement of the ground and so it may have adverse impacts on the adjacent area. In urban environments it is very common that such impacts need to be tightly controlled. One of the ways of economically and efficiently dealing with this kind of situation is likely to be by combining vacuum consolidation with the installation of soil-cement columns, as shown in Fig. 6.3. In this figure, the term CPVD refers to the installation of capped prefabricated vertical drains.

Clearly, if hybrid technologies such as the one suggested here, as well as any others that may emerge, are to gain greater use in engineering practice, then appropriate design methods will need to be developed. It is to be hoped that such methods will be firmly based on a robust theoretical underpinning combined with the vital experience that will be gained from well instrumented and documented field applications.

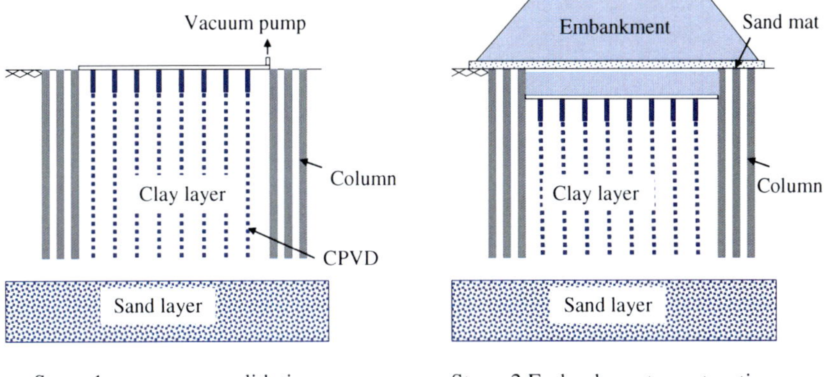

Fig. 6.3 Illustration of the principle of combining soil-cement columns and vacuum consolidation

References

Bergado DT, Chai JC, Miura N, Balasubramaniam AS (1998) PVD improvement of soft Bangkok clay with combined vacuum and reduced sand embankment preloading. Geotech Eng, Southeast Asian Geotech Soc 29(1):95–121

Chai JC, Miura N, Shen SL (2002) Performance of embankments with and without reinforcement on soft subsoil. Can Geotech J 39(8):838–848

Chai JC, Carter JP, Hayashi S (2006) Vacuum consolidation and its combination with embankment loading. Can Geotech J 43(1):985–996

Chew SH, Karunaratne GP, Kuma VM, Lim LH, Toh ML, Hee AM (2004) A field trial for soft clay consolidation using electric vertical drains. Geotext Geomembr 22(1–2):17–35

Igaya Y, Hino T, Chai JC (2010) Behavior of trial embankments and the results of environmental monitoring for the Ariake Sea Coastal Road Project. Proceedings of the 7th international symposium on lowland technology ISLT2010, Saga, Japan, pp 260–267

Jones CJFP, Glendinning S, Shim GSC (2002) Soil consolidation using electrically conductive geosynthetics. Proceedings of the 7th international conference on geosynthetics, vol 3, Nice, France, pp 1039–1042

Pothiraksanon C, Saowapakpiboon J, Bergado DT, Thann NM (2008) Reduction of smear effects around PVD using thermo-PVD. Ground Improvement 161:179–187

Poulos HG, Lee CY, Small JC (1989) Prediction of embankment performance on Malaysian marine clays. Proceedings of international symposium on trail embankment on Malaysian marine clays, vol 2, Kuala Lumpur, pp 4/1–4/10

Takakura A, Koga Y, Hamatake A, Koga H (2002) Strength properties of stabilized soil slab in road on soft ground. Research reports of Faculty of Science and Engineering, Saga University. 31(2):17–20

Index

J. Chai, J.P. Carter, *Deformation Analysis in Soft Ground Improvement*, Geotechnical, Geological and Earthquake Engineering 18, DOI 10.1007/978-94-007-1721-3,